WASTE AND WASTE MANAGEMENT

MANAGEMENT OF HAZARDOUS RESIDUES CONTAINING CR(VI)

WASTE AND WASTE MANAGEMENT

Additional books in this series can be found on Nova's website under the Series tab.

Additional E-books in this series can be found on Nova's website under the E-books tab.

WASTE AND WASTE MANAGEMENT

MANAGEMENT OF HAZARDOUS RESIDUES CONTAINING CR(VI)

MARIA JOSÉ BALART MURRIA
EDITOR

Nova Science Publishers, Inc.
New York

For permission to use material from this book please contact us:
Telephone 631-231-7269; Fax 631-231-8175
Web Site: http://www.novapublishers.com

LIBRARY OF CONGRESS CATALOGING-IN-PUBLICATION DATA
Management of hazardous residues containing Cr(VI) / editors, Marma Josi Balart Murria.
p. cm.
Includes bibliographical references and index.
ISBN 978-1-61470-637-3 (softcover)
1. Chromium--Environmental aspects. 2. Hazardous waste site remediation.
I. Balart Murria, Marma Josi.
TD196.C53M36 2009
628.4'2--dc22
2010014083

Published by Nova Science Publishers, Inc. † New York

CONTENTS

PREFACE

The first chapter is a case study of how untreated residues containing Cr (VI) dumped in landfields could get leached out Cr (VI) into the ground and ground water. This could occur in unsaturated soil, relatively close to the landfield having untreated residues containing Cr(VI), the degree of which is dependent on the specific interaction with the soil.

Chapter 2 is a short communication to the management of solid wastes containing Cr (VI), with a particular focus on two minimization routes of Chromite Ore Processing Residue (COPR) to generate "zero-waste". The versatility of one of the routes in that it could also be used for waste reprocessing from chromium chemicals production. The chapter also includes low lime and no-lime COPR processes. Management of COPR is described in detail in Chapter 14. Chapter 3 provides an introduction to the technologies for treating waste streams and sludges or solid wastes containing Cr (VI). The removal of Cr (VI) from waste streams involves physical processes such as adsorption on activated carbon and biosorption, and phytoremediation. Indirect reduction of Cr (VI) followed by adsorption, chemical precipitation or bioremediation is also described in this chapter. Of particular importance is the initial reduction which can be achieved by techniques such as bioreduction, treatment with zero valent iron (ZVI, Fe^0) or photocatalysis. More detail of the technologies for treating wastewaters and biological removal is given in the recovery section of the book. The removal of Cr (VI) from sludges or solid wastes involves stabilisation/solidification (S/S) treatments. Another key point introduced in this chapter is the degree of success of the S/S with additives dependent on the type of binding media and additives. Whereas the S/S techniques of encapsulation and vitrification are more successful.

Many metals and alloys such as steel, aluminium, zinc, tin, cadmium and their alloys are usually conversion coated. Chapter 4 explores non-Cr(VI) conversion coatings in current use that have become available since the early 1980's, namely chromium (III) based, permanganate based, phosphate based and synthetic polymer based conversion coatings. Chapter 5 in particular, shows how the biosorbent ectodermis of Opuntia pretreated with formaldehyde enhances Cr (VI) adsorption compared to that of the natural biosorbent without having being pretreated. The adsorption kinetics for the pretreated condition fitted best to the pseudo-second order kinetics and the mechanisms of adsorption fitted best to the Langmuir model. Thermal dewatering to reduce the waste sorbent mass and volume is also enhanced in the pretreated condition.

The following three chapters are all concerned with wastewater treatment for removal of Cr (VI). In Chapter 6 there is a detailed overview of the most important technologies for doing this, namely liquid-liquid extraction, liquid membranes technologies, adsorption and ion-exchange processing. Cancer and other health effects of chromium (VI) compounds are also discussed. Chapter 7 uses synthesized magnetic Fe-Mg oxide particles to recover Cr (VI) from wastewater. The results show the effect of calcinations temperature, pH, total dissolved substance, ionic strength, kinetics, chromate concentration and adsorption equilibrium isotherm. The innovative topic of the biological removal of Cr (VI) using activated sludges is much more recent. Chapter 8 discusses the effect of important factors concerning the bioreduction of Cr(VI) by activated sludge in batch reactors, such as biomass concentration, the presence, type, and concentration of carbonaceous substrates, the nitrogen source concentration, and Cr(VI) concentration. A mathematical model that describes the Cr(VI) bioreduction kinetics is defined, and the implementation and practical use of it are also discussed.

Once the residue has been directly generated or recovered, it is preferable to find an end-use. The ceramic and the cement industries have found beneficial uses of wastes containing Cr (VI). Practical applications have been examined thoroughly in the recycling chapters of this book. A comprehensive treatment of galvanic sludges containing Cr (VI) and their potential for industrial applications to produce glasses, bricks, tiles, pigments and glazes is described in Chapter 9. In Chapter 10 there is a detailed treatment of the susceptibility to incorporate chromium-tanned leather in mortars, using an ordinary Portland cement matrix, and an asphalt micro-surface layer. Chapter 11 covers the viability to incorporate industrial waste with chromium in concrete using the Portland cement matrix as well. The concrete samples are evaluated in terms of the chromium oxidation state —Cr (III) and Cr (VI)—, physicomechanical properties, durability and leachability tests. Chapter 12 introduces a new three-step hybrid technology for recovery of chromium from both secondary and tertiary wastes generated by the leather industry and proposes the application of the chromium-containing sludges in the ceramic industry. The chapter also includes the potential risk that Cr (VI) associated with leather goods could pose to the human health.

The versatility of the Portland cement is not only important in the recycling process but also in the S/S process. In parallel to Chapter 10, but this time from the viewpoint of the S/S process, Chapter 13 covers the treatment of tannery waste containing Cr (VI) using the Portland cement matrix, both with and without additions of sodium bentonite and organoclay; each of the two additives are tested for commercial use and are laboratory modified. Solid-state nuclear magnetic resonance spectroscopy and X ray diffraction are used to characterize the waste. Conventional and non-conventional differential thermal analysis, leaching and compressive strength tests are used to characterize the solidified samples and the solidification process. Chapter 14 is a comprehensive overview of COPR in terms of the mineralogy, evaluation of waste property and its treatment. For safe disposal and re-use of chromium rich waste materials, it is important to ascertain the environmental oxidizing or reducing conditions determining the two most common oxidation estates of chromium: trivalent Cr (III) and hexavalent Cr (VI). The analytical techniques: speciation analysis, fractionation procedures and leachability tests, for the investigation of the transformation and mobility of chromium in the environment are introduced in chapter 15. This chapter finalises with case studies of chromium wastes: safe disposal of tannery waste and sewage sludges and re-use of waste materials from steel production. Chapter 16 is entirely devoted to the fractionation procedure which is based on the Tessier's sequential extraction to determine the Cr (VI) content in waste coal fly ash. Although it is not discussed here, of particular interest is a technology to manufacture bricks from waste fly ash, which has been bought an patented by Australian manufacturer Vecor. Chapter 17 is a study of the susceptibility to oxidation of Cr (III) to Cr (VI) on an electric arc furnace slag during a long term leach test. This study is important from the point of view of waste management of materials containing Cr (III) because of their potential oxidation to Cr (VI). Therefore their waste management should be assessed accordingly. Peat generally has the capacity to act as hydraulic barrier layer, preventing the risk of pollution of the subsoil and ground water. Chapter 18 is a calculation of the rate of diffusion of Cr (VI) and Cu (II) through a compacted peat layer, taking into account the effect of adsorption on diffusion.

In: Management of Hazardous Residues Containing Cr(VI)
Ed: Maria José Balart Murria
ISBN: 978-1-61470-637-3

Chapter 1

THE TRANSFER AND SPECIATION TRANSFORMATION OF CR LEACHING FROM HAZARDOUS WASTES CONTAINING CR (VI) IN AN UNDERGROUND ENVIRONMENT

Xingmin Zhao[1,2], Deming Dong[1*], Xiuyi Hua[1], Yu Chen[1] and Wentao Wang[1]

[1] College of Environment and Resources, Jilin University, Changchun 130012, China;
[2] College of Resources and Environment, Jilin Agricultural University, Changchun 130118, China

ABSTRACT

Cr(VI) in hazardous residues disposed in the ground without anti-seepage or inertization measures could be leached out gradually by the action of precipitation or moisture in residues and moved down into the underground environment. Cr(VI) in soils often exist as anions (e.g., CrO_4^{2-}, $HCrO_4^-$ and $Cr_2O_7^{2-}$). Compared with Pb^{2+} and anions of As, soil adsorbed anions of Cr(VI) fast, but only a very limited amount was adsorbed. Cr(VI) not adsorbed by topsoil would be transported downward into the unsaturated zone and aquifers, and polluted the ground water. In this study, adsorption behavior of typical soil in northeast China to Cr(VI) and vertical transfer behavior of Cr(VI) in soils were investigated and compared with Pb^{2+} and anions of As. The changes of fractions and valences of Cr during the process of transfer were also examined. The results showed that adsorption amounts of the metals by three soils decreased in the order: Pb > As(V) > Cr(VI). The distribution coefficient (K) of Cr(VI) in black soil in northeast China is about 11 L/kg. It was found that Cr(VI) at the concentration of 10 mg/L would transport to 60 cm depth in the unsaturated zone within only 20 days. The mobility of Cr(VI) in the unsaturated zone is much greater than Pb^{2+} and anions of As because most of the Cr(VI)

* Corresponding author. Tel.: +86 431 85166401; fax: +86 431 85166401
E-mail address: dmdong@mail.jlu.edu.cn ((D. Dong).
Fund: The National Basic Research Program of China under Grant No. 2004CB418505.

was not adsorbed by soils. So Cr(VI) could transport into an aquifer after a certain period. The result of sandbox tests showed that when entering the aquifer, Cr(VI) was transported in each direction. So in places near the deposit site of hazardous residues containing Cr (VI), the groundwater was easily polluted. When Cr(VI) transported into the unsaturated zone and aquifer, significant transformation of the fraction and the valence of Cr were observed. Active factions were increased greatly, which indicated that once Cr(VI) was leached from hazardous residues, it was hazardous to soil and groundwater systems. The toxicity of Cr(VI) was changed significantly due to the change of valences. Therefore, there's remarkable risks of hazardous residues containing Cr(VI) disposed on ground without anti-seepage or inertization measures to ground water and agricultural products.

1. INTRODUCTION

Chromium contamination of soil and groundwater is a significant problem worldwide primarily owing to its use in numerous industrial processes (e.g., electroplating and manufacturing industries). Waste products from these processes normally contain Cr(VI) compounds, such as chromic acid and other oxidizing Cr(VI) cleaners. Cr(VI) comprises most of the wastes, and a smaller amount of Cr(VI) was reduced to Cr(III). Because of its toxic nature, Cr(VI) creates numerous environmental problems in waste products, mine wastes, and manufacturing slag piles. Cr(VI) is commonly spilled or leaked into the environment as a contaminant frequently with other associated wastes. Therefore, Cr(VI) may be transported into the environment as a contaminant through runoff and leaching of water, ultimately moving into the soil and groundwater. Cr(VI) in groundwater has also been documented in the vicinity of U.C. Davis in the Sacramento Valley. Concentrations of up to 0.2 mg/L have been reported from monitoring wells screened in alluvial deposits, and Cr(VI) is also present in off-site control wells and shallow upgradient wells [1].

Chromium commonly exists in two oxidation states in soils, Cr(III) and Cr(VI). Forms of Cr(VI) in soils are as hydrogen chromate ion ($HCrO_4^-$ predominant at pH < 6.5 or CrO_4^{2-} predominant at pH 6.5), and as dichromate($Cr_2O_7^{2-}$) predominant at higher concentrations and at pH 2 to 6. $Cr_2O_7^{2-}$ is more toxic to humans than CrO_4^{2-}. In neutral-to-alkaline soils, Cr(VI) is mostly soluble (e.g., Na_2CrO_4), but in moderately-to-sparingly soluble chromates (e.g., $CaCrO_4$, $BaCrO_4$, $PbCrO_4$) [2]. In more acidic soils (pH < 6), $HCrO_4^-$ becomes a dominant form. CrO_4^{2-} and $HCrO_4^-$ are the most mobile forms of Cr in soils. They can be taken up by plants and easily leached out into the deeper soil layers causing groundwater pollution [3]. Some minor amounts of Cr(VI) are bound in soils. This binding depends on the mineralogical composition and pH of the soil. The CrO_4^{2-} ion can be adsorbed by goethite, FeO(OH), aluminum oxides and other soil colloids with a positively charged surface [4]. $HCrO_4^-$, which occurs in more acidic soils, may also be held in soils or remain soluble [5]. Reviews of the processes that control the fate of Cr in soil and the effect these processes have on remediation are given in Bartlett [6] and Palmer and Wittbrodt [7]. In a study of the relative mobility of 11 different trace metals for a wide range of soils, Korte et al. [8] found that clayey soil, containing free Fe and Mn oxides, significantly retarded Cr(VI) migration. Cr(VI) was the only "metal" that was highly mobile in alkaline soils. The study also showed that free Fe oxides, total Mn, and soil pH influenced Cr(VI) immobilization, whereas soil properties such as cation exchange capacity, surface area, and clay had no significant influence on Cr(VI) mobility. The dominant Cr form depends strongly on pH; in acidic soils (pH < 4), it is $Cr(H_2O)_6^{3+}$, whereas at pH < 5.5 it is its hydrolysis products, mainly soluble $CrOH^{2+}$ [9]. Both these forms are easily adsorbed by clays. The process is intensified by an increase of pH, which can be interpreted in part due to an increase of negative charge on the clays. In contrast, mobile compounds such as citric acid ($H_3C_4H_5O_7$) and fulvic acid, form soluble Cr(III) complexes which control its oxidation to Cr(VI) in soils [2]. Both reduction of Cr(VI) to Cr(III) and the adsorption of Cr(VI) can occur in soil, even simultaneously. Therefore, this causes a problem in assigning one mechanism of observed attenuation of Cr(VI) in the subsurface [6].

Cr(VI)-contaminated sites are difficult to be delineated vertically and laterally, and may be complicated by naturally occurring Cr(VI) that has not been characterized. Cr(VI)-contaminated soil and groundwater resources should be assessed and remediated. For better understanding of the processes of soil-Cr interactions, in particular, the mobility and retention mechanism of Cr(VI) in soil, a study on the transport and fate of Cr(VI) in soil zones and aquifer was carried on by column leaching tests and sandbox tests in laboratory. Cationic Pb and anion As(V) were chosen as contrastive elements. To assess the environmental impact of Cr, the knowledge of the total concentration without considering Cr speciation is not sufficient, because the mobility and toxicity of heavy metals in contaminated soil are greatly controlled by its chemical fractions [10]. The mobility and bioavailability of heavy metal in solid samples were generally evaluated by selective sequential extraction, and metals were classified into five fractions: exchangeable (F1), bound to carbonates (F2), bound to Fe-Mn oxides (F3), bound to organic matter (F4) and residual (F5) [11]. Determination of valences of Cr is also very important, since their toxicity and mobility are associated with their oxidation/reduction state [12].

2. Collection and Preparation of Soil Samples

Soil samples were collected from three sites in moderately contaminated farmland in Northeast China. Site A ($42^0$04.001 N, $123^0$29.352 E) was about 60 m away from a pile of chromium wastes. Residual materials produced in industrial production containing chromium with great concentration were disposed for about fifty years. Leachate from the disposal had restrained the growth of corn plants covering the site. Site B ($43^0$53.503 N, 125^0 05.839 E) was about 30 m away from a river mainly polluted by industrial wastewater. Site C ($41^0$38.05 N, $123^0$4.35 E) was about 30 m away from another river seriously polluted by industrial wastewater and sewage. The farmland had been irrigated using the wastewater from the river for a long time; although wastewater irrigation was already prohibited, heavy metal pollution (especially Cd and Pb) of the agricultural soil had been serious. Soils were sampled from three layers: 0-20 cm, 20-40 cm and 40-60 cm in a profile. Soils were thoroughly mixed and air-dried after removing gravel and plant residues. The soils were crushed using an agate mortar and then passed through a 100 mesh nylon sieve. Some basic physicochemical properties of the soils are shown in Table 1.

Table 1. Physical and chemical properties of the soils used in this study

Soil	depth (cm)	pH	Sand (%)	Silt (%)	Clay (%)	BD (g/cm^3)	TOC (%)	EC (μs/cm)	CEC (mmol/kg)	pH_{pzc}
A	0-20	7.07	18	58	24	1.346	14.300	28	157.06	2.97
	20-40	7.21	26	52	22	1.336	6.940	42	160.00	
	40-60	7.20	38	45	17	1.326	5.235	51	137.24	
B	0-20	7.02	23	51	26	1.531	10.74	105	135.49	2.87
	20-40	6.95	31	48	21	1.583	9.077	75	119.65	
	40-60	6.99	41	41	18	1.594	1.563	77	100.55	
C	0-20	8.22	16	61	23	1.668	9.826	88	158	3.03
	20-40	8.37	21	59	20	1.503	8.904	102	153	
	40-60	8.16	35	48	17	1.412	10.74	128	165	

BD – Bulk density; TOC - Total organic carbon; EC - Electrical conductivity; CEC - Cation exchange capacity; pH_{pzc} –pH value at point of zero charge .

3. Thermodynamics of Cr(VI), As(V) and Pb Adsorption on the Soils

Three soils from 0-20 cm were used for adsorption experiment. The adsorption thermodynamics experiment of Cr(VI), As(V) and Pb was carried out for 24h with initial metal concentrations ranging from 0.05 to 100 mg/L by dilution of 1000 mg/L $K_2Cr_2O_7$, $Na_2HAsO_4 \cdot 7H_2O$ and $Pb(NO_3)_2$ (all analytical grade reagent) stock solution with distilled-deionized water (dd H_2O). Soil samples were weighed precisely and transferred into beakers and submerged in 500mL solution with different metal concentrations. The suspensions were stirred continuously with magnetic stirrers, maintaining pH value at 6.0 ± 0.1 by adding 1 mmol/L HNO_3 and 1 mmol/L NaOH. After 24h, the soil suspensions were centrifuged for 30 min at 5000 rpm, and the supernatants were collected for measurement of metal concentration. The concentrations of Pb in the extractants were measured using the flame atomic absorption method or the graphite furnace atomic adsorption method (AAS 700 equipped with GFAAS, Perkin Elmer, USA). Cr(VI) was measured by 1, 5-diphenyl carbazide spectrophotometric method [13] using a WFJ2-7200 Spectrophtometer (Unico (Shanghai) Instruments Co., Ltd, China). As(V) was determined by flow injection hydride generation atomic absorption spectrometry [14].

In this study, Langmuir [15], Freundlich [16] and linear adsorption isotherms were used to describe the data of adsorption thermodynamics. Langmuir isotherm is expressed as equation (1):

$$\Gamma = \frac{\Gamma_{\max} K_L [M]}{(1 + K[M])} \tag{1}$$

where Γ is the adsorption amount of metal on soil (mg/kg); Γ_{max} is the maximum adsorption amount of metal on soil (mg/kg); K_L is the Langmuir equilibrium coefficient (L/mg); and $[M]$ is the concentration of free metal ion in solution (mg/L).

Freundlich isotherm can be described as the following equation (2):

$$\Gamma = K_F[M]^{1/n} \tag{2}$$

where K_F is the Freundlich adsorption coefficient (L/kg); and n is a dimensionless parameter that varies between 0 and 1, depending on the character of adsorbent.

When partition of contaminant between solid and liquid phase is constant, adsorption is described by the following linear adsorption isotherm (3):

$$\Gamma = K_l[M] \tag{3}$$

where K_L is distribution coefficient (L/mg).

Sorption is important for the transport and fate of pollutants. Adsorption data and adsorption isotherms of Cr(VI), As(V) and Pb on the three soils are shown in Figure 1. Estimated Langmuir and Freundlich parameters for Pb and As(V) adsorption and linear parameters for Cr(VI) adsorption are listed in Table 2. Regression coefficients (R^2) showed that the data of Pb and As(V) adsorption followed Langmuir and Freundlich adsorption isotherms and the data of Cr(VI) adsorption followed linear isotherm in the concentration range investigated here.

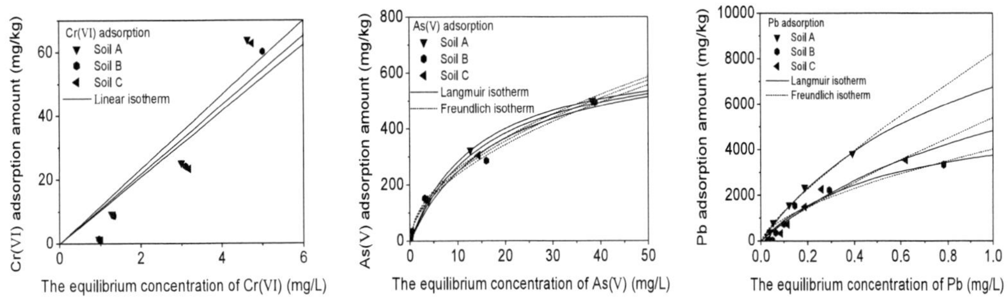

Figure 1. Langmuir and Freundlich isotherms for Pb, As(V) adsorption and linear isotherms for Cr (VI) adsorption on the three soils.

As shown in Figure 1, for the three soils, adsorption amounts of the three metals decreased in the order: Pb > As(V) > Cr(VI). From K in Table 2, it can be found that the adsorption amounts of Pb were about 2-3 magnitude grade of that of Cr(VI) and As(V). The adsorption amounts of As(V) were about 10 times of Cr(VI). Different charge characteristic led to different adsorption ability between Cr(VI)/As(V) and Pb. Positively charged Pb was adsorbed by negative sites on the soil surface. Anionic Cr(VI)/As(V) could be only adsorbed by goethite, FeO(OH), aluminum oxides and other soil colloids with positively charged surface sites [17]. The pH_{pzc} of the three soils were 2.97, 3.03, 2.87 respectively as listed in Table 1, which indicated that soil surfaces were negatively charged. Therefore, Pb^{2+} was adsorbed greatly, which might result in reduced threat of Pb contamination to groundwater. However, only limited amounts of Cr(VI) were adsorbed by soil, therefore, compared with Pb, the mobility of Cr(VI) in soil/groundwater system was much higher. The adsorption amounts of Cr(VI)/As(V) on the each soil were almost equivalent. The adsorption amounts of Pb on soil A were little greater than that on soil C and less Pb was adsorbed by soil B.

Table 2. Estimated Linear parameters for Cr(VI) adsorption and Langmuir, Freundlich parameters for As(V), Pb adsorption on the three soils

Equation	Parameter		Soil A		Soil B		Soil C	
			Estimated value	Error	Estimated value	Error	Estimated value	Error
Linear	Cr(VI)	K_l	11.674		10.403		10.870	
		R^2	0.9300		0.9420		0.9122	
Langmuir	As(V)	Γ_{max}	688	31	705	163	689	32
		K_L	0.07	0.008	0.054	0.030	0.069	0.008
		R^2	0.9980		0.9692		0.9980	
	Pb	Γ_{max}	12902	6022	5813	1549	10806	5910
		K_L	0.23	0.14	0.38	0.18	0.17	0.12
		R^2	0.9776		0.9426		0.9548	
Freundlich	As(V)	K_F	79.7	13.5	70.0	10.7	74.3	5.0
		n	0.51	0.05	0.53	0.05	0.52	0.02
		R^2	0.9897		0.9931		0.9985	
	Pb	K_F	1239	81	777	118	794	101
		n	0.86	0.10	0.67	0.13	0.84	0.14
		R^2	0.9688		0.9040		0.9401	

K_l is distribution coefficient;

K_L is the Langmuir adsorption coefficient (L/mg); Γ_{max} is the maximum adsorption amount of metal on soil (mg/kg);

K_F is the Freundlich adsorption coefficient (L/kg); n is a dimensionless parameter that varies between 0 and 1

4. TRANSPORT OF CR(VI), AS(V) AND PB IN SOIL UNSATURATED ZONES

Polyvinyl chloride (PVC) columns, with an inner diameter of 18.5 cm and a length of 60 cm, were packed with air-dried soil on the basis of original depth layer and bulk density in the field. The small holes for sampling effluents were drilled at depths of 20, 40, and 60 cm, respectively. Soil A was used for Cr(VI) and As(V) leaching tests and Soil C was used for Pb leaching tests. Before the leaching experiment, soils were saturated with dd H_2O at ionic strength of 0.01mol/L $CaCl_2$ for 24 h. Three soil columns were leached with the solution containing 10 mg/L Cr(VI), 10mg/L As(V) and 100mg/L Pb respectively (pH 6.0) at room temperature (18-20° C) and the influent flow was controlled at 2.8 mL/min. At certain intervals, the effluents were collected for the measurement of heavy metal concentration.

Successful description of column tests requires prediction of the breakthrough curves (BTC) for the effluents. The Thomas equation [18], as shown in eq 1 has been widely used for describing BTC.

$$\frac{C_t}{C_0} = \frac{1}{1 + e^{(K_T(q_0 m - c_0 V))/Q}} \tag{4}$$

where C_t is element concentration in the effluent (mg/L); C_0 is initial concentration of element (mg/L); K_T is Tomas rate constant (mL/(min• mg)); q_0 is maximum amount of element that can be absorbed by soil (mg/g); m is mass of the adsorbent (g); V is effluent volume (L); Q is flow rate (mL/min).

Yoon and Nelson [19] developed a relatively simple model to describe the adsorption and breakthrough of adsorbate vapors or gases with respect to activated charcoal. This model is not only less complicated than other models, but also requires no detailed data concerning the characteristics of adsorbate, the type of adsorbent, and the physical properties of adsorption bed.

The Yoon and Nelson equation in regards to a single component system is expressed as:

$$\frac{C_t}{C_0} = \frac{e^{(K_{YN}t - \tau K_{YN})}}{1 + e^{(K_{YN}t - \tau K_{YN})}} \tag{5}$$

where K_{YN} is rate constant (L/h); τ is time required for 50 % adsorbate breakthrough (h) and t is sampling time (h).

The results obtained from soil column leaching tests are shown in the form of BTCs fitted by the Thomas model and the Yoon-Nelson model (Figure 2 and Figure 3). Each regressed curves indicated that they were all well fitted with R^2 ranging from 0.828 to 0.992 (Table 3). The relative concentrations (C_t/C_0) of the three heavy metals in leachates were increased with leaching time and effluent volume. An apparent plateau of the BTC (C_t/C_0 0.8 ~ 1.0) appeared for each element in different sampling depth except Pb at 60 cm depth, which showed that the soils almost reached saturated adsorption to the heavy metals. At later periods of tests, the column for Pb leaching tests was blocked and leachates could not be collected, so Pb in effluents at 60 cm sampling depth had not reached the apparent plateau of BTCs.

As shown in Figure 2 and Figure 3, steep BTCs were obtained for Cr(VI) leaching because of its lesser adsorption capacity by soil. The breakthrough point for Cr(VI) leaching appeared at 5 h, 30 h and 60 h for the sampling depth of 20, 40, 60 cm respectively and maximum concentration in effluent appeared at 40, 96, 320 h respectively. In comparison to Cr(VI), the breakthrough point of As(V) was significantly delayed, which appeared at 23 h, 95 h and 140 h for the sampling depth of 20, 40, 60 cm respectively. The time needed to reach an apparent plateau of BTCs (C_t/C_0 ~1.0) was nearly 900 h, 1500 h and 3000 h longer for the sampling depth of 20, 40, 60 cm respectively. However, for Pb, the concentration in leachates increased more slowly and the breakthrough point appeared later (about several hundred to several thousand of hours) than that of Cr (VI) and As(V).

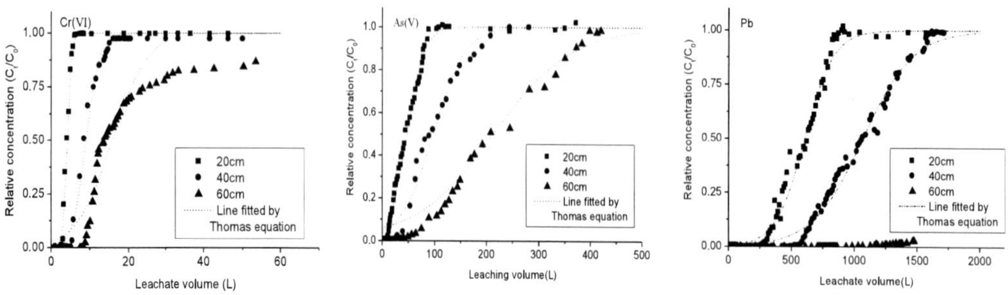

Figure 2. The predicted breakthrough curves obtained according to the Thomas model for Cr(VI), As(V) and Pb leaching tests.

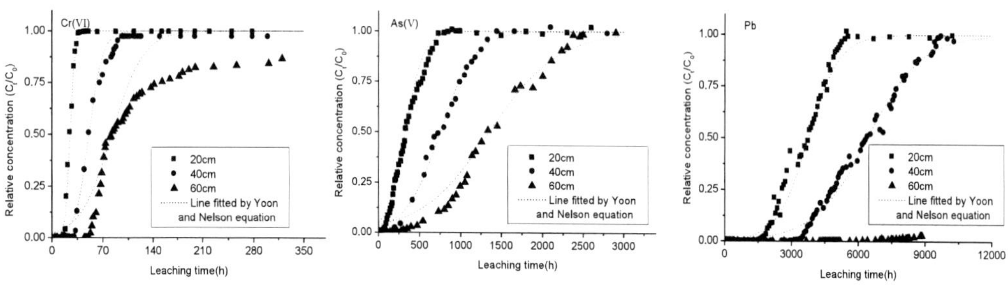

Figure 3. The predicted breakthrough curves obtained according to the Yoon and Nelson model for Cr(VI), As(V) and Pb leaching tests.

The fitted values of K_T and q_0 are presented in Table 3. With depth of unsaturated zone increased, the values of K_T decreased and q_0 was almost constant for each element. The predicted $q_{0,theo}$ and experimental $q_{0,exp}$ (soil capacity) obtained for each element showed negligible difference, respectively. The modeled values of $\tau_{0,theo}$ are also almost in agreement with the measured values $\tau_{0,exp}$. This indicated that data from column tests was well fitted by the Thomas and Yoon-Nelson model. According to the value of K_T (presenting the transport rate of pollutants in the soil zones), the rates of the three elements transport in the soil zones followed the order: Cr(VI) > As(V) > Pb. Differences in mobility and accumulation of the three elements in the soil zones were mainly related to soil adsorption character and metal ion properties [20].

Effluent Cr was entirely Cr(VI), and no Cr(III) was ever detected during leaching tests. The lack of effluent Cr(III) suggested that Cr(III) transport from hazardous wastes containing Cr in the soil zones was of minor importance, most likely because soil was an efficient sorbent for Cr(III), which was in agreement with the findings of Jardine et al. [21]. Likewise, no As(III) was detected in the effluents.

Table 3. Parameters predicted from the Thomas model and the Yoon-Nelson model for Cr(VI), As(V) and Pb leaching tests

Heavy metal		Cr (VI)			As (V)			Pb		
Depth(cm)		20	40	60	20	40	60	20	40	60
Thomas Parameters	K_T	537	165	62	17.3	9.16	3.71	0.24	0.14	–
	$q_{0,theo}$	0.0084	0.009	0.0105	0.095	0.097	0.142	12.13	10.68	–
	$q_{0,exp}$	0.0112	0.0122	0.01	0.179	0.181	0.183	10.48	10.36	–
	R^2	0.992	0.989	0.892	0.988	0.975	0.804	0.986	0.988	–
Yoon-Nelson Parameters	K_{YN}	0.326	0.108	0.047	0.0092	0.0048	0.0024	0.0014	0.0008	–
	$\tau_{0,theo}$	25	54	91	348	760	1351	3644	6423	–
	$\tau_{0,exp}$	24	50	82	325	780	1300	3750	6400	–
	R^2	0.992	0.985	0.849	0.982	0.992	0.828	0.985	0.987	–

K_T is Thomas rate constant (mL/(min• mg)); q_0 is maximum amount of element that can be absorbed by soil (mg/g);
K_{YN} is Yoon-Nelson rate constant (L/h); τ_0 is time required for 50 % adsorbate breakthrough (h).

5. Concentrations of Cr, As and Pb in Original Soils and Leached Soils

After the column leaching tests, the three columns were separated, extruded (~ 20 cm each). All soils from the specified depth range were thoroughly mixed, dried and ground. Original soils and leached soils were analyzed for element concentration and speciation.

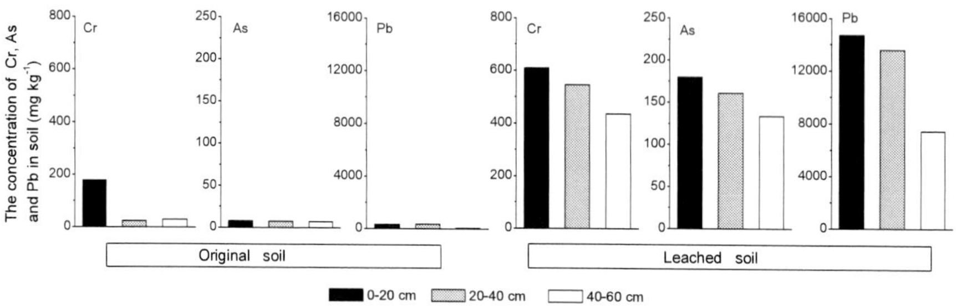

Figure 4. Total concentrations of Cr(VI), As(V) and Pb in soils with depth in original soils and leached soils.

The element concentrations in original soils and leached soils are presented in Figure 4. Element concentrations in the soils decreased with depth. For original soils, heavy metals had been accumulated in the soils due to the emissions of pollutants into the studied area in recent years. The concentrations of Cr, As and Pb increased dramatically after leaching tests. The concentration of each element in leached soils was slightly less in deeper soil zones, which might be due to the lower contents of clay particle, TOC and CEC in deeper soil zones. Concentrations of the three elements in leached soils were all remarkably higher than the third level of the Environmental Quality Standard for Soils [13]. Due to different enrichment ability of soil to different characteristic elements, there were great variations of each element concentration in soils, which followed the order of Pb > Cr > As.

6. Fractions of Cr, As and Pb in Original Soils and Leached Soils

Table 4. Fractions of the three heavy metals by selective sequential extraction

Fraction	Extractants
F1	Pb was extracted with 1 mol/L $MgCl_2$ (pH = 7.0), Cr and As were extracted with 1 mol/L KCl and 0.1 mol/L KH_2PO_4
F2	1 mol/L NaOAc adjusted to pH 5.0 with HOAc
F3	0.04 mol/L $NH_2OH{\cdot}HCl$ in 25 % (v/v) HOAc
F4	30 % H_2O_2+3.2 mol/L NH_4OAc in 20 % HNO_3
F5	HNO_3+$HClO_4$

A selective sequential extraction technique [11] was used to analyze the chemical fractions of Cr, Pb and As in original soils and leached soils. A minor modification of the selective sequential extraction technique was used for determination of exchangeable Cr and As in soils [22]. The method of selective sequential extraction was listed in Table 4.

For Cr, As and Pb in the original soils and leached soils, the relative concentrations of five chemical fractions are shown in Table 5. For the original soils, Table 5 showed that a significant fraction of Cr in the soils was bound in the residual fraction, and the other four fractions accounted < 3 % of the total Cr. Some studies showed that Cr was mostly present in residual fraction of soils at the horizon or profile scale [23] and are widely considered to have a very low geochemical mobility [24]. Pb and As were associated dominantly with the residual fraction, followed by the organic fraction. In general, the exchangeable and carbonate-bound Pb and As accounted < 7 % of the total Pb and As, respectively.

Table 5. The percentage distribution of Cr, As and Pb fractions in original soil and leached soil with depth

Heavy metal	Depth (cm)	Original soil					Leached soil				
		F1	F2	F3	F4	F5	F1	F2	F3	F4	F5
Cr	0-20	0.40	1.16	0.49	15.01	82.94	10.5	25.6	17	24.6	22.3
	20-40	0.51	1.16	0.86	14.27	83.19	9.2	13.6	19	28.5	29.7
	40-60	1.12	1.35	0.79	14.57	82.18	6.3	8.8	24	26.3	34.6
As	0-20	0.33	0.55	3.39	1.39	94.31	2.3	17.6	18.6	28.3	33.1
	20-40	0.40	0.72	5.06	2.24	91.57	1.0	9.3	29.5	31.3	28.9
	40-60	0.39	0.89	4.25	2.19	92.31	0.4	1.6	18.4	26.2	53.5
Pb	0-20	0.97	1.6	10.4	34.1	52.93	4.1	43.8	14	13.3	24.8
	20-40	0.03	0.45	13.8	13.1	72.62	6.0	20.1	15.6	22	36.3
	40-60	0.07	0.47	1.72	16.98	80.76	7.3	17.2	12.3	21.9	41.3

As the results show, there are some variations in the proportions of five fractions for the three elements after column leaching tests. The relative concentrations of other four fractions of the elements except residues were all increased. Because leaching time was greatly shorter than the time for soil-heavy metal interactions under the natural condition, it was very difficult for heavy metals from anthropogenic activities to enter in the firm mineral crystal lattice in soil to form the residual fraction in such a short time. After entering soils, a part of heavy metals were in the form of exchange fraction, and with the times, exchange faction would be adsorbed, mineralized by soil and transformed into other factions. Considering the absolute concentration, the five fractions of each element were all increased due to element addition into the soils. For Cr, the relative concentrations of the organic and Fe-Mn oxide fractions were significantly increased, especially the organic fraction (average 36 %). Likewise, through studying the fractions of heavy metals in sandy and loessial soils, Han and Banin [25] found that, after its addition to the soil, Cr was bound to the organic matter fraction. The organic matter, Fe-Mn oxide and carbonate fraction of As were increased, obviously. Lund and Fobian [26] found that As was retained by organic matter and Fe, Al and Mn oxides in contaminated industrial sites. For Pb, the carbonate fractions were increased greatly. The primary mechanism for Pb retention in clay soils was through precipitation of the metal ions with carbonates and amorphous oxides or hydroxides [27]. Yong and

Phadungchewit [28] proved that the presence of carbonates in a soil contributed significantly to the retention capacity of the soil.

The exchangeable fraction was the first to be brought into solution and was considered to be easily available for plant uptake. The carbonate fraction was susceptible to pH changes, and the Fe-Mn oxide fraction was unstable under low pH conditions. The organic fraction could be degraded under oxidizing conditions, and the residual fraction was not bioavailable since it was not expected to be solubilized over a reasonable period of time on natural conditions [11]. For Cr, As and Pb, the exchangeable fractions with high bioavailability and mobility were increased on average by 2.2%, 0.9%, 5.4% respectively, which indicated the increase of direct risk of Cr, As and Pb to soil/groundwater system. The amounts of non-residual fractions represent the amounts of potentially active trace elements [26]. In general, the high proportion of trace elements in non-residual fractions of the soils may suggest greater contribution of anthropogenic elements. The non-residual fractions of Cr, As and Pb in the leached soils after leaching tests were increased averagely by 37.57 %, 25.24 % and 34.64 % respectively. The increase of non-residual fractions elucidated the increase of potential risk of Cr, As and Pb to soil/groundwater system. From the results above we can see, if Cr(VI) was dissolved from hazardous residues , it would transport down with great speed. At the same time, its fractions also transformed, especially active factions were increased greatly, which indicated that once Cr(VI) was leached from hazardous residues, it was hazardous to soil and groundwater systems.

7. Valences of Cr and As in Original Soils and Leached Soils

Cr and As valences were determined using solvent extraction methods [12, 29]. For Cr(VI) extraction, 1.000 g soil and 50ml extractants (0.28 M Na_2CO_3 + 0.5 M NaOH) were added into a 200 ml beaker and mixed for 10 min. The beaker was heated to 95 °C for an hour. The extraction solution was acidified with nitric acid to a pH of 7-8 and then diluted to 100 ml for concentration measurement. Cr(VI) in extraction solution was measured by 1, 5-diphenyl carbazide spectrophotometric method [13] using a WFJ2-7200 Spectrophtometer (Unico (Shanghai) Instruments Co., Ltd, China). Total Cr were analyzed using an acid digestion method according to the Chinese standard, GB15618 [13] and measured by GFAAS.

As in soil was effectively extracted with concentrated hydrochloric acid (HCl) for about 30 min at 25 °C without changing its speciation [29]. As species (As(V), As(III)) was determined by flow injection hydride generation atomic adsorption spectrometry with variable concentrations of sodium tetrahydroborate [14].

Redox of Cr was governed by many factors including aeration reactions, soil moisture content, wetting and drying, iron and manganese status, microbial activity, organic matter, pH, and availability of electron donors and acceptors [30]. The organic matter present in soil is expected to reduce soluble chromate (CrO_4^{2-}) to insoluble Cr(III) oxide (Cr_2O_3) [31]. In this study, the proportions of Cr(III) in different depth of soil profiles were increased by about 6.0 % percent after column leaching tests. The proportions of As(III) in different depth of soil profiles after tests were increased by about 5.6 %. According to the standard electrode potential (φ^0) of redox reaction, φ^0 of Cr(VI) redox (+ 1.33 V) was greater than that of As(V)

(+ 0.56 V), so under the same condition, the degree of Cr (VI) reduction was significantly more than that of As(V). The increased Cr(III) and As(III) tended to be adsorbed by soils, which reduced the movement of them into groundwater.

Considering the absolute mass, the concentrations of Cr(III) were increased by about 400 mg/kg, which suggested that Cr(III) was potentially precipitating from solution after hundreds of hours of reaction. Cr(III) is considered to be an essential trace element for the proper functioning of living organisms, whereas Cr(VI) may exert toxic effects on biological systems. Also, Cr(VI) compounds are generally more soluble, mobile, and bioavailable in the environment compared with Cr(III) compounds. So, Cr(III) in hazardous residues was not easily dissolved, reversely, Cr(VI) was movable. In the process of transporting down, the reduction of Cr(VI) reduced its mobility and toxicity.

8. Transport and Speciation Transformation of Cr(VI) and As(V) in an Aquifer

A PVC box of 0.9 × 0.6 × 0.4 m size was drilled with 24 test holes at one side of: six horizontally, four vertically. Sand of 0.1-2 mm of depth was laid in the PVC box to simulate an aquifer with a depth of 28cm. At the top of aquifer, a 14 cm thick soil layer was set to simulate the unsaturated zone layer. Horizontal flow rate of groundwater was controlled at 2 cm/d. A simulative solution with 2 mg/L Cr(VI) and 2 mg/L As(V) was added to the sandbox with a flow rate of 500 mL/d. At certain intervals, the effluents from the 24 holes were collected for the measurement of Cr(VI) and As(V) concentration.

Cr(VI) can transport great distances in groundwater owing in part to its high solubility. The results of sanbox tests showed that Cr (VI) and As (V) concentrations in effluents increased as time increased. In comparison with As(V), Cr (VI) transported and diffused faster. Cr (VI) and As (V) transported in the aquifer not only in a vertical direction, but also a horizontal direction. So at the place near the deposit site of hazardous residues, the groundwater was easily polluted by Cr(VI). In the aquifer, speciation of Cr and As was adsorbed by soil and sand and transformed. As time increased, relative concentrations of Cr and As residue fraction were reduced gradually, but Cr bound to Fe-Mn oxides was increased and the other three fractions changed less. For As, relative concentration of exchangeable fraction was increased, the other three fractions remained almost unchanged. In terms of absolute concentration, each fraction of Cr and As was increased. Reduction reaction of Cr (VI) and As (V) occurred in box, but did not change significantly with time. It can be concluded from discussion above that Cr (VI) form hazardous residues can pass through soil unsaturated zone by leaching and enter an aquifer. In the aquifer, Cr (VI) transports faster, but speciation changes less.

Conclusion

Soil column leaching tests were carried out in order to investigate transport and fate of Cr(VI) in subsurface. The results showed that owing to lesser Cr(VI) adsorption by soil, if hazardous residues containing Cr(VI) were disposed on ground without anti-seepage or inertization measures, Cr(VI) would be easily leached out from hazardous residues and transported down into the underground environment. Compared with Pb^{2+} and anions of As, Cr(VI) transported in soil unsaturated zone the fastest. The transport rate of the three heavy metals decreased in the order of Cr(VI) >As(V) > Pb. The results of selective sequential extraction showed that bioavailable exchangeable-fraction and active non-residual fractions of Cr after leaching tests were increased, which elucidated the increased direct and indirect risk of Cr to soil/groundwater system. Compared with As and Pb, non-residual fractions of Cr in leached in soil was greatest, which indicated that Cr was active and bioavailable among the three elements. Under long-time water saturated conditions, the reduction reactions occurred for Cr(VI) and As(V), and the degree of Cr (VI) reduction was significantly more than that of As(V). The reduction reduced the transport of Cr(VI) into groundwater. The result of sandbox tests showed that when entering the aquifer, Cr(VI) transported great distances owing in part to its high solubility. In comparison with As(V), Cr(VI) transported and diffused faster, but speciation changed less. Through this study it can be summarized that Cr(VI) from hazardous residues can pass through soil unsaturated zone by leaching and can enter an aquifer. In the process of transportation, active and bioavailable fractions of Cr were increased. These factors together increased the threat of Cr(VI) to the soil and underground water system. So, it is suggested here that hazardous residues containing Cr(VI) must be dealt with by anti-seepage or inertization measures to prevent Cr(VI) being leached out.

REFERENCES

[1] Chung, J., Burau, R. G., & Zasoski, R. J. (2001). Chromate generation by chromate depleted subsurface materials. *Water, Air, Soil Pollut.*, 128, 407–417.

[2] James, B. R. (1996). The challenge of remediation chromium-contaminated soil. *Environ. Sci. Technol.*, 30, 248–251.

[3] Handa, B. K. (1988). Occurrence and distribution of chromium in natural waters of India, in *Chromium in Natural and Human Environments*, Nriagu, J.O. and Nieboer, E., Eds., Wiley Interscience, New York.

[4] Richard, F. C., & Bourg, A. C. M. (1991). Aqueous geochemistry of chromium: a review. *Water Res.*, 25, 807–816.

[5] James, B. R., & Bartlett, R. J. (1983). Behavior of chromium in soils: V. Fate of organically complexed Cr(III) added to soil; VI. Interaction between oxidation-reduction and organic complexation; VII. Adsorption and reduction of hexavalent forms. *J. Environ. Qual.*, 12, 169–181.

[6] Bartlett, R. J. (1991). Chromium cycling in soils and water: links, gaps and methods. *Environ. Health Prespective*, 92, 17–24.

[7] Palmer, C. D., & Wittbrodt, P. R. (1991). Processes affecting the remediation of chromium- contaminated sites. *Environ. Health Perspectives*, 92, 25–40.

[8] Korte, N. E., Skopp, J., Fuller, W. H., Niebla, E. E., & Aleshii, B. A. (1976). Trace element movement in soils: influence of soil physical and chemical properties. *Soil Sci.*, 122, 350–359.

[9] Ritchie, G. S. P. & Sposito, G. (1995). Speciation in soil, in *Chemical Speciation in the Environment* Ure, A.M. and Davidson, C.M., Eds., Blackie Academic and Professional, Glasgow, Scotland, 201–233.

[10] Peakall, D., & Burger, J. (2003). Methodologes for assessing exposure to metals: speciation, bioavailability of metals, and ecological nost factors. *Ecotoxicol. Environ. Safty,* 56, 110−121.

[11] Tessier, A., Campbell, P. G. C., & Bassoon, M. (1979). Sequential extraction procedure for the speciation of particulate trace metals. *Anal. Chem.,* 51, 844−851.

[12] Banks, M. K., Schwab, A. P., & Henderson, C. (2006). Leaching and reduction of chromium in soil as affected by soil organic content and plants. *Chemosphere,* 62, 255−264.

[13] National Environmental Protection Agency of China (1995). *Environmental Quality Standard for Soils* (GB 15618−1995).

[14] Sigrist, M. E., & Beldoménico, H. R. (2004). Determination of inorganic arsenic species by flow injection hydride generation atomic adsorption spectrometry with variable sodium tetrahydroborate concentrations. *Spectrochim. Acta Part B.,* 59, 1041−1045.

[15] Chiron, N., Guilet, R., & Deydier, E. (2003). Adsorption of Cu(II) and Pb(II) onto a grafted silica: isotherms and kinetic models. *Water Res*., 37, 3079–3086.

[16] Limousin, G., Gaudet,J.P., Charlet, L., Szenknect, S., Barthès, V., & Krimissa, M. (2007). Sorption isotherms: A review on physical bases, modeling and measurement. *Appl. Geochem*., 22, 249–275.

[17] Lund, U., & Fobian, A. (2004). Pollution of two soils by arsenic, chromium and copper. *Geoderma*, 49, 83−103.

[18] Thomas, H. C. (1994). Heterogeneous ion exchange in a flow system. *J. Am. Chem. Soc.,* 66, 1664−1666.

[19] Yoon, Y. H., & Nelson, J. H. (1984). Application of gas adsorption kinetics. I. A theoretical model for respirator cartridge service time. *Am. Ind. Hyg. Assoc. J.,* 45, 509−516.

[20] Alloway, B. J. (1990). *Heavy metals in soils*. Blackie & Son, New Jersey.

[21] Jardine, P. M., Fendorf, S. E., Mayes, M. A., Larsen, I. L., Brooks, S. C., & Bailey, W. B. (1999). Fate and transport of hexavalent chromium in undisturbed heterogeneous soil. *Environ. Sci. Technol.,* 33, 2939−2944.

[22] Barlett, R. J., & James, B. R. (1996). *Chromium. In: Spark, D.L. (Ed.), Methods of Soil Analysis: Part 3. Chemical Methods*. SSSA, ASA, Madison, WI, USA.

[23] Wilcke, W., & Amelung, W. (1996). Small-scale heterogeneity of aluminium and heavy metals in aggregates along a climatic transect. *Soil Sci. Soc. Am. J.,* 60, 1490−1495.

[24] Castro, M. T. T., Suarez, M. J. G., Gonzalez, A. P., & Vieira, R. S. (1998). *Heavy metal status of a soil developed over serpentine: statistical variability of total and extractable contents*. 16th World Congress of Soil Science. Montpellier, France.

[25] Han, F. X., & Banin, A. (1999). Long-term transformation and redistribution of potentially toxic heavy metals in arid-zones soils: II Incubation at the field capacity moisture content. *Water Air Soil Pollut.,* 114, 221−250.

[26] Lund, U., & Fobian, A. (1991). Pollution of two soils by arsenic, chromium and copper. *Geoderma,* 49, 83−103.

[27] Griffin, R. A., Shimp, J. D., Steele, J. D., Ruch, R. R., White, W. A., & Hughes, G. M. (1976) Attenuation of pollutants in municipal landfill leachate by passage through clay. *Environ. Sci. Technol.,* 10, 1262−1268.

[28] Yong, R. N., & Phadungchewit, Y. (1993). pH influence on selectivity and retention of heavy metals in some clay soil. *Can. Geotech. J.,* 30, 821−833.

[29] Chappell, J., Chiswell, B., & Olszowy, H. (1995). Speciation of arsenic in a contaminated soil by solvent extraction. *Talanta,* 4, 323−329.

[30] Kamaludeen, S. P. B., Megharaj, M., Juhasz, A. L., Sethunathan, N., & Naidu, R. (2003). Chromium-microorganisms interactions in soils: implications to remediation. *Rev. Environ. Contam. Toxicol.,* 178, 93−164.

[31] Bloomfield, C., & Pruden, G. (1980). The behavior of Cr(VI) in soil under aerobic and anaerobic conditions. *Environ. Pollut.,* 23, 103–114.

In: Management of Hazardous Residues Containing Cr(VI) ISBN: 978-1-61470-637-3
Ed: Maria José Balart Murria

Chapter 2

WASTE MANAGEMENT AND MINIMISATION OF THE RESIDUES CONTAINING CR(VI)

Tiangui Wang*[*] *and Yanmin Shen

School of Chemistry and Chemical Engineering, Henan University of Technology, Zhengzhou, China

ABSTRACT

The researches and practices for the minimization and management of the residues containing Cr(VI) was reviewed. The emphasis focuses on the minimization of COPR (chromite ore processing residues), mainly involving zero waste process or green manufacturing processes; low or no-lime processes; harmlessness and utilization. Management and utilization of ferrochrome slags, tanned leather wastes and galvanic sludges were also included.

1. INTRODUCTION

Chromium (VI) is now well known to be harmful to human beings and other organisms and has been designated as a carcinogen by The World Health Organization. Therefore, contamination of the environment caused by Cr (VI) has been of high concern all over the world. Unfortunately, a huge amount of wastes containing Cr (VI) is released into the environment each year, which include fumes containing Cr (VI) and Cr (VI) contained in wastewater and solids.

Chromium exists in nature mainly in the Cr(III) form. Hence, solids containing Cr (VI) come from various production activities, including processing of natural minerals containing chromium and use of chromium products. Two major sources of solids containing Cr (VI) are ferrochrome slag and chromite ore processing residues (COPR). Others include stainless steel slag, electroplate wastes, leather tanning wastes, municipal solid waste incinerator bottom ash and so on.

[*] E-mail: tgwang2006@126.com

2. Ferrochrome Slag

About 2.5 tons of chromite ore and 0.7 ton of reductant are consumed for every ton of FeCr produced. Fluxing agents, such as silica, are also used in the production process. Slags from different plants and processes have different chemical composition. According to Holappa and Xiao [1], chromium content of FeCr slags is typically 5% to 10%, of which a significant part is found as metallic particles dispersed in the slag. Therefore, recovery of metal from slags has been practiced [2]. Theoretically, Cr(VI) should not occur in ferrochrome slag because FeCr production is conducted in reduction circumstances with the existence of reductants. However, as Cr(III) is easily oxidized to Cr(VI) in alkaline mediums at elevated temperatures, Cr(VI) occurring in slag is inevitable during slag cooling and processing. If the slag experiences hexavalent chromium reduction before its returning to the slag dump, Cr(VI) content will be bearable and it will be safe to use the slag as construction material. Although some leaching tests of the ferrochrome slag show that the chromium content is high (1-3%), the leachability of Cr was only 0.03±1.7% of the total content; and a large part of which was in trivalent form, Cr(III) [3]. Currently, ferrochrome slag has been used for road construction. Whether the slag should be classified as waste, especially as hazardous waste, is arguable [4, 5].

Stainless steel slag has similar properties to ferrochrome slag. Slags produced by right methods can be considered environmentally friendly raw materials and can be used for many purposes [6].

3. COPR

Chromite ore processing residues typically contains 1-3% Cr(VI) and is the most hazardous Cr(VI) containing solid. Historically, a large amount of COPR has been discharged into the environment. Therefore, minimization and management of COPR is of concern in the countries where production of chromium chemicals result in the releases of COPR, such as China, the United States, Russia, India and Britain. Some technologies for minimizing COPR are discussed below.

3.1. Zero-Waste Technology

It is reported that a new 'zero-waste' process for chromium chemical and metal producers was developed at the University of Leeds [7]. It is said that the technique extracts 100% of the chromium and other valuable materials and can eliminate the need to dispose of hundreds of thousands of tonnes of chromium-containing waste in landfill sites, helping to meet increasingly stringent regulations worldwide and the technique also re-processes waste from chromium chemical's production, recovering all of the metal and other saleable by-products, but no technical details were given.

Zhang et al. [8] also reported a green manufacturing process of chromium compounds. Unlike the traditional high temperature roasting process, this process uses concentrated solutions of KOH or NaOH as reaction mediums and no other materials, except chromite and air, are needed. It claimed that all the elements in the raw materials can be separated and recovered, therefore, it achieves the zero emissions of chromium-containing residues and eliminates the pollution problem at the source. The conceptual design is shown in Figure 1 [9].

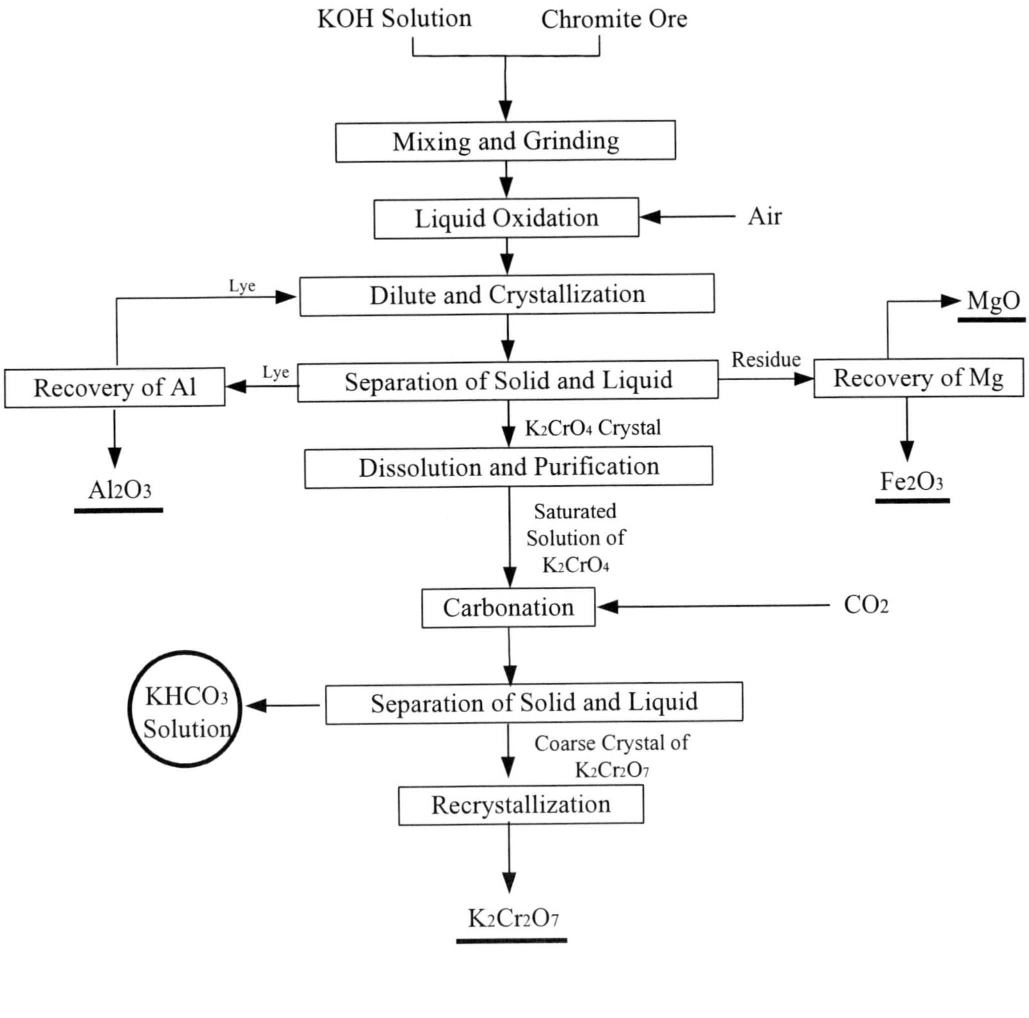

Figure 1. Schematic diagram of cleaner production process of chromates.

Yildiz and Sengil [10] also reported similar work. They reacted chromite with molten sodium hydroxide under oxidizing conditions to form sodium chromate. The fusion product was leached with methanol to remove the unreacted NaOH and then leached with water to obtain chromate solution. With this approach, both the amount and the Cr(VI) content of the residue can be greatly reduced.

3.2. Low Lime and No Lime COPR Processes

Lime-based processes, also called high-lime processes, remained exclusive as chromium extraction from chromite ore until the late 1950's/early 1960's when lime-free alternatives began to be developed. In the high-lime processes, limestone or dolomite was added to assist the operation so that up to 2.5-3.0 tons of COPR per ton of sodium dichromate were produced. Today virtually all producers in developed countries use lime-free extraction processes which produce much less COPR with much lower Cr(VI) content. However, most producers in developing countries, such as India, Pakistan and China, are still using high-lime processes [11]. Actually, apart from inclusion/exclusion of lime, details of the extraction process have not changed. It was, and still is, based on high temperature alkaline oxidation with the preferred alkali metal source being sodium carbonate. Nevertheless, due to much lower CaO and Cr(VI) content in COPR from no-lime processes, the treatment and disposal of the residues are much easier. In modern no-lime plants the water- insoluble post-extraction residue is processed further to reduce the residual Cr(VI) content to around 0.1–0.2% then passed to a chemical treatment stage prior to landfilling. Divalent iron- (e.g. $FeSO_4$) or sulphur-containing reducing agents are normally used and are capable of achieving residual Cr(VI) levels <1 ppm post-treatment prior to landfilling. Despite many years of researching alternatives, placement in landfills of the residues is still the world-wide industry standard today.

Lots of efforts have been taken to minimize, manage, utilize and treat COPR in China because China is the biggest COPR producer. The commercialization of the use of COPR to produce iron has been reported [12].

4. Other Solid Wastes

Chromium is used in electroplating to increase corrosion–resistance properties and improve wear qualities. Large amounts of chromium galvanic sludges come from the treatment of the galvanic process effluents. Although the sludges dominantly contain Cr(III), because there is a strong tendency and risk of Cr(III) being oxidized to Cr(VI), the management of the sludges is widely concerned. The methods of hydrometallurgical and pyrometallurgical have been used to recover chromium. Direct utilization of the sludges were also studied [13,14]. Another choice is to solidify and stabilize the chromium rich sludges by means of high temperature reactions [15–17]. In this case, the firing step is determinant to obtain a high immobilization degree of waste. The recycling of galvanic sludge as secondary raw material for the ceramic industry could help to reduce the amount of chromium containing waste.

Chromium salts are the most widely used chemical agents in the treatment of the skin in the leather industry. The skins treated with chromium salts exhibit good mechanical resistance, and a better hydrothermal resistance in comparison with the skins treated with vegetal substances. Unfortunately, processing one metric ton of raw hide only generates 200 kg of final leather product, but 250 kg of non-tanned solid waste, 200 kg of tanned waste, and 50,000 kg of wastewater. Thus, only 20% of the raw material is converted into leather, and more than 60% of the chromium is in the solid and liquid waste [18]. Similar to the electroplating, although the tanned wastes and the sludges from the liquid waste treatment hardly contain Cr(VI), because there is a strong tendency and risk of Cr(III) being oxidized to Cr(VI) [18-21], these wastes must be carefully managed [22,23]. Various utilization methods have been studied, such as incorporation of chromium-tanned leather residue into asphalt micro-surface layers to build road [24], using chrome-tanned leather shavings as a filler of butadiene–acrylonitrile rubber [25], producing protease, adsorbents, pigments and so on [26-34].

Conclusion

There are many kinds of residues containing Cr(VI). The fundamental measure to minimize these residues is to eliminate or substantially reduce the wastes from the original processes. Zero-waste technology has been developing and has been gaining emphasis in utilization. Various approaches to recycle or utilize the residues have also been developed. Currently, however, landfilling and road construction after detoxification are still the dominant practices to dispose the residues containing Cr(VI).

References

[1] Holappa, L.; Xiao, Y. J. S. Afr. Ins. Min. Metall. 2004, 429-438.
[2] Sripriya, R.; Murty, Ch. V. G. K. Int. J. Miner. Process. 2005, 75, 123– 134.
[3] Lind, B.B.; Fällman, A.-M.; Larsson, L.B. Waste Manage. 2001, 21, 255-264.
[4] Finnish Court Rules Ferrochrome Slag is not waste, http://www.icdachromium.com /industry_news.php?date=2006-05-09&s_date=May+2006#
[5] Oelofse, S. H. H.; Adlem, C. J. L. Hattingh, J. Overcoming bureaucratic obstacles to the re-use of metallurgical slag – A South African case study. http://www.infomine.com/publications/docs/Oelofse2007.pdf
[6] Ylipekkala, J. Quality management of chromium containing steel slag from melt phase to cooling, Thesis, Luleå University of Technology
[7] Zero-waste technology for chromium industry will reduce costs and protect the environment. http://www.albuw.ait.ac.th/Group_R/Chromium/report-4/abate.htm
[8] Zhang, Y.; Li, Z. H.; Qi, T.; Zheng, S. L.; Li, H.; Xu; H. Environ. Prog. 2005, 24, 44-50.
[9] Wang, T.; Xu, H. J. Hazard. Mater. 2005, B123, 176-180.
[10] Yildiz, K.; Sengil; I. A. Scand. J. Metall. 2004, 33, 251–256.
[11] Darrie, G. Environ. Geochem. Health, 2001, 23, 187–193.
[12] http://www.cn-fe.com/
[13] Andreola, F.; Barbieri, L.; Bondioli, F.; Cannio, M.; Ferrari, A. Lancellotti, M. I. J. Hazard. Mater. 2008, 156, 466–471.
[14] Costa, G.; Ribeiro, M. J.; Labrincha, J. A.; Dondi, M.; Matteucci, F.; Cruciani; G. Dyes Pigm. 2008, 78, 157-164.
[15] Magalhães, J. M.; Silva, J. E.; Castro, F. P.; Labrincha, J. A. J. Hazard. Mater. 2004, 106B, 169–176.

[16] Qian, G.; Yang, X.; Dong, Sh.; Zhou, J.; Sun, Y.; Xu, Y.; Liu, Q. J. Hazard. Mater. 2009, 165, 955-960.
[17] Magalhães, J. M.; Silva, J. E.; Castro, F. P.; Labrincha, J. A. J. Hazard. Mater. 2004, 106B, 139–147.
[18] Kolomaznik, K.; Adamek, M.; Andel, I.; Uhlirova, M. J. Hazard. Mater. 2008, 160, 514–520.
[19] Fernandes, H. R.; Ferreira, J.M.F. J. Eur. Ceram. Soc. 2007, 27, 4657–4663.
[20] Basegio, T.; Beck Leao, A. P.; Bernardes, A. M.; Bergmann, C. P. J. Hazard. Mater. 2009, 165, 604-611.
[21] Tahiri, S.; Albizane, A.; Messaoudi, A.; Azzi, M.; Bennazha, J.; Younssi, S. A.; Bouhria, M. Waste Manage. 2007, 27, 89–95.
[22] Erdem, M.; Ozverdi, A. J. Hazard. Mater. 2008, 156, 51–55.
[23] Bahillo, A.; Armesto, L.; Cabanillas, A.; Otero, J. Waste Manage. 2004, 24, 935–944.
[24] Krummenauer, K.; Andrade, J. J. O. Constr. Build. Mater. 2009, 23, 574–581.
[25] Przepiórkowska, A.; Chro´nska, K.; Zaborski, M. J. Hazard. Mater. 2007, 141, 252–257.
[26] Kumar, A. G.; Swarnalatha, S.; Sairam, B.; Sekaran,G. Bioresour. Technol. 2008, 99, 1939–1944.
[27] Oliveira, L. C. A.; Guerreiro, M. C.; Gonçalves, M.; Oliveira, D. Q. L.; Costa, L. C. M. Mater. Lett. 2008, 62, 3710–3712.
[28] Yılmaz, O. I.; Kantarli, C.; Yuksel, M.; Saglam, M.; Yanik, J. Resour. Conserv. Recycl. 2007, 49, 436–448.
[29] Lažau, R. I.; Păcurariu, C.; Becherescu, D.; Ianos, R. J. Eur. Ceram. Soc. 2007, 27, 1899–1903.
[30] Oliveira, L. C. A.; Goncalves, M.; Oliveira, D. Q. L.; Guerreiro, M. C.; Guilherme, L. R.; Dallago, G. R. M. J. Hazard. Mater. 2007, 141, 344–347.
[31] Basegio, T.; Haas, C.; Pokorny, A.; Bernardes, A. M.; Bergmann, C. P. J. Hazard. Mater. 2007, B137, 1156–1164.
[32] Rao, J. R.; Thanikaivelan, P.; Sreeram, K. J.; Nair, B. U. Environ. Sci. Technol. 2002, 36, 1372-1376.
[33] Song, J.; Tao, W.; Chen, W. J. Clean. Prod. 2008, 16, 591-597.
[34] Ferreira, D. F.; Almeida, P. D. Process of recovering leather residues and production of composite materials. WO 98018863, 1998-05-07.

In: Management of Hazardous Residues Containing Cr(VI)	ISBN: 978-1-61470-637-3
Ed: Maria José Balart Murria

Chapter 3

TECHNOLOGIES FOR TREATING CR(VI)-CONTAINING WASTES

Jason Scott[1]***, Yarong Li***[1,2]***, Gary Low***[2]***, Michelle Hone***[1] ***and Rose Amal***[1]

[1] School of Chemical Sciences and Engineering, The University of New South Wales, Sydney 2052, Australia. Email: jason.scott@unsw.edu.au

[2] Environmental Forensic and Analytical Science Section, Department of Environment, Climate Change and Water (NSW), Lidcombe 2141, Australia.

ABSTRACT

Within the environment, chromium is stable in either it's hexavalent (Cr(VI)) or trivalent (Cr(III)) oxidation states. Of these, Cr(VI) is more soluble, toxic, bioavailable and mobile than its trivalent counterpart. Anthropogenic activities such as electroplating and pigment manufacturing generate waste streams containing Cr(VI). The hazardous nature of Cr(VI)-containing wastes means they often require some form of pretreatment prior to disposal. This can include removal by physical processes such as adsorption on activated carbon, kaolinite, and flyash, by ion exchange, or by chemical, or more recently, biological treatment where it is typically reduced to the less soluble Cr(III). Cr(VI) may also be present in sludges or as a solid waste, for example, from the physical processing methods described above. In this instance, the waste may be subject to stabilisation/solidification treatment, designed to immobilise the Cr(VI) or Cr(III) in a solid matrix for subsequent disposal in landfill. This article highlights advances in techniques used to treat Cr(VI)-containing wastes, with an emphasis on biological treatment and stabilisation/solidification technologies.

1. INTRODUCTION

Chromium can exist in various oxidation states ranging from 0 to +6. Of these, the trivalent (Cr(III)) and hexavalent (Cr(VI)) forms are most often encountered as they are stable in the environment. Hexavalent chromium (either as chromate (CrO_4^{2-}) or as dichromate

($Cr_2O_7^{2-}$)) is the most soluble, toxic, bioavailable and mobile form of chromium. It is a potent and known carcinogen and mutagen and is toxic to many plants, aquatic animals and bacteria. Compared with Cr(VI), Cr(III) is less soluble and correspondingly less mobile in water and is a low level micronutrient for mammals.

Cr(VI) finds many industrial applications associated with electroplating, steel manufacturing, leather tanning, textile dying and photography. These processes generally produce waste streams containing elevated Cr(VI) concentrations and require subsequent treatment. Additionally, chromium ore processing residue (COPR), generated during the recovery of chromium from mineral reserves, can contain unacceptable Cr(VI) levels necessitating treatment. Soils may also possess naturally high Cr(VI) levels requiring treatment but in general anthropogenic activities are responsible for most Cr(VI) waste generation. The hazardous nature of these wastes necessitates some mode of pretreatment prior to disposal to ensure the Cr(VI) is not released into the environment.

Cr(VI)-containing wastes can be classed into two general groups: (1) aqueous wastes where the Cr(VI) is dissolved in a liquid stream; (2) solid wastes where the Cr(VI) is present on or in a solid matrix. The waste medium governs the method(s) required to treat the waste. In aqueous streams the Cr(VI) can be either directly treated in its hexavalent form or first reduced to the more manageable Cr(III) and then treated. Direct treatment of Cr(VI) may involve its adsorption on the surface of Activated Carbon (AC) [1], clays [2], fly ash [3], zeolites [4] or suitable biosorbents. Alternately the Cr(VI) can be extracted using anion exchange resins [5, 6]. Depending on system conditions, the adsorbent may convert the Cr(VI) to Cr(III), coupling capture and reduction in the one processing step. The Cr-loaded adsorbent can then be recovered for further treatment. Adsorbents may also be selected to facilitate recovery from the aqueous phase as demonstrated by Hu et al. [7] who employed the magnetic properties of maghemite to assist with particle recovery following Cr(VI) adsorption. Additionally, the magnetic properties allowed for nanoparticles to be utilised, increasing the surface area of the adsorbing material. Of the available sorbent materials available for Cr(VI) removal from aqueous streams, activated carbons (ACs) and biosorbents receive the most attention and accordingly will be further detailed.

2. Cr(VI) Adsorption

2.1. Activated Carbon

Current research into AC as a Cr(VI) adsorbent is primarily concerned with the effect of surface modifications to the material on its adsorption capacity. Cr(VI) adsorption on AC can be by either an ion exchange mechanism or oxidation mechanism with the favoured mechanism a function of Cr(VI) concentration and pH. Liu et al. [1] demonstrated ion exchange was the dominant mechanism for AC (source: coconut, commercial) adsorption of low Cr(VI) concentrations in water. They applied acid-base modifications (HNO_3-NaOH/NaCl) to the surface of their AC, favouring the ion exchange mechanism and considerably improving its adsorption capacity. Similarly, Babel and Kurniawan [8] reported modifying AC (source: coconut shell) with an oxidising agent improved its Cr(VI) adsorption capacity. In this instance they reported Cr(VI) adsorption increased in the order HNO_3-AC >

H_2SO_4-AC > chitosan-AC. Choi et al. [9] used a different approach to increase the Cr(VI) adsorption capacity of AC. They pre-adsorbed a cationic surfactant onto the AC which increased its adsorption capacity, the extent depending on the structure of the cationic head. The cationic surfactants promoted a net positive charge on the AC surface, enhancing Cr(VI) adsorption (as CrO_4^{2-}). Dubey and Gopal [10] found impregnating the surface of AC (source: groundnut husk) with $AgNO_3$ and then treating with HCl gave an Ag-loaded AC and improved its Cr(VI) adsorption capacity over the untreated AC. They described the improvement as due to the formation of insoluble silver chromate (Equation 1).

$$2Ag^+ + Cr_2O_7^-{}_{(soluble)} + H_2O \rightarrow 2AgCrO_{4(insoluble)}\downarrow + 2H^+ \qquad (1)$$

Stability of the surface Ag against leaching into solution was not raised in this or other related manuscripts by the authors.

In a slight variation to AC as a Cr(VI) adsorbent, multi-walled carbon nanotubes (MWCNTs) have also been investigated for this particular application with some success [11, 12]. For instance, Pillay et al. [11] reported a greater Cr(VI) uptake by MWCNTs compared with AC for Cr(VI) initially present at ppb levels.

2.2. Biosorbents

A cheaper alternative to AC as an adsorbent is biomass or biological wastes for removing Cr(VI) from wastewater streams. A broad variety of biomaterials have demonstrated a propensity for removing Cr(VI) and other heavy metals from aqueous streams predominantly due to their containing various functional groups (phenolics [13], amino, hydroxyl, carboxyl, sulfate [14]) suitable for heavy metal adsorption. A by no means comprehensive cross-section of biosorbents and their adsorption capacities at optimum pH values is given in Table 1. Cr(VI) adsorption capacities of other sorbents are included for comparison.

Table 1. Selected biosorbent and sorbent Cr(VI) adsorption capacities at optimum pH

Biosorbent/Sorbent	Cr(VI) ads. capacity (mg/g)	pH	Reference
Marine green alga	10.6	1	[15]
Marine brown alga (seaweed)	17.8	3	[16]
Fresh water alga	31	1-2	[14]
Grape waste (acidified)	99.3	4	[13]
Banana skin	249.6	1.5	[17]
Seed pods (*O. americanum* L)	83.3	1.5	[18]
Eucalyptus bark	45	2	[19]
Treated sawdust	9.6	3.5	[20]
Fungi (*Termitomyces clypeatus*)	54	3	[21]
Activated carbon	10.9	4	[8]
Ag-loaded activated carbon	11.4	3	[10]
Activated clay	1.4	2	[2]
Clinoptilolite (natural zeolite)	1.2	4	[4]
Maghemite	17.7	2.5	[7]

With the exception of the acidified grape waste, the Cr(VI) adsorption values in Table 1 are for the washed and dried biomaterials alone. Improvements in adsorption capacity can be achieved by washing the material with specific chemicals. For instance Basha et al. [16] were able to improve the Cr(VI) adsorption capacity of their marine brown alga by washing it either with the cross-linking reagent, epichlorohydrin (24.2 mg Cr(VI)/g adsorbent) or the oxidising potassium permanganate (20.1 mg Cr(VI)/g adsorbent). Similarly, Gupta and Rastogi [14] improved the Cr(VI) adorption capacity (35.2 mg Cr(VI)/g adsorbent) of their fresh water alga by acid treatment (HCl).

A drawback of biosorbents for Cr(VI) removal is they can leach substantial levels of soluble organic compounds during the adsorption process which may necessitate additional water treatment processes. Park et al. [17] observed the release of 826-872 mg/L of organics from their banana skin biosorbents during Cr(VI) removal. This may be addressed by modifying the surface of the biomass with an immobilising agent, as described by Yang and Chen [22].

3. Phytoremediation

Table 2. Cr(VI) phytoextraction and rhizofiltration capacities of selected plant species

Plant species	**Cr uptake (mg/kg DW)**		**Cr(VI) conc.**	**Soil**	**Reference**
Phytoextraction	**Roots**	**shoots/leaves**	**(mg/kg soil)**	**type**	
Indian Mustard (*Brassica juncea*)	390	20	100	Podzol	[23]
Fenugreek (*Trigonella foenumgraecum* L.)	12.3	3.52	10	Sandy	[24]
Spinach (*Spinacia oleracea* L.)	15.5	6.09	10	Sandy	[24]
Raya (*Brassica campestris* L.)	18.5	8.43	10	Sandy	[24]
Rhizofiltration			**Cr(VI) conc. (mg/L)**	**Nutrient solution**	
Sorghum (*Sorghum bicolor* L.)	6040	3130	80	Jensen	[25]
Alfalfa (*Medicago sativa* L.)	1150	910	10	Jensen	[25]
Water hyacinth (*Eichhornia crassipes*)	5265	89	10	Hoagland	[26]
Gramineae (*Leersia hexandra* Swartz)	3299	2164	30	Hoagland	[27]

Provided are initial Cr(VI) concentration and soil/nutrient solution. DW = dry weight

Another available Cr(VI) bioremediation technology is phytoremediation. Phytoremediation is a generic term encompassing: (1) phytoextraction or the capacity for living plant species (hyperaccumulators) to extract and accumulate heavy metals in their harvestable tissue from contaminated soils; (2) rhizofiltration or the capacity of aquatic plant species to uptake and store heavy metals in their roots from contaminated water; (3) phytodetoxification or the capacity of plant species to detoxify a metal (e.g. reduce Cr(VI) to Cr(III)) by chelation, reduction and/or oxidation. It represents a cheap and facile means of remediating tracts of contaminated land as well as water resources. Desirable properties of hyperaccumulators include tolerance to high levels of heavy metals, high heavy metal uptake capacity and high biomass production.

Various plant species have been demonstrated as potential candidates for the phytoremediation of Cr(VI) contaminated soils, as listed in Table 2. Included are the Cr concentrations in the roots and shoots/leaves following Cr(VI) exposure.

The values in Table 2 typically represent Cr uptake at the highest concentrations in each particular study. Additionally, in those studies where Cr speciation in the plant tissue following uptake was investigated [23, 26], the Cr was in a reduced state. Howe et al [28] similarly reported subterranean clover (*Trifolium brachycalcinum*) capable of completely reducing Cr(VI) at solution concentrations of 2.08 mg/L, but also observed Cr(VI) in the roots at elevated initial Cr(VI) concentrations (83.2 mg/L).

Table 2 demonstrates that, irrespective of the phytoremediation mode, Cr accumulation is higher in the roots than in the shoots or leaves for all plant species. The primary uptake route is via the root system with Cr translocation to stems, shoots and leaves restricted by plant physiological barriers and the generally low mobility of Cr(III). For phytoextraction this is a considerable limitation as ideally it is the aerial component of the plant (i.e. the shoots and leaves), recovered during harvesting, in which the highest chromium accumulation is desired. In the case of rhizofiltration, restricted chromium translocation is not as significant as the whole plant can be readily harvested.

4. Reduction Technologies

Indirect treatment involves initially reducing the Cr(VI) followed by adsorption, chemical precipitation or bioremediation. The initial reduction can be achieved by techniques such as bioreduction, treatment with zero valent iron (ZVI, Fe^0) or photocatalysis.

4.1. Zero Valent Iron (ZVI)

ZVI, in the form of a permeable reactive barrier, has been shown capable of effectively remediating groundwater contaminated with Cr(VI) [29]. The technology involves placing a barrier of Fe^0 in the path of a ground water flow and relies on reducing Cr(VI) to Cr(III). The Cr(III) can then precipitate as the hydroxide ($Cr(OH)_3$), mixed Fe(III)-Cr(III) hydroxide or oxyhydroxide species [30]. Cr(VI) reduction can be accomplished by Fe^0 or Fe^{2+} (Equations (2) and (3), respectively) after which it may form the mixed hydroxide and oxyhydroxide species mentioned above (Equations (4) and (5), respectively).

$$Cr_2O_7^{2-}{}_{(aq)} + 3Fe^0{}_{(s)} + 14H^+{}_{(aq)} \rightarrow 2Cr^{3+}{}_{(aq)} + 3Fe^{2+}{}_{(aq)} + 7H_2O \quad (2)$$

$$Cr_2O_7^{2-}{}_{(aq)} + 6Fe^{2+}{}_{(aq)} + 14H^+{}_{(aq)} \rightarrow 2Cr^{3+}{}_{(aq)} + 6Fe^{3+}{}_{(aq)} + 7H_2O \quad (3)$$

$$(1\text{-}x)Fe^{3+}{}_{(aq)} + xCr^{3+}{}_{(aq)} + 3H_2O \rightarrow Cr_xFe_{(1\text{-}x)}(OH)_{3(s)} + 3H^+{}_{(aq)} \quad (4)$$

$$(1\text{-}x)Fe^{3+}{}_{(aq)} + xCr^{3+}{}_{(aq)} + 2H_2O \rightarrow Cr_xFe_{(1\text{-}x)}(OOH)_{(s)} + 3H^+{}_{(aq)} \quad (5)$$

These precipitates may remain on the Fe^0 surface or conditions may be such that surface passivation may occur, for example by the formation of goethite (α-FeOOH) under high dissolved oxygen concentrations [31]. Additionally, within an actual groundwater environment, common ground water components such as hardness (Mg, Ca) and carbonates may further diminish long-term performance.

In a similar manner to that reported for AC, the capacity of ZVI to remove Cr(VI) can be altered by modifying Fe^0 morphology or its surface. Increased ZVI performance is directly related to increased surface area which has engendered the development of comparatively high surface area Fe^0 materials including hollow nanoporous structures [32] and chitosan-Fe^0 nanoparticle systems [33]. Surface modification may not always be beneficial as reported by Lai and Lo [34]. They found acid washing the Fe^0 was detrimental for Cr(VI) removal, particularly under simulated groundwater conditions, attributing this to more severe mineral precipitation, and ultimately passivation, by the initially more reactive acid washed surface.

4.2. Photocatalysis

Cr(VI) may also be reduced by photocatalysis. This technique involves catalytically activating a semiconductor material, such as titanium dioxide, with UV-light. The UV-light (photon) promotes an electron from a lower energy state (valence band, vb) to a higher energy state (conduction band, cb) in the semiconductor, leaving a 'positive hole' in the valence band.

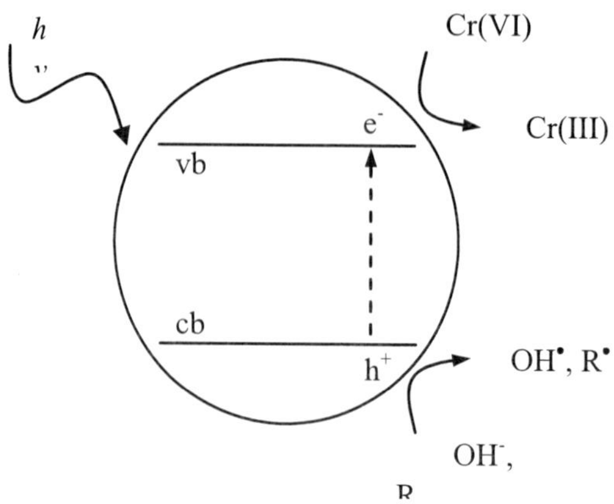

Figure 1. Cr(VI) photocatalytic reduction schematic. cb = conduction band, vb = valence band, R = organic species.

The photogenerated electrons and holes prompt redox reactions with the electrons capable of reducing surface-adsorbed species and the holes capable of oxidising surface-adsorbed hydroxyl groups or organics, in turn promoting free radical reaction processes as depicted in Figure 1.

In the presence of Cr(VI), the photogenerated electrons are capable of reducing it to its trivalent state (Figure 2). To avoid accumulation of the positive holes, and a subsequent reduction in efficiency due to elevated electron/hole recombination, organic species can be utilised as hole scavengers, with the choice of organic capable of influencing the Cr(VI) reduction rate [35]. The improved reductive performance in the presence of an organic will be beneficial in the case of mixed waste streams particularly as the organic itself is converted to the relatively benign CO_2 and water. The photocatalytic system can be modified for improved Cr(VI)-reductive performance as shown by Liu et al. [36] who coupled TiO_2 with AC. Alternately, photocatalytic materials capable of being activated by visible light [37] are available allowing the Cr(VI) reduction to be driven by sunlight, although Sun et al. [38] reported a synergetic system (TiO_2 (Degussa P25)/Cr(VI)/4-chlorophenol) which was able to be activated by visible light despite no modification to the TiO_2.

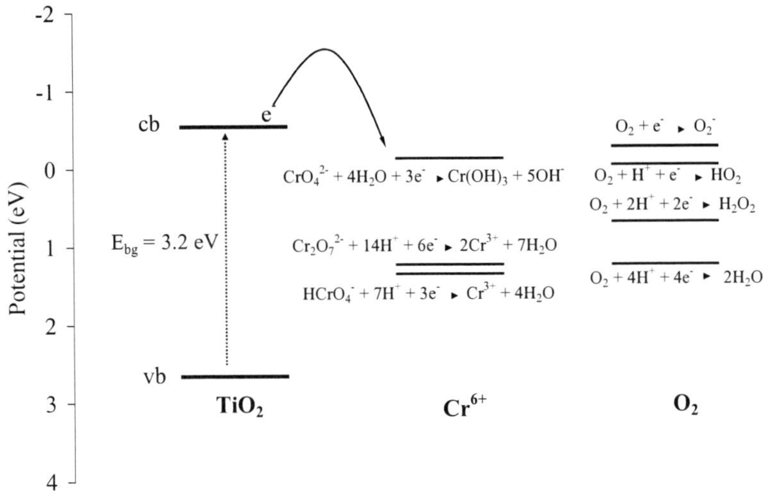

Figure 2. TiO_2 valence (vb) and conduction band (cb), Cr(VI) and O_2 redox potentials. E_{bg} = band gap energy (eV).

5. Bioreduction and Related Biological Treatments

The capacity for bioremediating Cr(VI) contaminated wastes has been known for over three decades following the identification of *Pseudomonas* strains by Romanenko and Koren'kov [39]. Biological treatment is more cost effective and has a lower environmental impact than the other techniques which suffer from the need for additional reagents and/or generate secondary waste products. Biological treatment may remove Cr(VI) by the mechanisms of bioaccumulation, bioadsorption or biotransformation [40, 41, 42]. Bioaccumulation is a metabolism-dependent process describing Cr(VI) uptake and storage by living bacterial cells. Bioadsorption is metabolism-independent and may be done through complexation/chelation, ion exchange, coordination, adsorption and microprecipitation by living or dead cells. Biotransformation includes direct reduction of Cr(VI) by enzymatic processes or indirect reduction by products of cell metabolism. Bioreduction has also been successfully coupled with other physical reduction technologies such as adsorption [43].

5.1. Bacteria Types

Various bacteria have been reported able to reduce Cr(VI), including sulfate-reducing (typically use sulfate as an oxidising agent) and Fe(III)-reducing (use Fe^{3+} as a terminal electron acceptor) microbes. A selection of these bacteria strains are listed in Table 3.

Table 3. Bacteria strains capable of reducing Cr(VI)

Cr-reducing bacteria	Sulfate-reducing bacteria	Fe(III)-reducing bacteria
Enterobacter cloaceae [40]	*Desulfovibrio vulgaris* [47]	*Thiobacillus ferrooxidans* [50]
Shewanella oneidensis [41]	*Desulfomicrobium norvegicum* [48]	*Shewanella alga* [51]
Acinetobacter haemolyticus [42]	*Desulfotomaculum reducens* [49]	
Pseudomonas CRB5 [44]		
Escherichia coli [45]		
Bacillus sphaericus [46]		

Bacterial consortiums have also been shown to reduce Cr(VI) [46, 52] with Quan et al. [53] reporting a synergetic effect between the strains in their particular consortium.

5.2. Bacteria Sources

Cr(VI) resistant and reducing bacteria and consortiums are typically sourced from industrial effluents [52, 54] or soils contaminated by anthropogenic activities [55, 56]. For instance, Zhu et al. [57] isolated Chromium-resistant bacteria (CRB) from a chromite ore processing residue (COPR) disposal site located in China. CRB have also been isolated from naturally occurring soils [46] and marine sediments [40, 41, 49]. Recently, we observed CRB in municipal landfill leachates (MLL) [58, 59].

5.2.1. Landfill Leachate

Bacteria in landfill leachate can reduce Cr(VI) to its trivalent form. Waste type can influence this as was illustrated in our work on Cr(VI) reduction by leachates collected from Municipal (MLL) and Non-Putresible (NPLL) landfills [59]. We observed that, despite the presence of biomass in both leachates, Cr(VI) was reduced by MLL but not by NPLL. The lack of Cr(VI) reduction in NPLL could be attributed to dissimilar bacteria species, organic type and/or the lower organic loading compared with MLL.

The presence of putrescible waste in a landfill does not automatically imply the leachate will be capable of reducing Cr(VI) as was indicated by our recent studies involving municipal landfills located throughout the metropolitan Sydney region (New South Wales, Australia). As illustrated in Figure 3, samples collected from four separate landfills (MLL1-4), as well as samples taken from the same landfill (MLL4) but from old (MLL4(o)) and new (MLL4(n)) sections, exhibited markedly different Cr(VI) reduction capabilities. MLL1 and the newer section of MLL4 demonstrated no Cr(VI)-reducing capabilities, MLL2 and MLL3 demonstrated an essentially first order Cr(VI) reduction rate and the old section of MLL4 exhibited a second order Cr(VI) reduction rate. Comparing the (initial) Cr(VI) reduction rates with the various characteristics of each leachate we found the rate to be proportional to the BOD_5 content but not to any other characteristics. The fact no Cr(VI) reduction is observed in sterilised samples of the leachates confirms the reduction is bacteria-mediated.

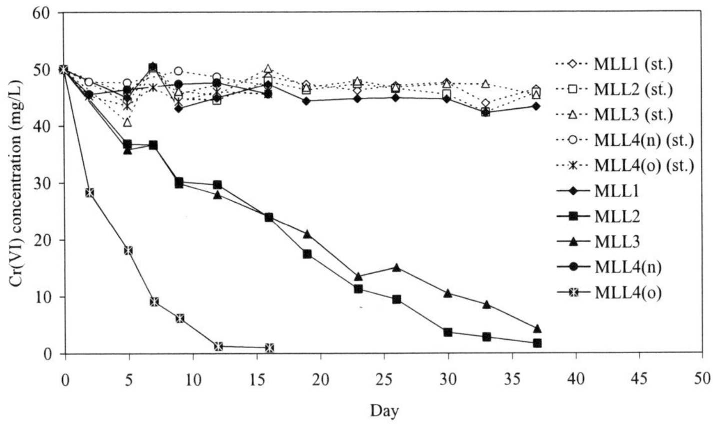

Figure 3. Cr(VI) reduction profiles for sterilised (st.) and non-sterilised landfill leachates. Initial Cr(VI) concentration = 50 mg/L. Leachate concentration = 75 (v/v%).

5.3. Reduction Mechanisms

Cr(VI)-reducing bacteria are either directly or indirectly involved in the Cr(VI) reduction mechanism. Direct microbial reduction has been ascribed as predominantly due to enzymatic soluble Cr(VI) reductase activity [60, 61]. Hydrogenases, nitroreductases and quinone reductases have been implicated in the Cr(VI) reductase process. Direct microbial reduction may also occur via non-enzymatic processes such as dissimaltory electron reduction where the Cr(VI) acts as a terminal electron acceptor, accepting electrons from donor species such as acetate or hydrogen [62].

Indirect microbial Cr(VI) reduction occurs via a biotic-abiotic reduction mechanism where a concomitant species in solution (e.g. a metal cation) is microbially reduced with the product in turn chemically reducing Cr(VI) to Cr(III). This mechanism was initially reported for Fe by Wielinga et al. [63], elaborated on by Hansel et al. [64] and more recently demonstrated to occur in MLL by our group [65]. In this mechanism, bacteria reduces Fe(III) to Fe(II) which then reduces the Cr(VI), producing Cr(III) species including its hydroxide and mixed Fe/Cr hydroxides. In fact, the Cr(VI) reduction by Fe(II) and mixed hydroxide formation reactions are similar to the processes occurring during ZVI treatment and accordingly can be described by Equations (3) and (4), respectively. Figure 4 portrays a simplified version of this electron shuttle mechanism.

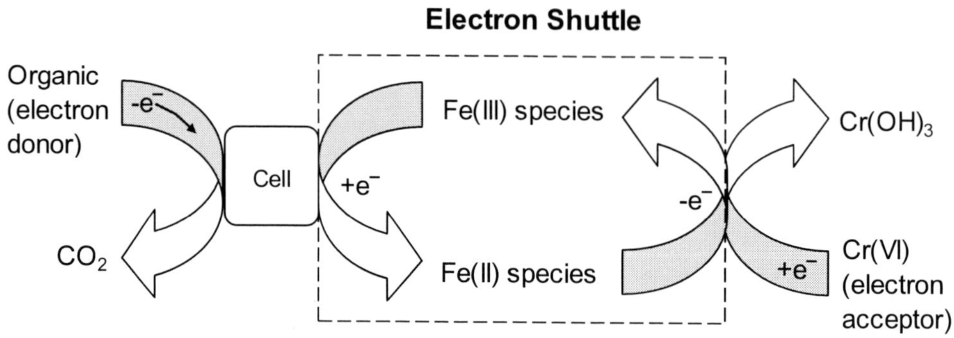

Reprinted from Journal of Hazardous Materials, Vol 161, Y. Li, G. K. C. Low, J. A. Scott, R. Amal, The role of iron in hexavalent chromium reduction by municipal landfill leachate, pp. 657-662, Copyright (2008), with permission from Elsevier.

Figure 4. Simplified mechanism for Cr(VI) reduction in MLL using Fe(II)/Fe(III) as an electron shuttle.

6. Stabilisation/Solidification

Whereas the previously described techniques are generally applied to Cr(VI)-contaminated aqueous streams, stabilisation/solidification (S/S) is typically applied to solid waste systems. This can include secondary wastes from treatment of aqueous streams, contaminated soils or COPR. In S/S treatment, an additive is applied to the waste which assists in binding the Cr(VI) within a solid matrix. This may be by directly binding the Cr(VI) or by reducing and binding it as Cr(III). Ordinary Portland Cement (OPC) is perhaps the most traditional S/S additive for heavy metals, relying on the highly alkaline environment it provides to precipitate heavy metals as their hydroxides and other species. In the instance of Cr(VI) the mechanism is untenable as the Cr(VI) remains soluble under alkaline conditions. This was demonstrated in both our experimental [66] and modeling studies [67]. Experimentally, we observed Cr(VI) to be released from OPC when leached with acetic acid over the (final) pH range 5-12, indicating a high level of solubility when present in cement under any conditions. Simulating the system using the PHREEQC geochemical modeling package indicated Cr was highly soluble and readily leached in the presence of acetic acid, both initially from the calcium-silicate-hydrate (C-S-H) surface and then from the pores of the C-S-H matrix as it dissolved. When leached by MLL the model indicated Cr(VI) was predominantly reduced to Cr(III), which precipitated, in the early stages of the leaching process. The reduction was governed by the redox and pH properties of the leachate. As leaching progressed the redox and pH values of the system increased leading to oxidation of the Cr(III) to Cr(VI) and an increase in its solubility.

Alternative binders, used stand-alone or as additives to OPC, can improve heavy metal S/S. Waste materials such as fly ash [68, 69] and slag [70] or other compounds such as lime [68], soluble silicates [69], superplasticisers [71], barium [72] and ferrous sulfates [73] have been considered, although they are not always effective for Cr(VI) immobilisation. For instance, Zhang et al. [69] when investigating geopolymers (formed by activating 'Class F' fly ash with sodium silicate solutions) as possible heavy metal binders, reported them suitable for Pb and Cd but essentially ineffective for Cr(VI). On the other hand, Dermatas and Moon [68] found a mixture of quicklime and 'Class C' fly ash capable of immobilising Cr(VI) despite the individual components being unable to do so. They suggested the pozzolanic compound, $3CaO \cdot Al_2O_3 \cdot 0.5CaCrO_4 \cdot 0.5CaSO_4$, as responsible for stabilisation in this instance.

Ferrous sulfate has been considered as a possible stabiliser of COPR [73, 74] with limited success. The logic behind COPR treatment with $FeSO_4$ is the Cr(VI) will be reduced to the more stable Cr(III) by the Fe(II). However, the highly alkaline conditions promoted by the COPR favor competing reactions, forming ferrous hydroxides and carbonates, which scavenge the Fe(II) before it can react with Cr(VI). This may be overcome by increased $FeSO_4$ addition or pH control (e.g. H_2SO_4 addition). The role of the sulfate ions is to compete with chromate ions in the COPR, forcing them into solution and facilitating their interaction with Fe(II). If the pH is not closely controlled the sulfate can react with the COPR matrix to form ettringite ($Ca_6Al_2(SO_4)_3(OH)_{12}.26H_2O$) which can lead to problems associated with material swelling.

Other heavy metal S/S techniques include encapsulation and vitrification. The addition of a cationic polysufide, CaS_x, to COPR was found by Tinjum et al. [73] to effectively stabilise Cr(VI) against leaching in laboratory column tests. Characterisation of the treated waste revealed the COPR particles were encapsulated by a sulfur-containing compound. Xu et al. [75] successfully used vitrification to S/S Cr(VI)-contaminated sewerage sludge. Their vitrification process involved mixing clay, sodium silicate and dried sewerage sludge to up to 1000°C to form the ceramic material ceramsite. Similarly, Wei et al. [76] heat treated Cr(VI)-sorbed kaolin and found the Cr(VI) could be reduced and stabilised as the lowly soluble Cr_2O_3 between 500-900°C while at temperatures of 1100°C a chromium-silicate species was observed.

CONCLUSION

Cr(VI)-contaminated wastes are generally present either as an aqueous stream or solid matrix, which will govern the type of treatment selected. Aqueous Cr(VI) wastes have traditionally been commercially treated by adsorption using inorganic media such as Activated Carbon or clays. Receiving recent considerable interest are biosorbents as alternatives to these established materials, which have distinct advantages in they are cheap, readily available and in some instances, waste products themselves. Given some biosorbents have Cr(VI) adsorption capacities superior to their inorganic counterparts the potential for this technology in Cr(VI) waste management is significant. Phytoremediation represents another cheap and facile means of treating Cr(VI)-contaminated wastes (both aqueous and solid), although restricted translocation of Cr(VI) from plant roots to aerial tissue limits its application for treating solid wastes. Bacteria, tolerant to Cr(VI) and capable of directly or indirectly mediating its reduction, represent another cheap and facile biological treatment technology for remediating Cr(VI) wastes. Phytoremediation and bacterial reduction hold the advantage over other technologies in that they can be applied to both contaminated aqueous (e.g. water) and solid (e.g. soil) systems. Other non-biological technologies utilising zero valent iron and photocatalysis are capable of reducing Cr(VI) but higher technology and implementation costs suggest they are not as tenable as bioremediation for large-scale application. Overall, it is apparent bioremediation will play an increasing role in remediating Cr(VI)-containing wastes as it represents a group of cheap, straightforward and environmentally amenable technologies.

Most of the technologies for treating aqueous wastes generate secondary Cr-contaminated waste products. These can require additional treatment which is usually in the form of stabilisation/solidification (S/S). S/S may also be used on contaminated sites or for treating COPR and appears to have mixed success for, in particular, Cr(VI) wastes. Success is based on the choice of binding media and additives as they are critical in governing the ability of the matrix to effectively secure the Cr. The greatest assets of S/S are it is simple and reasonably cheap and may incorporate other process technology wastes such as slags or fly ash as binding agents.

REFERENCES

[1] Liu, S. X.; Chen, X.; Chen, X. Y.; Liu, Z. F.; Wang, H. L. *J. Hazard. Mater.* 2007, *141,* 315-319.
[2] Weng, C. H.; Sharma, Y. C.; Chu, S. H. *J. Hazard. Mater.* 2008, *155,* 65-75.
[3] Rao, M.; Parwate, A. V.; Bhole, A. G. *Waste Manage.* 2002, *22,* 821-830.
[4] Ouki, K.; Kavannagh, M. *Waste Manage. Res.* 1997, *15,* 383-394.
[5] Shi, T.; Wang, S.; Liu, Y.; Jia, S.; Changming, D. *J. Hazard. Mater.* 2009, *161,* 900-906.
[6] Xing, Y.; Chen, X.; Wang, D. *Environ. Sci. Technol.* 2007, *41,* 1439-1443.
[7] Hu, J.; Chen, G.; Lo, I. M. C. *Water Res.* 2005, *39,* 4528-4536.
[8] Babel, S.; Kurniawan, T. A. *Chemosphere* 2004, *54,* 951-967.
[9] Choi, H. D.; Jung, W. S.; Cho, J. M.; Ryu, B. G.; Yang, J. S.; Baek, K. *J. Hazard. Mater.* 2009, *166,* 642-646.
[10] Dubey, S. P.; Gopal, K., *J. Hazard. Mater.* 2007, *145,* 465-470.
[11] Pillay, K.; Cukrowska, E. M.; Coville, N. J. *J. Hazard. Mater.* 2009, *166,* 107-1075.
[12] Hu, J.; Chen, C.; Zhu, X.; Wang, X. *J. Hazard. Mater.* 2009, *162,* 1542-1550.
[13] Chand, R.; Narimura, K.; Kawakita, H.; Ohto, K.; Watari, T.; Inoue, K. *J. Hazard. Mater.* 2009, *163,* 245-250.
[14] Gupta, V. K.; Rastogi, A. *J. Hazard. Mater.* 2009, *163,* 396-402.
[15] El-Sikaily, A.; Nemr, A. E.; Khaled, A.; Abdelwehab, O. *J. Hazard. Mater.* 2007, *148,* 216-228.
[16] Basha, S.; Murthy, Z. V. P.; Jha, B. *Chem. Eng. J.* 2008, *137,* 480-488.
[17] Park, D.; Lim, S. R.; Yun, Y. S.; Park, J. M. *Bioresource Technol.* 2008, *99,* 8810-8818.

[18] Levankumar, L.; Muthikumaran, V.; Gobinath, M. B. *J. Hazard. Mater.* 2009, *161,* 709-713.
[19] Sarin, V.; Pant, K. K. *Bioresource Technol.* 2006, *97,* 15-20.
[20] Baral, S. S.; Das, S. N.; Rath, P. *Biochem. Eng. J.* 2006, *31,* 216-222.
[21] Das, S. K.; Guha, A. K. *Colloid. Surface B* 2007, *60,* 46-54.
[22] Yang, L.; Chen, J. P. *Bioresource Technol.* 2008, *99,* 297-307.
[23] Bluskov. S.; Arocena, J. M.; Omotoso, O. O.; Young, J. P. *Int. J. Phytoremediat.* 2005, *7,* 153-165.
[24] Dheri, G. S.; Brar, M. S.; Malhi, S. S. *Commun. Soil Sci. Plan.* 2007, *38,* 1655-1672.
[25] Bonfranceschi, B. A.; Flocco, C. G., Donati, E. R. *J. Hazard. Mater.* 2009, *165,* 366-371.
[26] Lytle, C. M.; Lytle, F. W.; Yang, N.; Qian, J. H.; Hansen, D.; Zayed, A.; Terry, N. *Environ. Sci. Technol.* 1998, *32,* 3087-3093.
[27] Zhang, X. H.; Liu, J.; Huang, H. T.; Chen, J.; Zhu, Y. N.; Wang, D. Q. *Chemosphere* 2007, *67,* 1138-1143.
[28] Howe, J. A.; Loeppert, R. H.; Derose, V. J.; Hunter, D. B.; Bertsch, P. M. *Environ. Sci. Technol.* 2003, *37,* 4091-4097.
[29] Wilkin, R. T.; Su, C.; Ford, R. G.; Paul, C. *J. Environ. Sci. Technol.* 2005, *39,* 4599-4605.
[30] Hou, M.; Wan, H.; Liu, T.; Fan, Y.; Liu, X.; Wang, X. *Appl. Catal. B: Environ.* 2008, *84,* 170-175.
[31] Lo, I. M. C.; Lam, C. S. C.; Lai, K. C. K. *Water Res.* 2006, *40,* 595-605.
[32] Cao, J.; Clasen, P.; Zhang, W. X. *J. Mater. Res.* 2005, *20,* 3238-3243.
[33] Geng, B.; Jin, Z.; Li, T.; Qi, X. *Chemosphere* 2009, *75,* 825-830.
[34] Lai, K. C. K.; Lo, I. M. C. *Environ. Sci. Technol.* 2008, *42,* 1238-1244.
[35] Wang, L.; Wang, N.; Zhu, L.; Yu, H.; Tang, H. *J. Hazard. Mater.* 2008, *152,* 93-99.
[36] Liu, S. X.; Chen, X. Y.; Chen, X. *J. Hazard. Mater.* 2007, *143,* 257-263.
[37] Das, D. P.; Parida, K. M. *Catal. Surv. Asia* 2008, *12,* 203-213.
[38] Sun, B.; Reddy, E. P.; Smirniotis, P. G. *Environ. Sci. Technol.* 2005, *39,* 6251-6259.
[39] Romanenko, V. I.; Koren'kov, V. N. *Mikrobiologiya* 1977, *46,* 414-41.
[40] Iyer, A.; Mody, K.; Jha, B. *Mar. Pollut. Bull.* 2004, *49,* 974-977.
[41] Daulton, T. L.; Little, B. J.; Jones-Meehan, J.; Blom, D. A.; Allard, L. F. *Geochim. Cosmochim. Acta* 2007, *71,* 556-565.
[42] Zakaria, Z. A.; Zakaria, Z.; Surif, S.; Ahmad, W. A. *J. Hazard. Mater.* 2007, *146,* 30-38.
[43] Silva, B.; Figueiredo, H.; Qunitelas, C.; Neves, I. C.; Tavares, T. *Micropor. Mesopor. Mater.* 2008, *116,* 555-560.
[44] McLean, J.; Beveridge, T. J. *Appl. Environ. Microbiol.* 2001, *67,* 1076-1084.
[45] Shen, H.; Wang, Y. T. *Appl. Environ. Microbiol.* 1993, *59,* 3771-3777.
[46] Pal, A.; Paul, A. K. *Microbiol. Res.* 2004, *159,* 347-354.
[47] Lovely, D. R.; Phillips, E. J. P. *Appl. Environ. Miocrobiol.* 1994, *60,* 726-728.
[48] Michel, C.; Brugna, M.; Aubert, C.; Bernadac, A.; Bruschi, M. *Appl. Microbiol. Biotechnol.* 2001, 55, 95-100.
[49] Tebo, B. M.; Obraztsova, A. Y. *FEMS Microbiol. Lett.* 1998, *162,* 193-198.
[50] QuiIntana, M.; Curutchet, G.; Donati, E. *Biochem. Eng. J.* 2001, *9,* 11-15.

[51] Vázquez-Morillas, A.; Vaca-Mier, M.; Alvarez, P. J. *Europ. J. Soil Biol.* 2006, *42,* 99-106.
[52] Cetin, D.; Donmez, S.; Donmez, G. *J. Environ. Manag.* 2008, *88,* 76-82.
[53] Quan, X.; Tan, H.; Zhao, Y.; Hu, Y. *J. Hazard. Mater.* 2006, *B137,* 836-841.
[54] Campos, J.; Martinez-Pacheco, M.; Cervantes, C. *Antonie van Leeuwenhoek* 1995, *68,* 203-208.
[55] Megharaj, M.; Avudainayagam, S.; Naidu, R. *Curr. Microbiol.* 2003, *47,* 51-54.
[56] Thacker, U.; Parikh, R.; Shouche, Y.; Madamwar, D. *Process Biochem.* 2006, *41,* 1332-1337.
[57] Zhu, W.; Chai, L.; Ma, Z.; Wang, Y.; Xiao, H.; Zhao, K. *Microbiol. Res.* 2008, *163,* 616-623.
[58] Li, Y.; Low, G.; Lei, Y.; Halim, C.; Amal, R. *Aust. J. Chem.* 2004, *57,* 967-970.
[59] Li, Y.; Low, G. K. C.; Scott, J. A.; Amal, R. *J. Hazard. Mater.* 2007, 142, 153-159.
[60] Shen, H.; Wang, Y. T. *Appl. Environ. Microbiol.* 1993, *59,* 3771-3777.
[61] Desai, C.; Jain, K.; Madamwar, D. *Process Biochem.* 2008, *43,* 713-721.
[62] Francis, C. A.; Obraztsova, A. Y.; Tebo, B. M. *Appl. Environ. Microbiol.* 2000, *66,* 543-548.
[63] Wielinga, B.; Mizuba, M. M.; Hansel, C. M.; Fendorf, S. *Environ. Sci. Technol.* 2001, *35,* 522-527.
[64] Hansel, C. M.; Wielinga, B. W.; Fendorf, S. *Geochim. Cosmochim. Acta.* 2003, *67,* 401-412.
[65] Li, Y.; Low, G. K. C.; Scott, J. A.; Amal, R. *J. Hazard. Mater.* 2009, *161,* 657-662.
[66] Halim, C. E.; Amal, R.; Beydoun, D.; Scott, J. A.; Low, G. *Cement Concrete Res.* 2004, *34,* 1093-1102.
[67] Halim, C. E.; Short, S. A.; Scott, J. A.; Amal, R.; Low, G. *J. Hazard. Mater. A125* 2005, 45-61.
[68] Dermatas, D.; Moon, D. H. *Environ. Engin. Sci.* 2006, *23,* 77-87.
[69] Zhang, J.; Provis, J. L.; Feng, D.; van Deventer, J. S. J. *J. Hazard. Mater.* 2008, *157,* 587-598.
[70] Shi, C.; Fernández-Jiménez, A. *J. Hazard. Mater.* 2006, *B137,* 1656-1663.
[71] Shi, H. S.; Kan, L. L. *J. Hazard. Mater.* 2009, *162,* 913-919.
[72] Chyrsochoou, M.; Dermatas, D.; Moon, D.H.; Christodoulatos, C.; Wazne, M.; French, C.; Morris, J.; Kaouris, M. *ASTM Special Technical Publication* 2006, *1482,* 165-175.
[73] Tinjum, J. M.; Benson, C. H.; Edil, T. B. *J. Geotech. Geoenviron. Engin.* 2008, *134,* 1791-1803.
[74] Dermatas, D.; Chrysochoou, M.; Moon, D. H.; Grubb, D. G.; Wazne, M.; Christodoulatos, C. *Environ. Sci. Technol.* 2006, *40,* 5786-5792.
[75] Xu, G. R.; Zou, J. L.; Li, G. B. *J. Hazard. Mater.* 2008, *153,* 1031-1035.
[76] Wei, Y. L.; Chiu, S. Y.; Tsai, H. N.; Yang, Y. W.; Lee, J. F. *Environ. Sci. Technol.* 2002, *36,* 4633-4641.

In: Management of Hazardous Residues Containing Cr(VI) ISBN: 978-1-61470-637-3
Ed: Maria José Balart Murria

Chapter 4

AN OVERVIEW OF CHROMIUM (VI) FREE CONVERSION COATINGS IN CURRENT USE

John W. Bibber*

Laboratory Director, Sanchem, Inc., Chicago, IL. U.S.A.

ABSTRACT

The expression “conversion coating,” as used in the metal finishing industry, makes reference to a chemical or electrochemical process that converts the surface of any given metal into a surface that is more corrosion resistant and / or more easily accepts applied secondary coatings (i.e. paint) [1]. They are rather thin (generally not more than 6000 nm thick), quickly and easily formed, easily scratched and if used to improve adhesion, are generally coated shortly after being formed to prevent any degradation of the conversion coated surface. Since the early 1920’s the conversion coatings of choice have almost all contained some form of chromium (VI) due in large part to the corrosion resistance the chromium (VI) imparts to the conversion coated surface generated. Their first major application was in the conversion of zinc and zinc alloy surfaces. The use of chromium (VI) then quickly spread to aluminum and a host of other metals and their alloys. This chapter will explore the alternatives that have become available since the early 1980’s in terms of their relative effectiveness, cost and ease of use.

1. INTRODUCTION

The two most widely used industrial metals on the planet are iron and aluminum and their alloys. Iron and / or steel (an alloy of iron and carbon) are almost always conversion coated before use or electroplated with zinc, tin or cadmium. In some cases aluminum is electroplated, but in general it is used as the pure metal or in the form of an alloy. Conversion coatings for use on iron and steel, zinc, tin and / or cadmium surfaces may be traced back many hundreds of years but widespread industrial conversion coating of these metals began in

* Tel: 312 733 6100; Fax: 312 733 7432
E-mails: jbibber_sanchem@yahoo.com; Jo10bi12@aol.com

the late 19th century. Aluminum and its alloys were not available in large industrial quantities until the early 20th century and the conversion coating of aluminum and its alloys began in earnest in the 1920's. The four primary methods of conversion coating in current use are: (1) the generation of a film of mixed metal chromium hydroxides and / or oxides; (2) the generation of a film of mixed metal manganese hydroxides and / or oxides; (3) the generation of a precipitated film of mixed metal phosphates and (4) the use of various synthetic organic and / or inorganic polymers with or without associated metal oxides and / or hydroxides. This chapter contains the following 5 sections: (1) cleaning and deoxidization; and the four primary alternatives to chromium (VI) methods of conversion coating in current use (which have become available since the early 1980's): (2) chromium (III), (3) permanganate, (4) phosphate and (5) synthetic polymers.

2. Cleaning and Deoxidization

The most important part of any metal processing procedure (conversion coating, electroplating, painting, etc.) is the proper cleaning and deoxidization of the metal's surface. Cleaning is the removal of surface oils and other organic compounds, while deoxidization is the industrial expression for the proper removal of unwanted inorganic compounds (i.e. metal oxides), which may interfere with further processing of the metal without any significant attack on the surface of the metal being processed. The actual method used on any given metal or metal alloy will be dependent upon the metal or metal alloy being processed, its specific chemical reactivity and the type of soil being removed. As such, cleaning and deoxidization is very much a part of the conversion coating process itself and in many cases some of the same materials used in the conversion coating process, are used in the cleaning and deoxidization process. Starting around 1980, one of the first uses of chromium (VI) to be curtailed was the use of dichromates, or chromic acid itself, as deoxidization agents on one or more of the metals under discussion. The use of chromium (VI) as a deoxidization agent enhanced the performance of chromium (VI) conversion coating systems applied later, but harmed the performance of most non-chromium (VI) conversion coating systems as they left traces of chromium in various oxidation states on the surface of the metal or metal alloy being processed.

Iron and / or steel products are generally cleaned by the use of a highly alkaline (pH of 12 or above) cleaner as iron and / or steel are quite stable towards oxidation in an alkaline environment. Deoxidization is generally accomplished with the use of a mild hydrochloric acid solution as hydrochloric acid readily dissolves iron oxides, activates the surface of the iron and is rather inexpensive. Zinc or zinc electroplated articles are rather reactive towards oxidation and in point of fact, zinc is plated on steel as it will preferentially oxidize and in the process protect the steel from oxidation. Zinc plated articles are generally decreased, if necessary, and then cleaned with a simple non-ionic cleaner [2]. The zinc is then deoxidized with a 30 to 60 second dip in a strongly alkaline (pH of 12 or above) solution (zinc will dissolve in strongly alkaline or strongly acid solutions), rinsed and then given a 30 to 60 second dip in a mild acid solution to remove any zinc hydroxides, rinsed again in mineral free water and immediately conversion coated. Tin and cadmium are rather noble metals and somewhat less reactive than the steel to which they are plated and they are generally applied for other reasons than simply protecting the steel or iron articles from corrosion. The metals are soft and provide a degree of lubricity and in both cases they present a pleasing bright and reflective surface (i.e. tin cans). Strong alkaline cleaners will in both cases (for cadmium aqueous ammonia based cleaners are often used) serve to clean and deoxidize the metals and in some cases they are cleaned by being made the cathode in an electrolytic cell containing an alkaline solution. In many cases these metals are conversion coated shortly after being deposited on a steel or iron surface itself [3]. Freshly electroplated metals have a rather open, porous and more reactive surface (Figure 1) which is desirable when the articles are to be coated. In many cases the deposited metal is heated to the melting point in a process referred to as "reflowing," to better seal the steel or iron surface, give a more bright and reflective surface and reduce the reactivity of the electroplated metal (Figure 2).

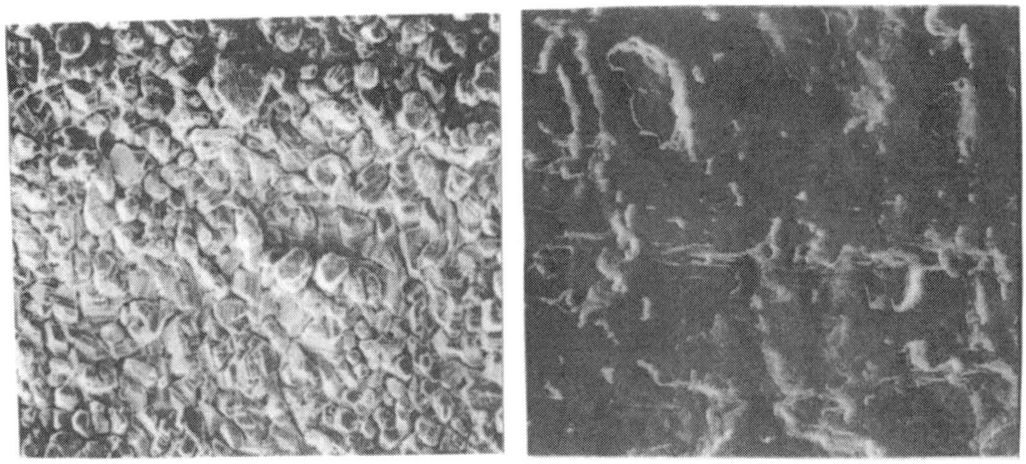

Figure 1. Typical electroplated surface without "reflowing" at 1000 magnification (a).

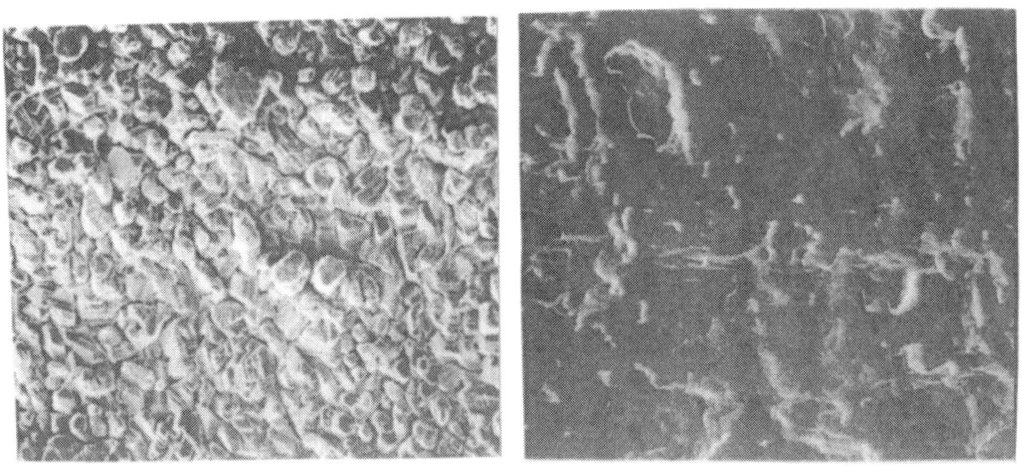

Figure 2. Typical electroplated and "reflowing" surface at 1000 magnification (b).

As in the case of iron and / or steel or steel plated articles, the cleaning and deoxidization of the aluminum and its alloys are processes that are very much a part of the conversion coating process itself. Mildly alkaline cleaners (normally not more than a pH of 11.0) are generally used and they should be non-etching as an alkali etched aluminum surface will generally leave difficult to remove alloyed elements such as elemental silicon or heavy metal oxides (i.e. copper and zinc oxides) (Figure 3). To prevent excessive etching, chromate's (prior to about 1980) were at one time added. Additives now in use include silicates, nitrates and / or phosphates. Silicates are now used to a lesser extent as they tend to be too effective and in many cases will leave difficult to remove deposits of various silicates [4]. As in the case of aluminum cleaners, deoxidization of aluminum surfaces was at one time carried out in chromic acid or nitric acid / dichromate based products. They were very effective in preventing the pitting (Figures 4-6) of aluminum surfaces by maintaining a thin film of aluminum oxides on the aluminum or aluminum alloy surface. Since about 1980, various other oxidizing agents have come into widespread use to prevent the pitting of aluminum surfaces without leaving deposits of chromium oxides or other heavy metals [5] and as a result making it difficult to apply non chromium (VI) conversion coating systems.

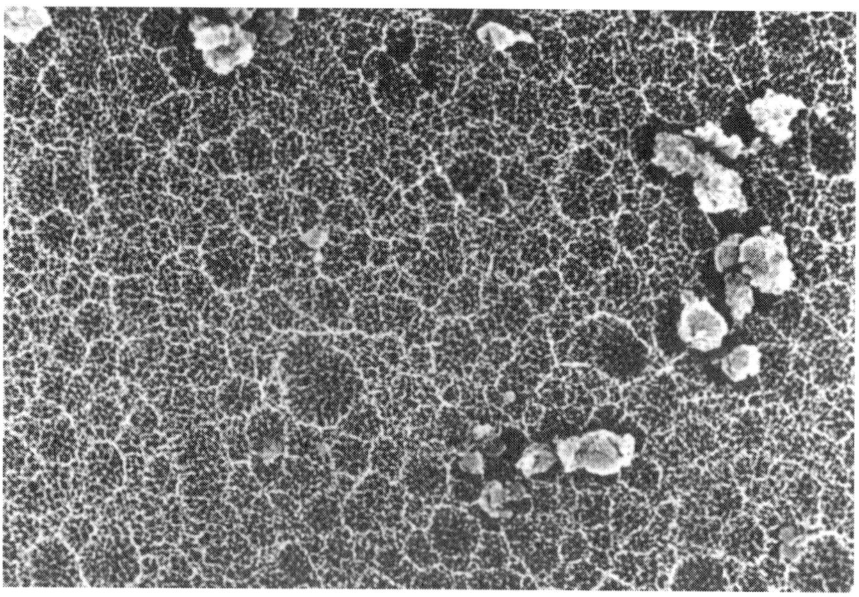

Figure 3. Alkaline etched high silicon aluminum alloy at 1000 magnification. Note the dark black deposits of elemental silicon and the lighter metal oxide deposits.

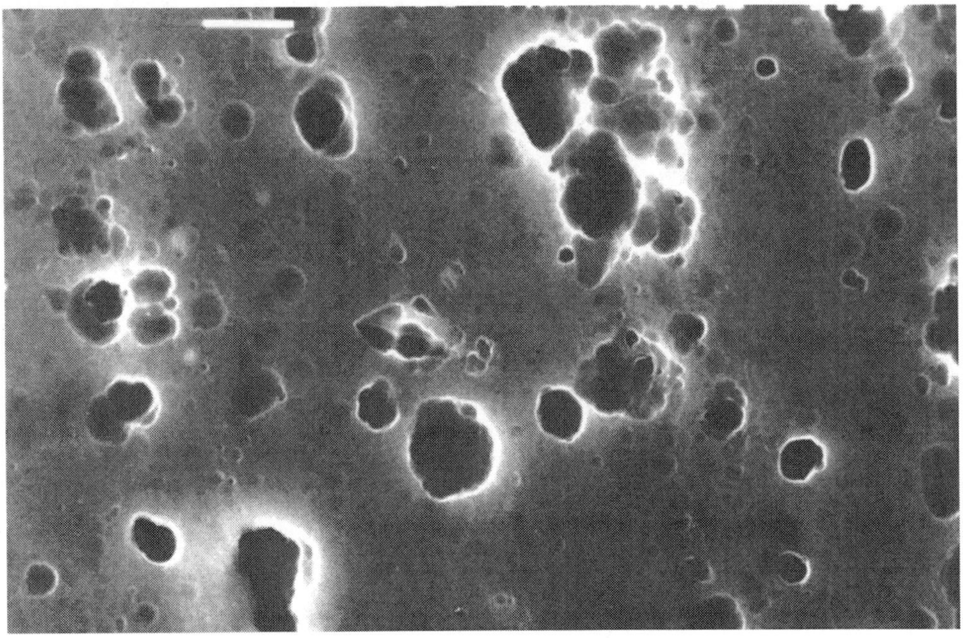

Figure 4. Acid deoxidized aluminum with no inhibitors at 1000 magnification. Note the extensive pitting.

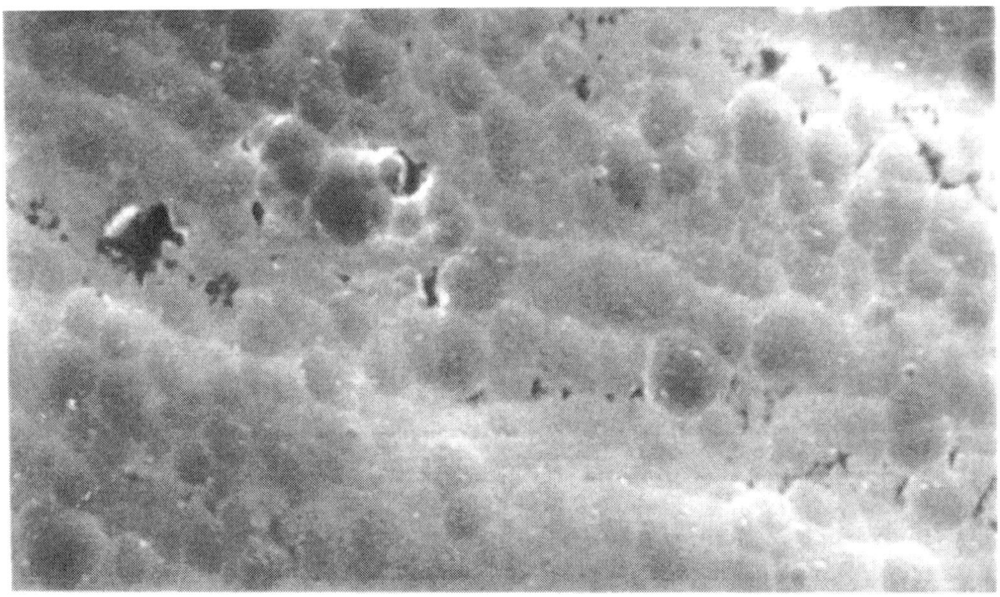

Figure 5. Inhibited acid deoxidized aluminum surface at 1000 magnification.

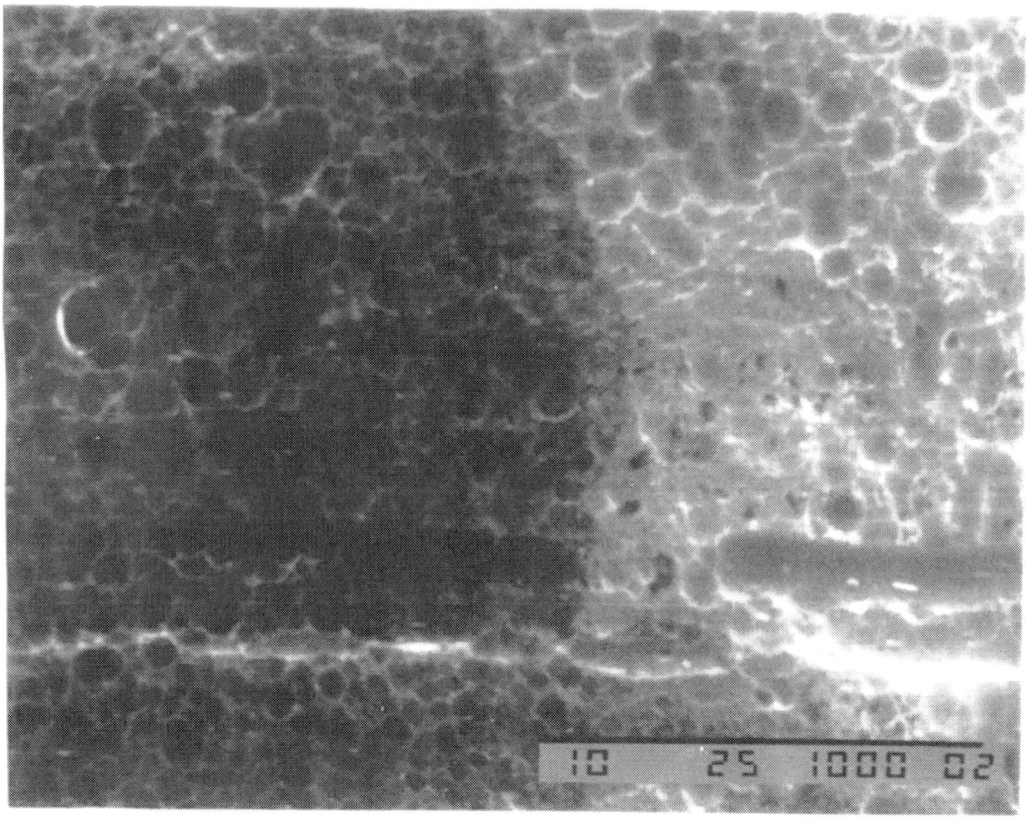

Figure 6. Aluminum surface deoxidized in a heavy metal containing deoxidizer at 1000 magnification. Note the dark colored layer of heavy metal deposit.

3. Alternatives to Chromium (VI) based Conversion Coatings in Current Use

3.1. Chromium (III) based Conversion Coatings

All conversion coating processes use chromium (VI) based conversion coating systems as a baseline for judging the performance of other systems. Were it not for the issue of toxicity they would still be in wide spread use in the metal finishing industry. Chromium (III) based systems began to be used in military applications [6] by about 1990 to decrease the amount of chromium (VI) used in metal processing facilities. They are included here as chromium (III) will be stable as chromium (III) under rather specific conditions of pH and as such not considered to be toxic. However, any conversion coatings produced by the use of chromium (III), will revert to the plus six state once they have been applied to a metal's surface and allowed to sit in the open air for any extended period of time (Figure 9) [7,8] and will not meet the current European RoHS (Restrictions of Hazardous Substances) requirements for chromium (VI). In addition they are rather difficult to apply as there is no driving force (oxidation or reduction) directing them to attach themselves to the metal being conversion coated. Chromium based conversion coatings are produced through the reduction of chromium (VI) and corresponding oxidation of the metal being conversion coated to form a jell-like mixture of various low oxidation state chromium compounds (mostly chromium (III) hydroxide) mixed with the oxides and / or hydroxides of the metal or metal alloy they were applied to. When freshly formed, they have an open and porous structure that will readily accept secondary coatings (i.e. paint). It is generally recommended that the secondary coating be applied as soon as possible and generally not more then eight hours after the application of the conversion coating [9]. After approximately eight hours these conversion coatings will complete the formation of a rather hydrophobic inorganic polymer system not too different than that formed by "Portland Cement" and held together by water or hydration. It is this polymer that is primarily responsible the extremely good corrosion resistance of these conversion coating systems. It is assumed that excess chromium (VI), trapped in the polymer, act to "self seal" any breaks in the coating. These deposits of not yet reacted chromium (VI) are also responsible for the various shades of light yellow to brown color observed in these conversion coating systems on aluminum, zinc and cadmium surfaces. These colors are a rather useful feature in that they serve as an indicator of the relative thickness of the applied coating and also indicate if the coating has properly formed. The various colors also serve to indicate the degree of corrosion resistance and / or paint adhesion a particular coating has. Relatively light or "clear" coatings will only give a bare minimum of corrosion resistance and are generally used for surfaces that will receive a secondary coating or for surfaces that need a degree of electrical conductivity. The "clear" coatings are generally about 100 to 120 nm thick with a mass of about 161 to 269 mg / m^2. An iridescent to light yellow coating will be about 200 to 250 nm thick and have a mass of about 430 to 646 mg / m^2. A yellow to light brown coating will range from about 300 to 600 nm in thickness with a mass of about 861 mg to 2150 mg / m^2. One major drawback to the use of these coatings is that they will decompose when heated above 70°C due to the loss of the water that holds the inorganic polymer together [10] (Figure 7]. This feature restricts their use in powder coating applications as they require extensive heating. The cost of these systems is primarily driven by the fact that they are quite toxic and must be disposed of properly. Chromate based conversion coatings have been used on tin plated steel for over 70 years now to provide for a bright reflective surface that accepts secondary coatings and will not stain when exposed to high sulfur content food.

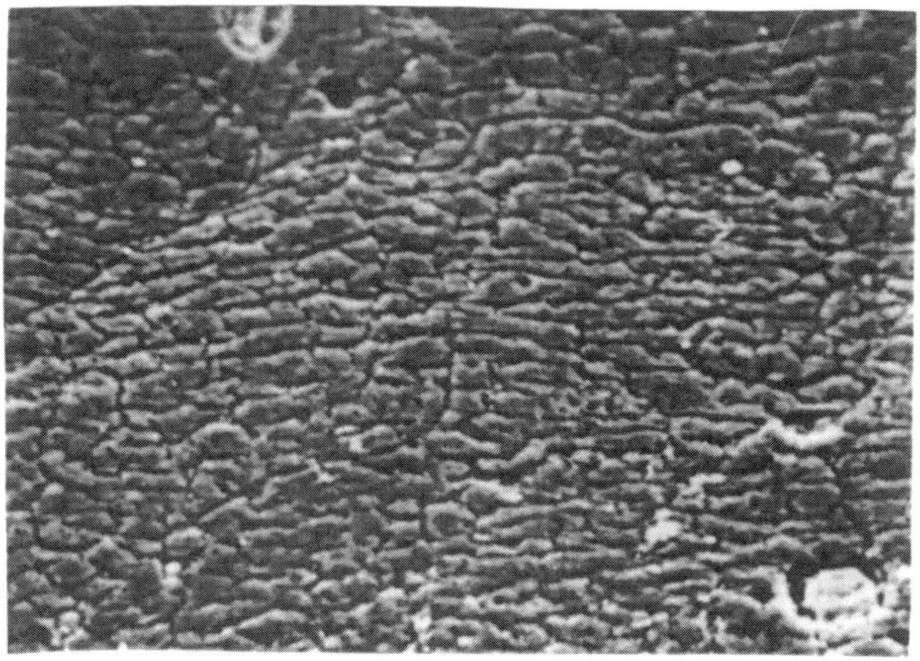

Figure 7. Decomposition of chromate based conversion coating due to the loss of water of hydration after being heated to 93°C at 1000 magnification.

3.2. Permanganate Based Conversion Coatings

Going across the third row transition metals, from left to right, there are a great many similarities between the elements in their respective chemical and physical properties until you reach iron. Thus, the conversion coatings produced by chromium (VI) compounds are, as expected, quite similar to the one's generated by the manganese (VII) found in permanganate compounds [11]. Manganese generated conversion coating systems produce a jell-like structure similar to that produced by chromium, but not similar enough to give the same level of corrosion resistance. They are much more stable towards heat and as a result may be used in powder coating applications. With aluminum you will obtain essentially the same range of colors and corrosion resistance, with all but the more corrosion susceptible high copper and / or zinc alloys. The more corrosion susceptible alloys will have the same level of corrosion resistance when used with a secondary inorganic or organic seal. With zinc, cadmium and steel, the permanganate based system will have a quite similar characteristic color pattern and will match all of the corrosion resistant characteristics of the chromium based systems when used with an inorganic silicate based seal or a secondary organic seal while still allowing for electrical conductivity. Its characteristic color changes provide for a quick and easy way to determine if the conversion coating has properly formed negating the necessity for expensive re-painting and unexpected corrosion failures. If the conversion has not formed properly, it may be quickly and easily reprocessed. A "clear" permanganate conversion coating for maximum electrical conductivity may be produced, if so desired, and the presence of the coating may be confirmed by a simple and easy to use chemical process. The permanganate based system does not require an excessively clean surface as it actually assists in the cleaning process by converting organic residue to carbon dioxide and water. Permanganates have been in use for over 90 years in the treatment of potable water systems and they are on the list of materials allowed in drinking water [12]. Due to increased use and production of permanganates, their cost had been decreasing.

3.3. Phosphate Based Conversion Coatings

The use of various heavy metal phosphates (iron, manganese, chromium, etc.) is the oldest conversion coating method in use and still produces one of the best conversion coating methods for adhesion on zinc surfaces available (Figure 8). The corrosion resistance of phosphate generated conversion coating systems is quite poor with the exception of those coatings sealed with chromates, but even then chromate's used by themselves will give better results. Zinc phosphated surfaces are always generated from baths that contain a significant amount of zinc as it acts to slow down the activity of the bath [13] to produce the proper degree of porosity in the coating. This results in a relative high concentration of zinc in the bath, a high chemical usage rate and heavy metal disposal problems which serves to drive up the cost of the process. Iron phosphating systems need the addition of oxidizing agents or more noble metals (generally copper salts) to increase the rate of phosphating and obtain the proper degree of porosity, but they are in general relatively inexpensive and very little has been done to improve on these systems over the last thirty years. Improvements in corrosion resistance are still obtained with a final rinse in a chromium (VI) based solution.

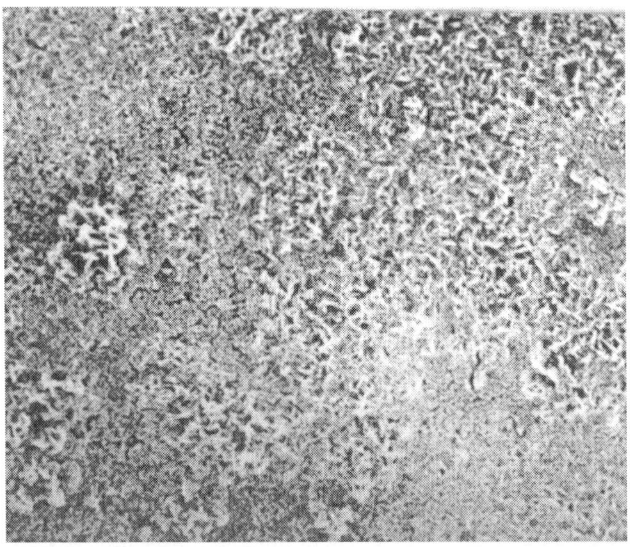

Figure 8. Relatively thin zinc phosphate conversion coating on zinc plated steel surface at 1000 magnification. Note huge number of attachment points for the application of a secondary coating - paint.

The most effective phosphating systems for the treatment of aluminum are those that contain chromium (VI) compounds and this continues to be the case. In many processing shops attempts will be made to use iron and / or zinc phosphating baths to process aluminum. This will result in deposits of zinc or iron on the surface of the aluminum giving poor adhesion and galvanic corrosion due to the presence of dissimilar metals on the surface of the aluminum which will result in the pitting of the aluminum. In addition, aluminum is a negative catalysis to the phosphating of zinc and iron [14] with 0.76 g/L being enough to poison the bath toward zinc or iron phosphating. Phosphating processes are quite similar in composition with the exception of relatively minor variations in the oxidizing agents used and / or the heavy metals added to the bath. They require constant monitoring to maintain them within a narrow pH and concentration range, but are relatively inexpensive to operate with the exception of the waste disposal cost. Chromium (VI) phosphate based coatings on aluminum are generally 200 to 300 nm thick and a light green in color due to extensive reduction to chromium (III) during the phosphating process which serves to identify them as being chromium (VI) based coatings.

3.4. Synthetic Polymer Based Conversion Coatings

Synthetic organic based polymers for improving the adhesion of water based paints were developed in the early 1970's and are quite similar in composition to that of tannic acids which have been used since the 19th century. In this application, they work quite well and provide some corrosion resistance to humid conditions on zinc and in particular iron and steel. They will not provide for any salt spray corrosion resistance unless first mixed with chromium (VI) salts, as is often done when they are used on aluminum. As they are very effective wetting agents, they will in many cases mask the presence of organic soils by giving a "water break-free" surface on soiled metals parts which then leads to adhesion failures. Like tannic acid, they are only effective at certain pH and concentration levels and as a result require constant monitoring. As a general rule they will begin to decompose at about 165 °C, leaving a brown stain on parts being power coated. As a such they do not work well in many powder coating applications. Polyamide and / or polyamine based polymer systems have recently come into use as a replacement for chromate's to prevent the staining of tin plated steel and silver [15] that will not decompose at temperatures up to 276 °C.

Since the late 19th century, various polymeric networks of metal and silicon oxides have been used to prepare glass and ceramic surfaces to accept applied coatings [16]. Starting in the 1970's, the process was used to improve the adhesion characteristics of metals using a low temperature application of various inorganic jells of transition metal alkoxides and silicon alkoxides. This has become known as the "sol – gel" process as a colloidal suspension ("sol") is first formed and then caused to undergo gelation ("gel") to form a unique inorganic network of transition metal oxides and silicon oxides [17]. If properly prepared and properly applied, these coatings will provide for excellent adhesion, but little corrosion resistance. However, the degree to which the metal surface needs to be cleaned is quite extreme and only obtained on a consistent basis in a laboratory setting. The films are quite thin (about 40 to 100 nm) and difficult to detect and determine if they have been properly applied until one has an adhesion failure as they give no indicating color change. Corrosion resistance, in terms of salt spray resistance will only be obtained with a secondary dip in a chromate containing bath.

Solutions of various rare earth salts of cerium, lanthanum, or yttrium will produce a conversion coating consisting of the various hydroxides of these metals and the metal to which they are applied in a manor similar to that of chromium (VI) conversion coating systems [18]. That is to say an inorganic polymer held together with water of hydration. As in the case of the chromium (VI) based conversion coatings, this will not stand up to the temperatures used in powder coating operations due to the loss of the water of hydration and it is not held together nearly as strongly as the chromium based conversion coating system. The rare earth metals used are quite rare and quite expensive although not at the moment considered to be toxic. Used by themselves, they will not meet the corrosion requirements expected for aluminum as outlined in United States military specification "MIL-DTL-5541F," but will show a significant improvement in corrosion resistance. The coatings will range in color from "clear" (about 323 mg / m^2) to yellow (about 645 mg / m^2) and as such will allow a quick visual determination of presence of the coating. These coatings may be sealed with chromium (VI) or other inorganic compounds (silicates) to improve upon their adhesion and / or corrosion resistant properties.

Cobalt compounds which have been complexed with ammonia will generate a conversion coating of various cobalt hydroxides and / or oxides on metal surfaces [19]. This is once again a polymer based upon water of hydration which will decompose when used in powder coating applications. Thin films (about 30 nm) will be essentially "clear" while thicker films (about 200 nm) will be yellow in color. The metals to which it is applied must of necessity be quite clean and the cleaning stages are quite extensive. Without an additional seal in a chromium (VI) solution they will not provide sufficient corrosion resistance, but the adhesion properties are acceptable. Cobalt has been banned from use in California and is known to be toxic. In addition, it is relativity rare and rather expensive.

CONCLUSION

Chromium (VI) compounds will produce the most corrosion resistant conversion coatings available due to their ability to form a unique inorganic polymer system that very effectively seals the surface of the metal to which they are applied an as a result enjoy a rather unique position in metal finishing similar to that of silver in photography. For that reason, they have been in use for almost a hundred years in one way or another and would still be widely used were it not for their toxicity. Any replacement must of necessity be as closely related to chromium and its chemistry as possible while at the same time being non – toxic, readily available and inexpensive. Manganese sits next to chromium in the periodic chart and as expected, the chemistry of the manganese (VII) oxidation state is the most closely related to that of chromium (VI) as are the mixed metal oxides they both form with other metals such as zinc. The zinc hydroxide, manganese hydroxide inorganic polymer system being the most closely matched to that of the corresponding chromium / zinc hydroxide based system and indeed the corrosion resistance of the manganese / zinc system is every bit as good as the chromium / zinc inorganic system. Cobalt is relatively rare and is now on the list of toxic materials for use in California and many other areas of the world. In addition the inorganic polymer system it forms requires a degree of metal preparation and process control very difficult to maintain outside of a laboratory. The rare earth metals are quite rare, expensive and once again require very specific conditions of surface preparation and process control. "Sol-Gel" chemistry will produce excellent adhesion characteristics, but little or no corrosion resistance and once again, need very specific and difficult to achieve degrees of surface preparation in "real world" metal working facilities. Organic polymer systems are quite good when used to enhance the adhesion of water based coatings, but provide little or no corrosion resistance and will decompose when heated to the temperatures required in many power coating applications. Phosphate based systems are quite good for adhesion, but offer little or no corrosion resistance, will often decompose due to the loss of water of hydration when heated for use in power coating applications and actively increase the degree of unwanted plant growth. They are, however, quite inexpensive.

Figure 9. Zinc electroplated bolts conversion coated with chromium (III) (left) convert to chromium (VI) after exposure to 100% humidity at 49°C for 14 days.

REFERENCES

[1] Definitions Committee, Federation of Societies for Coating Technology, Philadelphia, 1978, p.177.

[2] Spring, S. Industrial Cleaning; Prism Press, Melbourne, 1974, p. 20.

[3] Mason, S. R. *U.S. Patent*, 1,827,204, U.S.A., 1927.

[4] Spring, S. Industrial Cleaning, Prism Press, Melbourne, 1974, p. 144.

[5] Tadros, M. E. *U.S. Patent*, 4,883,541, U.S.A., 1989.

[6] Agarawala, V. S. *U.S. Patent*, 5,304,257, U.S.A., 1993.

[7] Bibber, J. W. Non-Chrome Containing Conversion Coatings for Zinc and Zinc Alloys; *Metal Finishing*, Vol. 106, No. 4, 2008, p. 41-46.

[8] Rochester, T., Kennedy, Z. W. Unexpected Results from Corrosion Testing of Trivalent Passivates; *Plating and Surface Finishing*, Vol. 10, No. 14, 2008, p.94.

[9] British Military Specification: DEF STAN 03-18/1 (2000).

[10] Wernick, S., Pinner, R. The Surface Finishing and Treatment of Aluminum and its Alloys, ASM International, Metals Park, OH. 1987, p.264.

[11] Danilidis, I., Hunter, J., Scamans, G. M., Sykes, M., Manganese - Based Conversion Treatment; University of Oxford, OX13PH, England, Alcan International Ltd., Banbury, Oxon, OX1675P, England, 1998.
[12] Drinking Water Systems Components, Component Materials and Treatment Additives in Accordance with ASNI/NSF 60 and 61", Underwaters Laboratories, Inc.,1997.
[13] Biestek, T., Weber, T. Electrolytic and Chemical Conversion Coatings; Portcullis Press, Ltd., Redhill, Surrey, U.K. 1976, p. 200.
[14] Biestek, T., Weber, T. Electrolytic and Chemical Conversion Coatings; Portcullis Press, Ltd., Redhill, Surrey, U.K. 1976, p. 198.
[15] Bibber, J. W. *U.S. Patent*, 6,830,821, U.S.A. 2002.
[16] Brinker, C.J., C. Scereer Sol-Gel Science; Academic Press, New York; 1990.
[17] Filliagi, M .J. et. Al. *J. Biomed. Mat.*, Vol. 33 p.225,1996.
[18] Mansfeld, F., et. Al. *Corrosion*, Vol. 45, No..8, 1989, p. 615.
[19] Carson, L. R. Third Annual Workshop on Chromate Replacements in Light Metal Finishing, Albuquerque, N.M.; Sept. 15 - 16, 1993.

In: Management of Hazardous Residues Containing Cr(VI) ISBN: 978-1-61470-637-3
Ed: Maria José Balart Murria

Chapter 5

VOLUME REDUCTION OF CR(VI)-BEARING SORBENT MATERIALS

C. Barrera-Díaz[1*], B. Bilyeu[2], L. A. Bernal-Martínez[1] and J. Cruz-Olivares[1]

[1] Universidad Autónoma del Estado de México. Facultad de Química. Paseo Colón intersección Paseo Tollocan S/N. C.P. 50120, Toluca, Estado de México, México
[2] Xavier University of Louisiana, Department of Chemistry, 1 Drexel Drive, New Orleans, Louisiana 70125, USA

ABSTRACT

Natural and formaldehyde-treated ectodermis of Opuntia were evaluated as Cr(VI) sorbents, in terms of both effectiveness and consolidation or mass reduction before disposal. Under optimum conditions at pH 4, natural ectodermis of *Opuntia* (NE) reaches adsorption equilibrium in 60 minutes, whereas that pretreated with formaldehyde (PF) only took 5 minutes. The rate of the Cr(VI) adsorption by the PF followed pseudo-second order kinetics with rate constants of K_2 = 15.97 g mg/min. The mechanism of the adsorption fit best to the Langmuir model with constants of K_L = 0.257 L/mg and q_{max} = 54.971 mg/g, and a regression correlation of 0.981. The maximum removal of Cr(VI) was 79%. SEM and EDS confirm that Cr(VI) is adsorbed on the surface of the sorbent. In both sorbents, the drying process takes place in three stages: a short induction period, a rapid drying stage in which the drying rate is proportional to the free moisture of the product, and a slower third phase in which the mechanism responsible for the water outlet of the product is the molecular diffusion of moisture to reach equilibrium.

Higher drying temperatures increase the rate and decrease the time for drying in both, but the PF material dried faster. Thus, the formaldehyde treatment improves drying in addition to the sorption improvement.

* To whom correspondence should be addressed.
Tel.: + (52)-(722)-2173890; FAX: + (52)-(722)-2175109
E-mail address cbarrera@uaemex.mx

1. Removal of Cr(VI) From Wastewater Using Biosorbent

Hexavalent chromium Cr(VI) is carcinogenic and mutagenic, a strong oxidizing agent, and irritating to plant and animal tissue in low doses, as well as capable of diffusing readily through soil, water and even skin [1,2]. Unlike many metals, Cr(VI) is usually found as oxyanions in aqueous solution, specifically $Cr_2O_7^{2-}$, CrO_4^{2-} and $HCrO_4^-$, the relative distribution of which depends on the solution pH and concentration [3,4]. However, none of them form insoluble species, such that separation is not feasible from wastewater through a direct precipitation method [5].

Adsorption is an effective physiochemical wastewater treatment process for removing metal ions which produces high quality effluents [6]. The development of inexpensive adsorbents for the treatment of wastewaters is an important area in environmental science [7,8]. The removal of Cr(VI) from wastewater can be achieved via sorption-based processes, including synthetic resins [9,10], activated carbon [11], inorganic sorbent material [12], crab shell [13], brown seaweed [14], polyacrylamide-grafted coconut coir pith [15], wheat bran extract [16], roots of *Typha latifolia* [17], eucalyptus bark [18], carrot residue [19] or other biosorbents derived from biomass. Biosorbents are cheap, abundant, and environmentally friendly [20].

Although adsorption is effective, disposal of chromium-laden exhausted sorbent requires care and expense. Therefore, in this study, we evaluate dewatering and thermal processes that can be used to reduce the waste sorbent mass and volume.

2. Biosorbent Preparation

The ectodermis of *Opuntia* sp. was collected in the northern part of the Estado de México. It was rinsed with distilled water to remove dirt, then sun dried for 7 days. The dry biomass was crushed, milled and sieved through a No. 20 mesh, then stored in a desiccator.

2.1. Formaldehyde Treatment

Twenty grams of natural ectodermis of *Opuntia* sp. was treated in one liter of 0.2% (v/v) formaldehyde solution for 24 hours with continuous stirring. The solid biomass was filtered, rinsed with distilled water, and dried at 18 °C for 48 hours, then stored in desiccators [21,22]. Natural ectodermis of *Opuntia* sp. leaches brown or yellow components into aqueous solutions, so the biomass was treated with formaldehyde to extract the organics. The pretreatment with formaldehyde solution (0.2% v/v) resulted in a 4% weight loss due to removal of organic compounds but increased the drying rate.

2.2. Drying Procedure

The dehydration of the natural ectodermis of *Opuntia* sp. and the pretreated material was performed in a dryer at temperatures of 25, 50, 100 and 150 °C. Samples of approximately 10 g were individually weighed and distributed in metal trays of known weights. The sets (trays and samples) were placed in the drying chamber and the loss of water in time was measured by discontinuous weighing with ±0.01 g precision to a constant weight [23-25].

3. Characterization Methods

3.1. Scanning Electron Microscopy (SEM)

Topography and elemental analysis of certain features were performed on a JSM-6510LV microscope with energy dispersive x-ray spectroscopy (EDS).

3.2. Metal Detection in Aqueous Solution

The colorimetric method was used to measure the concentrations of the chromium species. A solution of 1–5 diphenylcarbazide in acid forms a pink complex with Cr(VI), which can be spectrophotometrically analyzed at 540 nm (Perkin-Elmer Lambda 25). All calibrations and procedures were carried out in accordance with AWWA standards [26].

3.3. Thermogravimetric Analysis (TGA)

The thermogravimetric analyses were carried out in a SDT Q600 TA Instruments analyzer, which was operated under a nitrogen atmosphere at a heating rate of 25 °C/min from 25 to 1000 °C.

4. Application of Kinetic and Isotherm Models to the Removal of Cr (VI)

4. Adsorption Kinetics

The adsorption equilibrium experiments were conducted to determine the adsorption capacity for Cr(VI) under a given set of conditions. Adsorption studies were carried out by batch technique to obtain rate and equilibrium date. The batch equilibrium tests were conducted at a constant temperature (18 ºC). A series of test tubes with 10 mL of 50 mg/L Cr(VI) solution and 120 mg of sorbent were agitated for specific times ranging from 1 min to 6 h, to determine the kinetics. The pH was maintained at 4.0±0.1 during the process. At specific time intervals, the solutions were centrifuged to separate the solid adsorbent from the supernatant. Hexavalent chromium in the supernatant liquid phase was analyzed by means of atomic absorption, and the adsorbent was dried and characterized using SEM. Triplicate experiments were done in order to average the results. Kinetic data was analyzed with the pseudo-first-order and pseudo-second-order equations [27-30]. The pseudo-first-order equation is as follows:

$$\frac{1}{q_t} = \frac{K_1}{q_1}\left(\frac{1}{t}\right) + \left(\frac{1}{q_1}\right) \tag{1}$$

The pseudo-second-order equation can be written as:

$$\frac{t}{q_t} = \frac{1}{K_2 q_2^{\ 2}} + \frac{t}{q_2} \tag{2}$$

where, q_t is the sorption capacity (mg/g) of Cr(VI) at equilibrium and at a time t, respectively, q_1 is the maximum adsorption capacity (mg/g) for pseudo first-order adsorption, K_1 is the pseudo-first order rate constant for the Cr(VI) adsorption process (min^{-1}), q_2 is the maximum adsorption capacity for the pseudo-second order adsorption (mg/g), K_2 is the pseudo second order rate constant (g/mg min).

4.1. Adsorption Isotherms

Langmuir and Freundlich models were used to fit the experimental data [31-33]. The Langmuir isotherm model assumes uniform energies of adsorption onto the surface with no transmigration of adsorbate in the plane of the surface. The Langmuir equation can be written in the following form:

$$q_e = \frac{Q_o K_L C_e}{1 + K_L C_e} \tag{3}$$

where q_e is the solid phase equilibrium concentration (mg/g), C_e the liquid phase equilibrium concentration (mg/L), K_L the Langmuir constant (L/mg), and Q_o is the maximum adsorption capacity (mg/g).

The Freundlich isotherm describes adsorption where the adsorbate has a heterogeneous surface with adsorption sites that have different energies of adsorption and which are not always available. The energy of adsorption varies as a function of the surface coverage and is represented by the Freundlich constant K_F (mg/g):

$$q_e = K_F C_e^{\ 1/n} \tag{4}$$

where q_e is the amount adsorbed (mg/g), C_e is the equilibrium concentration of the adsorbate (mg/L), and K_F (mg/g) and n (dimensionless) are Freundlich constants related to the adsorption capacity and adsorption intensity is the heterogeneity factor which has a lower value for more heterogeneous surfaces, respectively [34,35]. The adsorption isotherms for the biomasses were obtained using 10, 30, 50, 100, 150, 200, 250, 300, 350, and 400 mg/L of Cr(VI) concentrations.

5. Optimization on a Laboratory Scale of the Removal Process of Cr (VI) From Wastewater

Sorption kinetics is one of the most important characteristics of biosorption and is a very important factor in design and process optimization in industry [36, 37]. The effect of contact time on the adsorption of Cr(VI) using natural (NE) and pretreated with formaldehyde (PF) biosorbents are shown in Figure 1. The sorption kinetics has been extensively studied, and it has been observed that the sorption rate is rapid at the beginning of the process and then becomes slower as equilibrium is approached. In this study, the equilibrium time for NE was 60 min, whereas PF attained equilibrium within the first 5 min of contact. The kinetics models of pseudo-first-order and pseudo-second-order were applied and the kinetics constants for the biosorption of Cr(VI) are presented in Table 1.

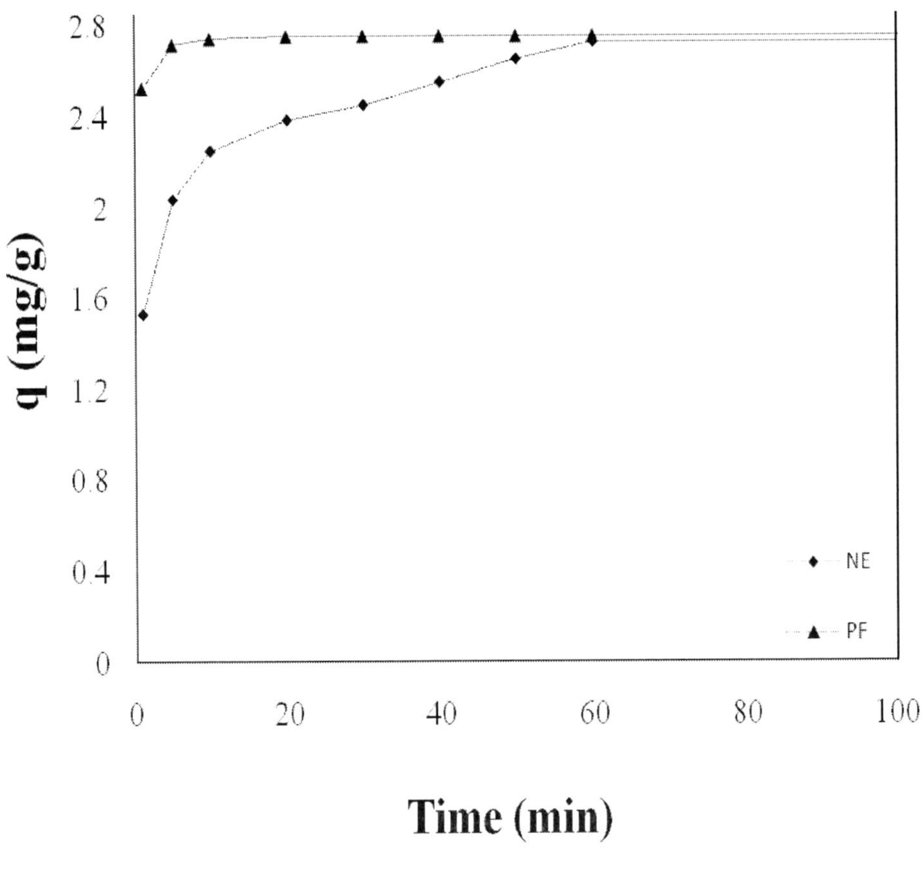

Figure 1. Kinetics of Cr(VI) biosorption from aqueous solution as function of time. The Cr(VI) concentration in the aqueous solution was 50 mg/L, the pH of the solution was 4.0 ± 0.1 at room temperature.

Table 1. Kinetics constants for the biosorption of Cr(VI) in pseudo-first and second order models

	Eq. (1) Pseudo-first order			Eq. (2) Pseudo-second order		
	K_1 (min^{-1})	q_1 (mg/g)	r^2	K_2 (g mg/min)	q_2 (mg/g)	r^2
NE	0.057	1.354	0.852	0.189	2.816	0.968
PF	0.827	1.306	0.869	15.97	2.696	0.990

Adsorption Isotherms. The adsorption isotherm studies are of fundamental importance in determining the adsorption capacity of Cr(VI) onto the biomass and to determine the mechanistic parameters associated with Cr(VI) adsorption, The results of the experiment were fitted according to the Langmuir and Freundlich models [28, 32]. Figure 2 shows the biosorption isotherm for Cr(VI) onto PF. Table 2 shows the Langmuir and Freundlich constants and correlation regression coefficients of Cr (VI) adsorption for NE and PF biosorbents. The adsorption isotherms of Cr(VI) uptake by PF at pH 4.0 ± 0.1 were fit to Langmuir and Freundlich models, with the Langmuir model better fitting the experiment data. The Langmuir biosorption isotherm assumes that biosorption takes place at specific homogeneous sites within the biosorbent and has found successful application in many monolayer biosorption processes [27].

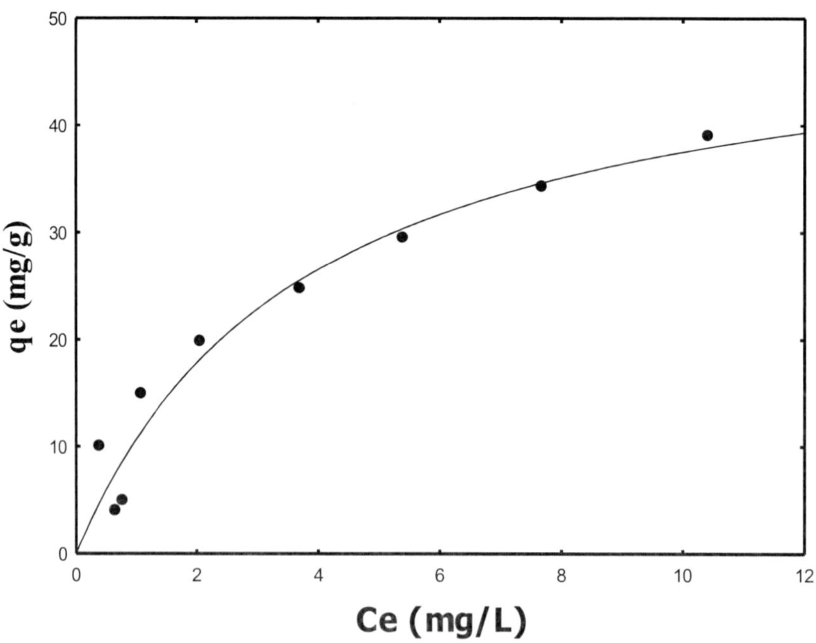

Figure 2. Adsorption isotherms for Cr(VI) onto PF at room temperature and pH 4.0 fit to the Langmuir model.

Table 2. Langmuir and Freundlich constants for the removal of Cr(VI)

Eq. (3)	**Langmuir**			**Eq. (4)**	**Freundlich**		
	K_L (L/mg)	**$q_{e\,max}$ (mg/g)**	**r^2**	**K_F (mg/g)**	**$1/n$**	**n**	**r^2**
NE	0.045	76.610	0.836	2.341	2.357	0.434	0.910
PF	0.257	54.971	0.981	12.093	0.543	1.869	0.950

6. Physical-Chemical Characterization Of Biosorbent

6.1. Scanning Electron Microscopy (SEM)

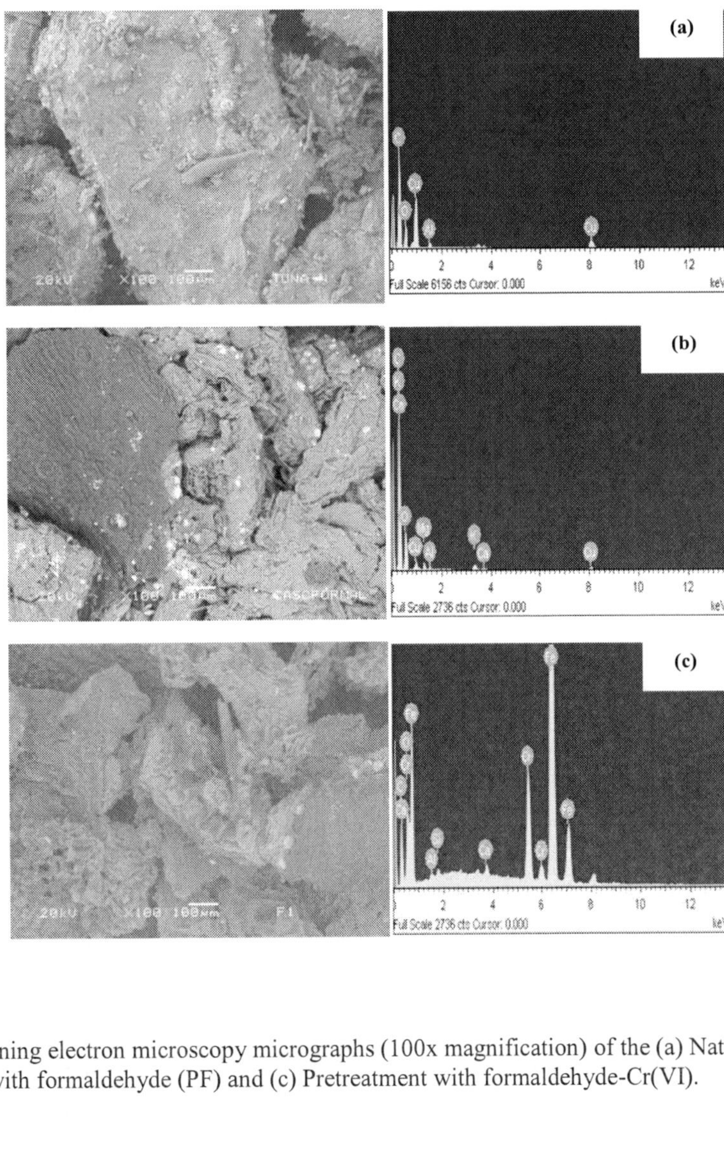

Figure 3. Scanning electron microscopy micrographs (100x magnification) of the (a) Natural (NE), (b) Pretreatment with formaldehyde (PF) and (c) Pretreatment with formaldehyde-Cr(VI).

SEM provides information about the morphology, the surface texture and elemental composition of the *Ectodermis* of *Opuntia* in its natural, pretreatment, and post-sorption states. The micrographs in Figure 3 show the surface morphologies. The surface of the natural in Figure 3a shows a heterogeneous structure with inorganic particles irregularly dispersed in the organic matrix. After pretreatment with formaldehyde, the morphology is more homogeneous as a consequence of the reaction between the hydroxyl groups of the cellulose with formaldehyde, Figure 3b. Elemental analysis can also be done during SEM by EDS analysis. As shown in Figure 3a, carbon, oxygen, copper and aluminum are the main elements present in the natural *Ectodermis* of *Opuntia*. After the pretreatment, the peak of Ca, K, Al, Mg and Cu are detected as shown in Figure 3b and in Figure 3c shows the presence of Cr over the surface of the control biomass was confirmed by EDX analysis.

6.2. Thermogravimetric Analysis (TGA)

The thermogravimetric behavior of the natural ectodermis of *Opuntia* and that pretreated with formaldehyde is shown in Figure 4. The TGA curves indicate that the pretreated material has a higher thermal stability than the natural one. The decomposition onset temperature of the pretreated material was 225 °C, while the natural ectodermis of *Opuntia* began at 219 °C. Both samples showed a moisture loss around 150 °C: the natural material lost about 6.5%, while the pretreated lost about 15%.

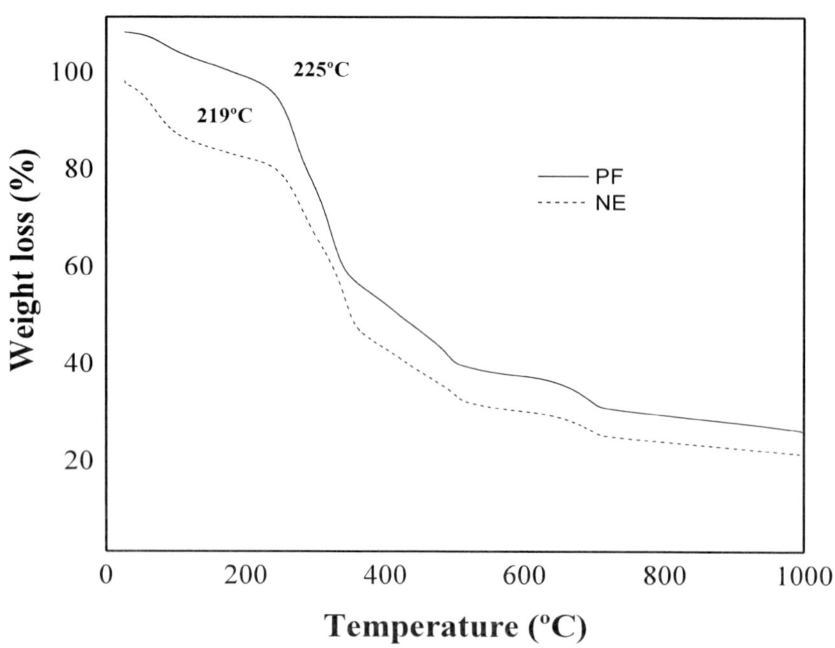

Figure 4. Thermogravimetric analysis of the natural (NE) and pretreated with formaldehyde (PF) ectodermis of *Opuntia*.

7. Drying Results of Biosorbent (Sludge)

The temperature effect on the dehydration of NE and PF, is shown in Figures 5 and 6. Higher drying temperatures increase the drying rate and decrease the drying time. The relationship indicates that diffusion is the main mechanism of the migration of moisture in the drying process of NE and PF. The greatest drying speed was obtained at 150°C, whereas it was lowest at 25°C. This behavior was described in studies of the drying of mint leaves [38] and onions [39].

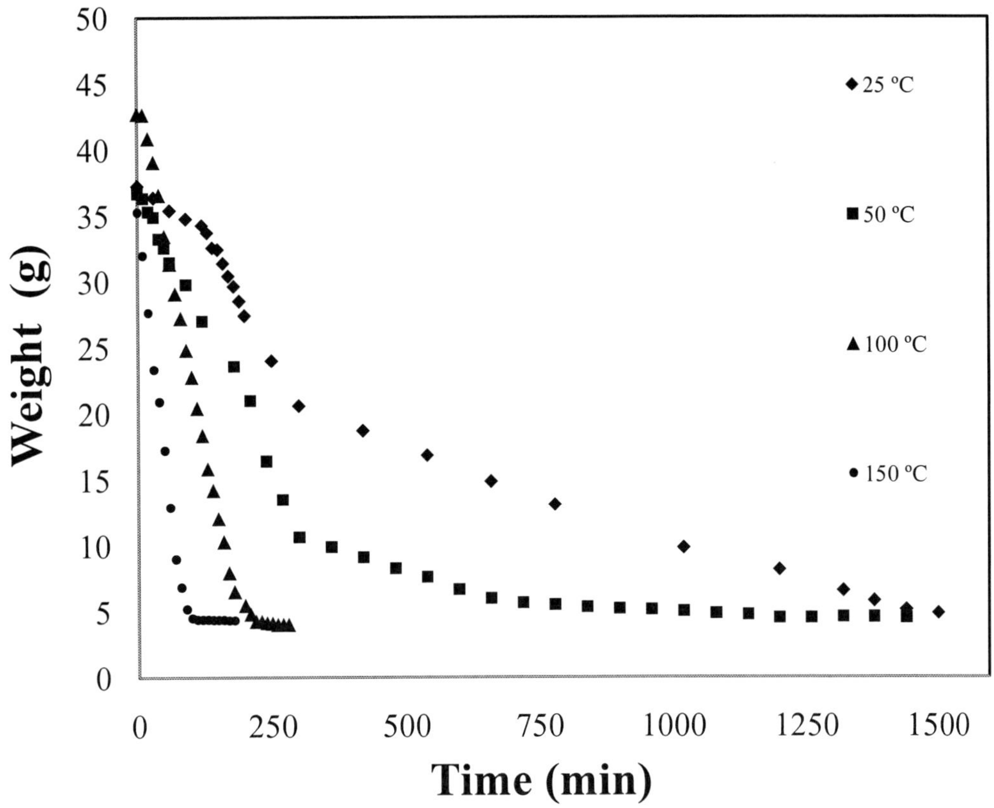

Figure 5. Weight loss as a function of time for natural ectodermis of *Opuntia* sp. at four temperatures.

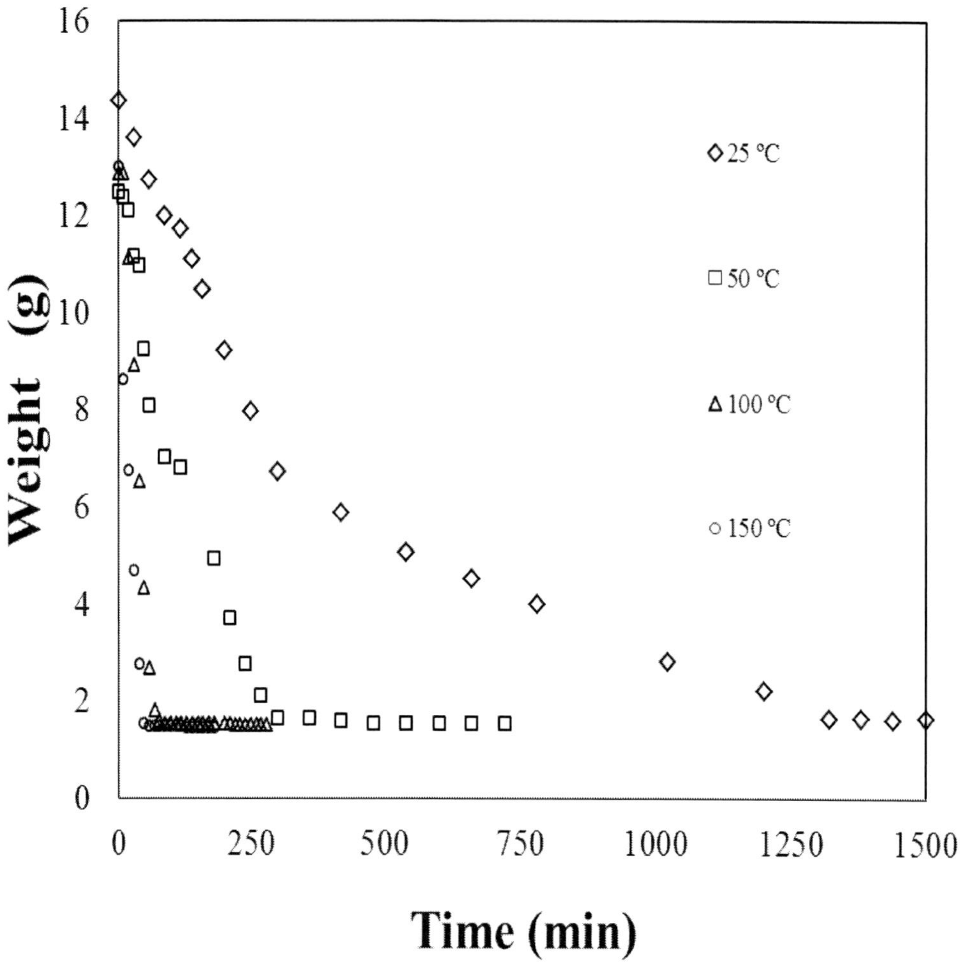

Figure 6. Weight loss as a function of time for pretreated ectodermis of *Opuntia* sp. at four temperatures.

The drying rate curves are shown in Figures 7 and 8, where the periods of induction, constant speed, and low speed are evident. The first phase (I) shows a short induction period where the material is adapted to the rapidly drying conditions. In the second stage (II), the drying rate varies linearly with moisture. Finally, in the third drying stage (III), the relationship between the moisture and dry weight tends to remain constant.

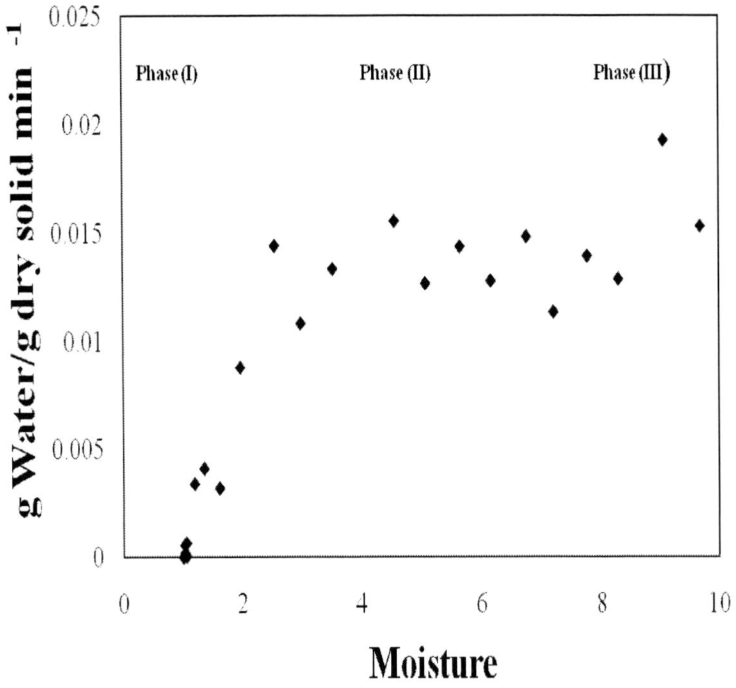

Figure 7. Drying rate curves of natural ectodermis of *Opuntia*.

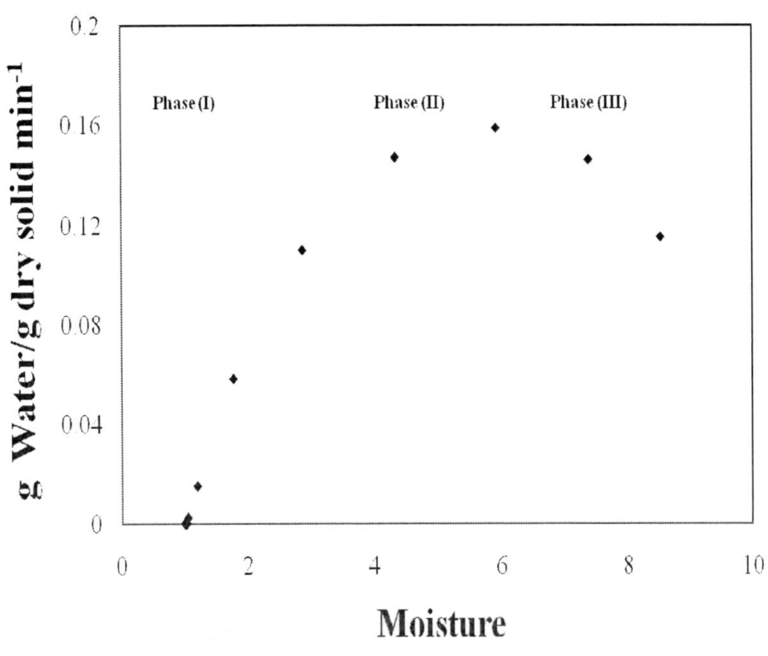

Figure 8. Drying rate curves of the pretreated ectodermis of *Opuntia*.

CONCLUSION

Ectodermis of *Opuntia* is an effective Cr(VI) sorbent and is enhanced by a formaldehyde pretreatment. The formaldehyde extracts organic components which tend to leach during adsorption and also interacts with the active hydroxyl groups on the sorbent. Under optimum conditions at pH 4, natural ectodermis of *Opuntia* (NE) reaches adsorption equilibrium in 60 minutes, whereas that pretreated with formaldehyde (PF) only took 5 minutes. The rate of the Cr(VI) adsorption by the PF followed pseudo-second order kinetics with rate constants of K_2 = 15.97 g mg/min. The mechanism of the adsorption fit best to the Langmuir model with constants of K_L = 0.257 L/mg and q_{max} = 54.971 mg/g, and a regression correlation of 0.9812. The maximum removal of Cr(VI) was 79%. SEM and EDS confirm that Cr(VI) is adsorbed on the surface of the sorbent.

In both sorbents, the drying process takes place in three stages: a short induction period, a rapid drying stage in which the drying rate is proportional to the free moisture of the product, and a slower third phase in which the mechanism responsible for the water outlet of the product is the molecular diffusion of moisture to reach equilibrium. Higher drying temperatures increase the rate and decrease the time for drying in both, but the PF material dried faster. Thus, the formaldehyde treatment improves drying in addition to the sorption improvement.

REFERENCES

[1] Stoepler, M., Hazardous Metals in the Environment Techniques and Instruments in Analytical Chemistry Elsevier, New York, 1992, 73.

[2] Gaulhofer, J.; Bianchi, V., Chromium, metals and their compounds in the environment VCH Whenheimer, 1991, 30.

[3] Dean, R. J.; Butlerb, O.; Fisherc, A.; Gardend, M. L.; Cressere, S. M.; Watkinsf, P.; Caveg, M. *J. Analytical Atomic Spectrometry*. 13.

[4] Barrera, D. C.; Palomar, P. M.; Romero, R. M.; Martínez, S. *J. Appl. Electrochem. 33*, 61-71.

[5] Nriagu, J.; Nioeber, E., Chromium in the Natural and Human Environments Wiley Series, New York, 82.

[6] Choy, K. K. H.; Porter, J. F.; McKay, G. *Langmuir. 20,* 9646–9656.

[7] Wagner, K.; Schulz, S. *J. Chem. Eng. Data. 46,* 322–330.

[8] Jain, A. K.; Gupta, V. K.; Jain, S.; Suhas. *Environ. Sci. Technol. 38,* 1195–1200.

[9] Sengupta, A. K.; Clifford, D. *Ing. Eng. Chem. Fundam. 25,* 249–258.

[10] Ureña, N. F.; Díaz, J. P.; Barrera, D. C.; Romero, R. M.; Palomar, P. M. *Rad. Phys. Chem. 68*, 819–825.

[11] Pérez, C. M.; Martín, M. J. M.; Torregrosa, M. R. *Water Res. 29,* 2174–2180.

[12] Lehmann, M.; Zouboulis, A. I.; Matis, K. A. *Chemosphere. 39,* 881–892.

[13] An, H. K.; Park, B. Y.; Kim, D. S. *Water Res. 35,* 3551–3556.

[14] Park, D.; Yun, Y.; Cho, H. Y.; Park, J. M. *Ind. Eng. Chem. Res. 43*, 8226–8232.

[15] Unnithan, M. R.; Vinod, V. P.; Anirudhan , T. S. *Ind. End. Chem. Res. 43*, 2247–2255.

[16] Dupont, L.; Guillon, E. *Environ. Sci. Technol. 37,* 4235–4241.

[17] Barrera, D. C.; Colín, C. A.; Ureña, N. F.; Romero, R. M.; Palomar, P. M. *Environ. Technol. 25,* 907–917.
[18] Kratochvil, D.; Pimentel, P.; Volesky, B. *Environ. Sci. Technol. 32*, 2693–2698.
[19] Sarin, V.; Pant, K. K. *Bioresource Technol. 97*, 15–20.
[20] Nasernejad, B.; Esslam, Z. T.; Bonakdar, P. B.; Esmaail, B. M.; Zamani, A. *Process Biochem. 40,* 1319–1322.
[21] Chen, P. J.; Yang, L. *Ind. Eng. Chem. Res. 44*, 9931.
[22] Bernal, M. L. A.; Hernández, L. S.; Barrera, D. C.; Ureña, N. F.; Bilyeu, B. *Ind. Eng. Chem. Res. 47*, 1026-1034.
[23] Silva, A. S.; Almeida, F. A. C.; Lima, E. E.; Silva, F. L. H.; Gomes, J. P. *Cienc. Tecnol. Aliment.* 6, 13-19.
[24] Krokida, M. K.; Karathanos, V. T.; Maroulis, Z. B.; Marinos, K. D. *J. Food Engineering. 59,* 391-403.
[25] Akpinar, E. K. *J. Food Engineering. 73*, 75-84.
[26] APHA, AWWA, WEF, Standard Methods for the Examination of Water and Wastewater; American Public Health Association, Washington, D.C. 19th Ed., 1995, 3-101-102.
[27] Ho, Y. S.; McKay G. *Adsorption Science and Technology*. 20, 8.
[28] Barrera, H.; Ureña, N. F.; Bilye, B.; Barrera, D. C. *J. Hazardous Materials. 136,* 846-853.
[29] Park, D.; Yun, Y. S.; Park, J. M. *Environ. Sci. Technol. 38,* 4860-4864.
[30] Shubo, D.; Yen, P. T. Environ. Sci. Technol. *39,* 8490–8496.
[31] Unnithan, M. R.; Vinod, V. P.; Anirudhan, T. S. *Ind. Eng. Chem. Res.* 43, 2247.
[32] Park, D.; Yun, Y. S. Y.; Park, J. M. *Chemosphere.* 60, 1356.
[33] Dupont, L.; Guillon, E. *Environ Sci. Techn. 37*, 4235-4241.
[34] Gupta, V. K.; Suhas, I.; Ali, V. K.; Saini. *Ind. Eng. Chem. Res.* 43, 1740-1747.
[35] Barrera, D. C.; Almaraz, C. C.; Olguín, G. Ma. T.; Romero, R. M.; Palomar, P. M. *Environmental technology*. 26, 821-829.
[36] Kurniawan, T. A.; Chan, G. Y. S.; Lo, W. H.; Babel S. *Science of the Total Environment.* 366, 409.
[37] Tunali, S.; Akar, T.; Özcan, A. S.; Kiran, I.; Özcan, A. *Separation and Purification Technology.* 47, 105-112.
[38] Doymaz, I. *J. Food Engineering*. 74, 370-375.
[39] Sharma, G. P.; Verma, R. C.; Pathare, P. B. *J. Food Engineering. 37,* 361-366.

In: Management of Hazardous Residues Containing Cr(VI) ISBN: 978-1-61470-637-3
Ed: Maria José Balart Murria © 2011 Nova Science Publishers, Inc.

Chapter 6

ADVANCES IN CHROMIUM (VI) SEPARATIONS TECHNOLOGY FROM WASTEWATERS

F.J. Alguacil, M. Alonso, F.A. López, A. López-Delgado* and I. Padilla

Centro Nacional de Investigaciones Metalúrgicas (C.S.I.C.), Avda. Gregorio del Amo 8, 28040 Madrid (Spain)

1. OVERVIEW

Because chromium is widely used in a number of industries, chromium-contained wastewaters can originate from myriad sources. Despite their presence, Cr(VI) compounds are regardless as highly toxic due to their various health and carcinogenic effects. Several technologies can be used to remove this toxic element from liquid elements, thus in this chapter, three separation technologies are considered, all of which can be used to improve the selective removal and concentration of the element from wastewaters or contaminated waters prior to its final disposal.

Liquid-liquid extraction is a proven technology for the recovery of heavy metals and other pollutants and is useful in medium to large scale operations, and when the solute concentration is high. Extractants are becoming increasingly selective, allowing specific compounds to be separated from the aqueous phase while others are retained in it. The capital outlay for such equipment can be expensive, large volumes of organic phases are required and performance would be limited by hydrodynamic constrains such as flooding and entrainment. Sometimes, the potential cross-contamination of the aqueous stream with the organic solution must be considered. In the operation, a water insoluble extractant dissolved in a suitable diluent is contacted with the aqueous phase and the metal or solute transfers into the organic phase. This loaded organic phase is then further contacted with a second or strip aqueous phase. The extracted species transfer back into the strip phase, from which it is finally processed. The organic phase is also regenerated and is returned to further extraction

* E-mail corresponding author: alopezdelgado@cenim.csic.es

operation. Due to the characteristics of chromium (VI) compounds two types of extractants are considered: basic (or anionic) and neutral or solvating reagents.

Liquid membrane technologies using organic phases extended the range of conditions under which liquid-liquid extraction processes are viable. The organic phase used in these technologies resembled very closely those used in the liquid-liquid extraction operation. The term liquid membrane is given to a system in which the membrane that separates the feed and stripping phases consists of a thin film or organic solution. Two types of liquid membranes are considered. The first, a surfactant or emulsion liquid membrane, consists of a water-in-oil emulsion formed from droplets of the aqueous strip solution contained in the organic phase. The latter consists of an organic diluent, an organic extractant or carrier to combine to the species to be transported and a surfactant to stabilize the emulsion. Once formed, this emulsion is put into contact with the aqueous feed solution in a suitable contactor. The solute of interest in the feed, reacts with the carrier phase at the feed-organic interface and migrate across the organic solution to the inner aqueous strip interface where it is stripped. The organic extractant which is regenerated, then migrates back to the feed interface and the extraction process continues. When the extraction is complete, the loaded emulsion phase is separated from the aqueous raffinate and it is broken to separate the aqueous strip and the organic phase. Then, all the above process is repeated. In the second type of liquid membranes considered here, the organic phase, now without the surfactant, is immobilized into the pores of a porous solid polymer, resulting in the so-called supported liquid membranes. The polymeric support, which is generally of a hydrophobic nature, consists of microporous membranes in the form of a sheet or hollow fiber. The polymer is wet by the organic solution which migrates (and fills) into the pores to produce the liquid membrane. This then forms a physical barrier between the aqueous feed and strip solutions. The mechanism of transport is somewhat the same as in the surfactant liquid membranes. Recent modifications or improvements to the supported liquid membranes operation mode included: non-dispersive solvent extraction (NDSX), pseudo-emulsion hollow fiber strip dispersion (PEHFSD) and hollow fiber renewal liquid membrane (HFRLM).

Adsorption and ion exchange processing are useful to remove species from non-clarified wastewaters and when the concentration of the pollutant is extremely low. In these technologies, a solid matrix replaces the extractant or carrier as the medium for mass transfer. The apparent lack of selectivity of various adsorbants, i.e. activated carbon, can be overcome by the use of byosorbants, i.e. chitosan, algaes, fungi, etc. Ion exchange technology consists in the exchange of ions in solution with the counterions of the same charge contained in insoluble synthetic polymers. Coordination and chelating ion exchange resins are polymers with covalently bound side chains containing donor atoms that are able to form complexes with selected metal ions or with their complexes. Since some of the properties of such exchangers are not good enough, especially when referring to selectivity, the concept of solid polymeric extractants (TVEX and Levextrel resins) and solvent impregnated resins (SIRs) have been developed.

2. Chromium (VI): Occurrence and Uses

The only chromium-bearing raw material of economic importance is chromite, though this mineral has a variable chemical composition, the ideal chemical formula for it is $FeO{\cdot}Cr_2O_3$. Major known chromium deposits occur in South Africa, Zimbabwe and Finland.

Since the industrial demand mostly calls for the intermediate compounds of chromium, only a little fraction of the chromite is processed to yield metal chromium. Besides the metal, various industrial applications call for differents forms of chromium, such as chromite itself, ferrochromium, chromite refractory bricks, chromite foundry sand, chromic acid, etc.

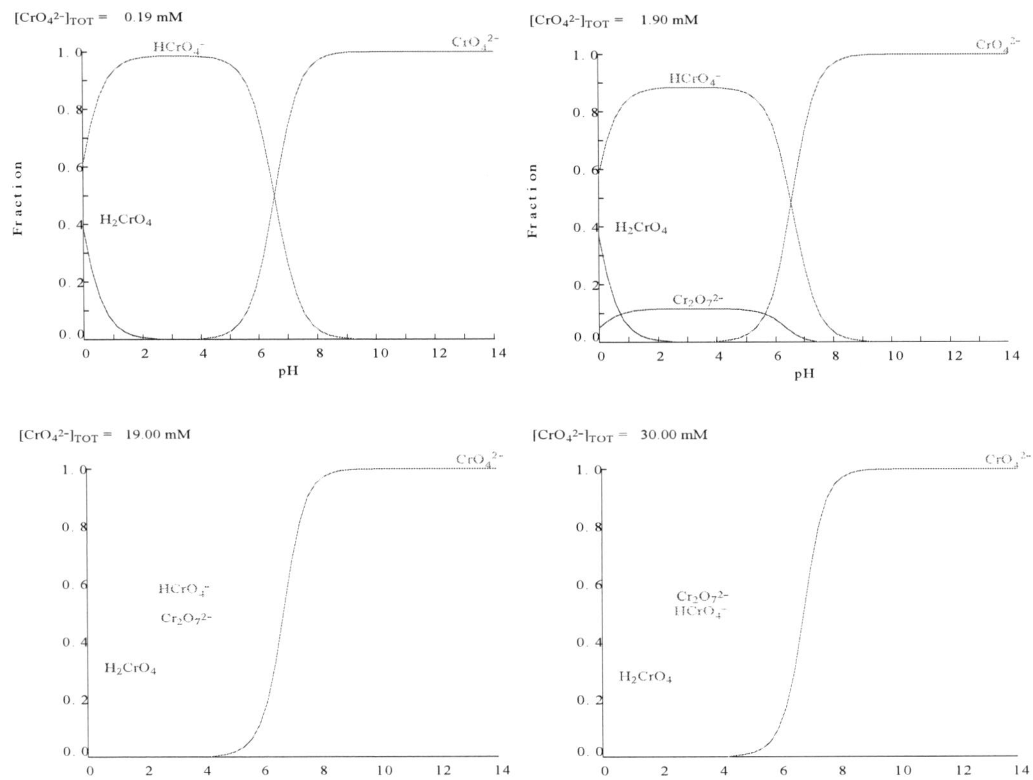

Figure 1. Chromium (VI) speciation at various pH values and total chromium concentration in the aqueous phase.

Being that chromium is widely used in a number of industries, chromium-contaminated liquid effluents or waters can originate from various sources. Consequently, these effluents may contain a wide range of concentrations (from the ppb to the hundreds of mg/L ranges) of either Cr(VI), Cr(III) or both.

In aqueous solutions, Cr(VI) exists as $HCrO_4^-$ and $Cr_2O_7^{2-}$ species at acidic-near neutral pH values, whereas at alkaline pH values CrO_4^{2-} is the only predominant species (Figure 1). Moreover, the predominancy of $HCrO_4^-$ and $Cr_2O_7^{2-}$ is dependent upon the total concentration of chromium (VI) in the aqueous media, being $HCrO_4^-$ the predominant species up to 10^{-2} M total Cr(VI) concentration.

3. Cancer and Other Health Effects of Chromium (VI) Compounds

A number of metals are of concern because of their toxic properties, whereas some metals are essential for survival and health of humans (and animals). Thus, metals have been classified as essential, beneficial or detrimental.

Table 1. Target sites of Cr(VI) compounds accepted as carcinogens in humans

TARGET SITE	REMARKS
Lung	Definitive
Sinonasal cavity	Definitive
Liver	Possible
Esophagus	Possible
Prostate	Suggested
Stomach	Suggested
Kidney	Suggested

Included in trace elements recognized as essential for human health is chromium. Paradoxically, severe exposure and intake of chromium compounds (specifically Cr(VI) compounds) lead to the consideration of the element as highly toxic for humans. Though cancer related to the respiratory system is the most common effect of Cr(VI), other cancers have also been associated with exposure to the various Cr(VI) compounds (Table 1).

Non-cancer diseases upon Cr(VI) intake are related to irritation of skin and nasal tissues and gastric, hepatic and renal distresses. Experimental data also suggests that concern about the potential adverse effects of chromium compounds on human reproduction and development are justified, even though there is no direct human evidence.

4. The Need For Chromium (VI) Removal From Liquid Effluents and Waters

Besides the concerns about the relationship between chromium (VI) and cancer and non-cancer diseases, industries are being forced either by regulatory, cost pressures and social pressures to minimize the amount and environmental sensitivity of the effluents they produced.

Anthropogenic activities are responsible for the presence of chromium in the environment. The presence of chromium in water is the result of mineral-weathering processes, though soluble organic chromium compounds and the mobilization of chromium compounds from sediments also contributed to the presence of the element, in whatever form, in waters or liquid effluents. Also, waters can be contaminated with liquid wastes from various industries (see Section 2). Additionally, solid wastes from these (chromium) industrial processes or municipal incineration can find its way and contaminated waters if not properly disposed of.

Liquid effluents have traditionally been treated by a combination of physico-chemical processes:

flocculation,
precipitation,
filtration,
and biological processes:
activated sludge,
biofilm,

which are usually effective in the removal of organic pollutants but fail to reduce the concentration of heavy metals to levels suitable for their safe disposal in whatever form. Also, these technologies usually do not provide the selectivity necessary for built-up streams with a value for their recycle or reuse, even in the same industry.

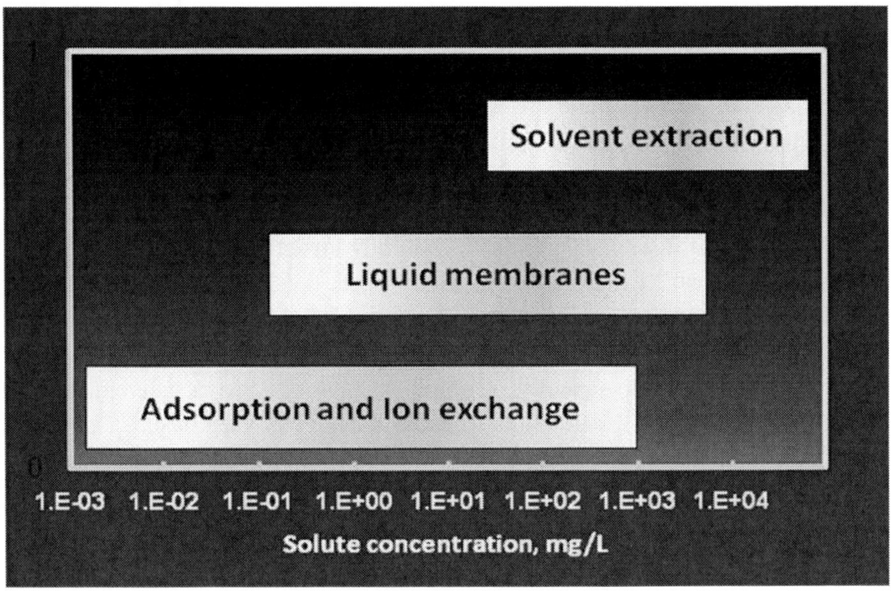

Figure 2. Approximate range of solute concentrations for liquid-liquid extraction, liquid membranes and adsorption and ion-exchange processes usefulness.

Three separations technologies (liquid-liquid extraction, liquid membranes and ion exchange and adsorption) can provide the means to split a given liquid effluent into two or more liquid phases that are either recyclable, saleable or waste.

By a first consideration of the chromium (VI) (or whatever solute) concentration in the liquid solution, each one of the above technologies are useful in a determined range of solute concentrations (Figure 2).

Liquid-liquid extraction can be used where the concentration of the solute is high, whereas ion exchange and adsorption processes are useful at very dilute solute concentrations. Liquid membrane technologies work appropriately between both of the above.

5. LIQUID-LIQUID EXTRACTION

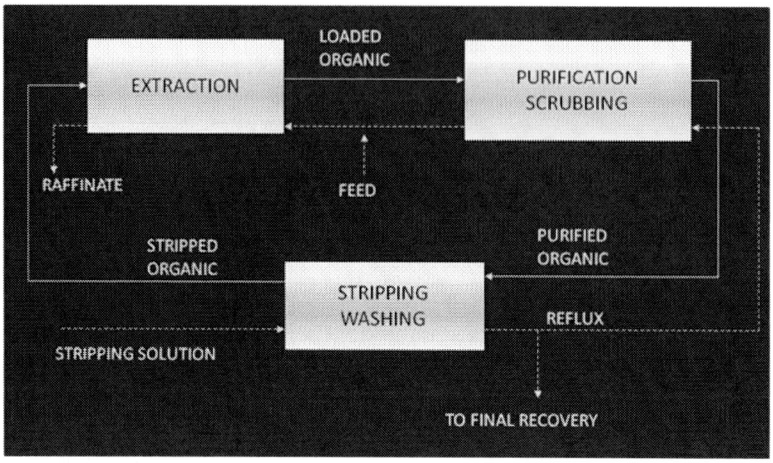

Figure 3. The process of liquid-liquid extraction.

The large amount of continuous research and development work, carried out over past decades, on both extractants and operational systems have resulted in the adoption of liquid-liquid extraction (LLE onwards) as a unit process in a number of hydrometallurgical operations for the recovery of heavy metals. As an extension of its applicability, LLE is considered the most established separations technology for liquid effluents recycling.

A scheme of a typical LLE circuit is shown in Figure 3. The metal-bearing liquid solution (i.e. a liquid effluent) and an organic solution are fed, usually in a counter-current mode, into a suitable contactor in which the two phases are mixed. As a consequence, the metal of interest is transferred from the aqueous to the organic solution. After convenient settling, the aqueous phase or raffinate is separated and treated for the recovery of other metals, recycled to some point upstream or downstream of the general process treatment of the given effluent, or waste.

The loaded organic solution goes to another stage in which it is scrubbed, with a suitable aqueous solution, to remove some co-extracted impurities from the above step, and then, the loaded organic solution passes to a further stage in which the metal is stripped from the organic solution by a suitable aqueous (stripping) solution producing a concentrated and purified solution of the metal, which is processed to metal (or metal compound) recovery, dumping, etc.

The now free-metal organic solution is back recycled to another extraction stage, since in the stripping stage it is conveniently regenerated, if not, the organic solution is further treated in another stage (not shown in the figure) prior to entering the extraction operation.

For the success of the whole operation, very often each one of the above steps may involve several contactors in each step.

The key to the operation is evidently the organic solution, this solution is normally composed of an extractant or mixed extractants and a diluent; sometimes a third component or modifier is also needed.

During all these years, extractants (an organic compound in itself) are becoming increasingly selective, allowing the specific separation of some solutes from another. In LLE practice three main types of extractants are considered:

Acidic and chelating reagents.
Basic or anionic reagents.
Neutral or solvation reagents.

Acidic and chelating reagents extracted metals by a cation- related exchange reaction:

$$nM^{+}_{aq} + nHR_{aq} \Leftrightarrow MR_{n_{org}} + nH^{+}_{aq} \tag{1}$$

where M represents the metal and HR the extractant. Since chromium (VI) only forms anionic compounds, this type of extractants can not be used for Cr(VI) removal from liquid effluents or wastes and they will not be further considered in this chapter.

Basic or anionic extractants, extracted anionic metal complexes by an anion exchange reaction, that in the particular case of a Cr(VI) compound is:

$$HCrO^{-}_{4\,aq} + R_3NH^{+}X^{-}_{org} \Leftrightarrow R_3NH^{+}HCrO^{-}_{4\,org} + X^{-}_{aq} \tag{2}$$

from the above, in the extraction of the metal compound the extraction reaction is shifted to the right;, aq and org subscripts refereed to the aqueous and organic solutions, respectively.

In order to achieve this exchange, the amine (R_3N) must be first converted to an appropriate amine salt to provide the anion (X^-) to exchange with the metal compound:

$$HX_{aq} + R_3N_{org} \Leftrightarrow R_3NH^{+}X^{-}_{org} \tag{3}$$

that is, the amine reacts with a i.e. mineral acid (HX) to form the amine salt in the organic solution. This reaction can occurs in the same extraction step of the metal itself, if suitable acidity to form the amine salt is available in the aqueous solution, or in a step prior to metal extraction. The above reaction means that amines can be used to remove inorganic acids, i.e. chromic acid, from a given effluent.

Table 2. Basic extractants

Type	Formula	Name	Manufacturer
Primary amine	RNH_2	Primene 81R	Rohm and Haas
		Primene JMT	Rohm and Hass
Secondary amine	R_2NH	Amberlite LA2	Rohm and Haas
Tertary amine	R_3N	Alamine 304	Cognis
		Alamine 336	Cognis
		Hostarex A324	Hoechst
		Hostarex A327	Hoechst
Quaternary amine salt	$R_3NCH_3^{+}Cl^{-}$	Aliquat 336	Cognis

R represents alkyl chains.

In any case, the amine salt should be considered as being the active extractant and not the free amine (though there are some exceptions to this general rule).

Basic reagents comprised amines and quaternary ammonium salts, some of these industrial available extractants are shown in Table 2.

It is clear, that if a quaternary ammonium salt is used, the reaction of the formation of the amine salt is not needed to extract the metal, thus these compounds can be directly used to extract metals even at alkaline pH values.

The stripping reaction can involve a variety of chemical reactions which basically depends upon the metals to be stripped.

The third type of extractant, neutral or solvating reagents, are based on the solvation of neutral inorganic molecules or complexes by electron-donor extractants. By means of this solvation, the solubility of inorganic species in the organic solution is increased.

Table 3. Neutral derivatives of organophosphorous acids and phosphine

Type	Active substance	Name	Manufacturer
Phosphate	$(RO)_3PO$	TBP	Various
Phosphonate	$(RO)_2RPO$	DBBP	Various
Phosphine oxide	R_3PO	TOPO	Various
		Cyanex 921	Cytec
		Cyanex 923	Cytec

R represents alkyl chains.

There are two main groups of extractants belonging to this type of reagents; extractants containing oxygen bonded to carbon, such as ethers, esters, alcohols and ketones, and those containing oxygen or sulphur bonded to phosphorous. Some examples of extractants with P=O bond are shown in Table 3.

Though it is very difficult to give an overall or general extraction reaction for these extractants, the extraction mechanism is mainly the same for all of them and it is the oxygen of the phosphoryl group responsible for the coordination bond with the metal species. These reagents can also extract acids from aqueous solutions, i.e:

$$H_2CrO_{4_{aq}} + nL_{org} \Leftrightarrow H_2CrO_4 \cdot L_{n_{org}} \tag{4}$$

where L may represents any of the compounds shown in Table 3, and n is the number of molecules of the solvation reagent involved in the reaction, this number must be determined from experimental results.

In the case of phosphates and phosphonates derivatives, there is the possibility of more than one coordination bonds formation through their other oxygen atoms, this influenced the extraction as well the stripping performance of a given system.

As in the case of amines, the stripping of neutral reagents is characteristic of each system, and a general rule may not be given; however, in a number of extraction systems involving these neutral extractants, water is an effective stripping reagent.

One of the features of these solvation reagents is that they can be used without diluent, that is in concentrations up to 100%, and thus improving their extraction abilities. The exception to this rule are TOPO, Cyanex 921 and Cyanex 471X, (alkyl thiophosphate derivative) which are solids.

Either of the extractants considered for LLE practice should have the following characteristics:

- be relatively inexpensive,
- have a low solubility in water (aqueous solutions) and high solubility in aliphatic or aromatic diluents,
- have good stability,
- will not form stable emulsions with the aqueous solutions, and should have good coalescing properties when mixed with diluents or modifiers (if necessary),
  - o have high metal (solute) loading capacity, and the most possible selectivity against other metals (solutes),
- be non-flammable (high flash point), non-volatile and non-toxic,
- have good kinetics of extraction and stripping,
- be easy to be stripped from metal (solute) loaded.

The second most important part of the organic phase in LLE operation is the diluent, which refers to the organic liquid in which the extractant and modifier are dissolved to form the organic solution. Often, the diluent comprises the major component of the organic solution. Diluents are usually relatively inexpensive crude oil fractions, and can be aliphatics, aromatics or mixtures of both types of oils. The major properties of the diluents in LLE practice are basically the same than that of the extractants, and they are used to:

- decrease the viscosity of the extractant,
- decrease emulsion-forming tendencies of extractants, and improving dispersion and coalescence properties of the organic phase,
- provide a suitable concentration of the extractant as required for each particular system.

Figure 4. Third phase formation in liquid-liquid extraction.

Diluents are generally considered inert and do not enter into the extraction mechanism, but this rule is sometimes far from reality, and in fact the properties of the diluent can greatly vary the results and performance of a given organic solution. Broadly speaking, the performance, and thus the choice, of a diluent for a particular process can not be predicted before experimentation.

A third component of the organic phase is the modifier. Sometimes, a third phase or a secondary organic phase, appeared in the settler either after extraction or stripping operation and after phases settling. The formation of this third phase can visualize in the left-hand side vessel of Figure 4, whereas in the right-hand side the correct appearance of the common two phases (organic and aqueous) can be also seen. The formation of this phase must be avoided, in operation a system with this third phase is unpracticable, and the problem is generally solved by changing the diluent of the organic solution or by adding a modifier. The modifiers are organic compounds, with the same properties of the extractants, and the most used are: alcohols (2-ethylhexanol, n-decanol, p-nonylphenol) and tri-n-butyl phosphate (TBP), some of these are in fact extractants by themselves. The required amount of modifier to be added to an organic phase to avoid the problem of third phase formation must be experimentally determined.

Figure 5. Mixer-settler units in liquid-liquid extraction operation.

Equipment in LLE extraction practice is related to mixer-settlers (Figure 5, in which two mixer-settler are shown under operation). The smaller vessels are the mixers in which the aqueous and organic phases are intimately mixed, and the emulsion passed continuously to the settlers (the bigger vessels) where emulsion splits into two phases, the upper corresponding to the organic phase, and the lower corresponding the organic phase. Also in the Figure, the principle of continuous counter-current operation can be seen, the organic phase from the left-hand settler unit passes continuously to the mixer of the right-hand unit, whereas the aqueous phase from the right-hand settler unit passes continuously to the mixer of the left-hand unit. The sequence is repeated as many units (or stages) are necessary and connecting the extraction, scrubbing and stripping operations. In continuous co-current operation mode, all the phases flows in the direction.

In industry, most common equipment is box-type mixer-settler units, though columns are not uncommon and also centrifugal separators can be found, especially if mud or a solid product is formed during the operation.

For liquid effluents with a low dilute concentration of the metal to be separated, the technology is limited by the need for high aqueous to organic ratios. Other odds in relation with the use of this technology in the treatment of a number of liquid effluents are summarized as:

capital outlay for equipment can be expensive,
relative large volumes of organic solutions are required,
performance may be limited by flooding and entrainment,
potential for cross-contamination of the aqueous stream with the organic solution,
aqueous solutions may be as well clarified as possible.

Recent advances (2004-2008 years) in chromium (VI) LLE included few references using alcohols, calyx[4]arene diamide derivatives, amines, quaternary ammonium salts, TBP and Cyanex 923.

6. Liquid Membranes

A membrane is a semi-permeable barrier through which selected species may diffuse. Normal membrane technologies, such as nanofiltration, ultrafiltration, etc, operate primarily through size exclusion, thus, the selectivity for specific solutes is somewhat limited.

Thus, combining the physically selective permeability of the above membrane operations with the chemically selective separation provided by the LLE process, greatly enhanced the performance of the membrane technology, and then the concept of liquid membranes (LMs) arises.

Two important phenomena must be immediately considered in LMs practice: the facilitated transport and the coupled transport.

The LM is located between a liquid effluent and a stripping solution, then the carrier phase, similar to the organic phase used in LLE operation, binds to the diffusing species on one side of the membrane, carrying it across to the other side and then, owing to the alternate chemical environment, releases it in the strip phase. This is the so-called facilitated transport.

The phenomena of coupled transport in LMs are due to that species are pumped "uphill" against its own concentration gradient as a consequence of an existing concentration gradient of a second species.

In LMs both types of transport occur simultaneously resulting in the facilitated coupled transport. Two types of facilitated coupled transport must be considered in LMs operations:

Countertransport. Metal (solute) and counter ion diffuse in opposite directions. This type of transport is typical of cationic/chelating extractants, and thus, due to the anionic nature of the chromium (VI) species in the aqueous solutions will not be further considered here.

Cotransport. Metal (solute) and counter ion diffuse in same direction. This is typical of basic and solvation extractants.

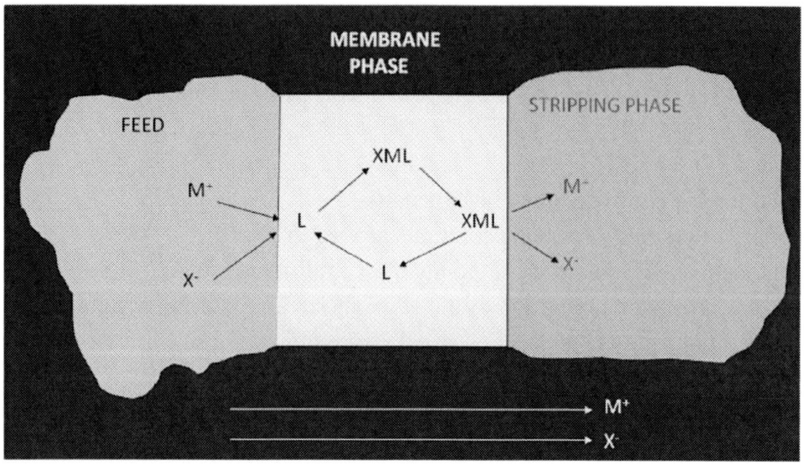

Figure 6. Mechanism of cotransport.

Figure 6 shows the principle of the cotransport mode using a solvation extractant (L), it can be seen that a neutral species is formed in the membrane phase, and both the metal (M^+) and the counter ion (X^-) flows in the same direction. Very often the metal forms an anionic complex in the aqueous feed solution and the counter ion is a cation.

Two types of LMs can be considered, unsupported and supported liquid membranes. In the first case, only emulsion liquid membranes will be described, since bulk liquid membranes has only a limited utility in lab and batch scale.

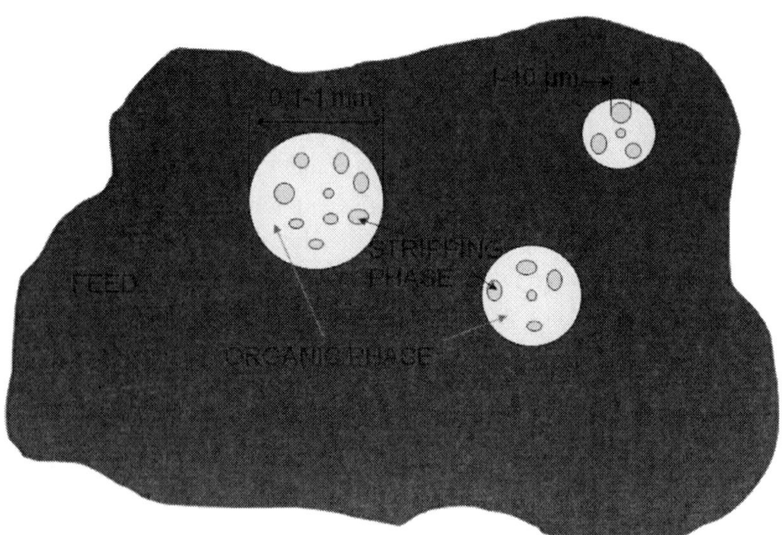

Figure 7. Schematic view of emulsion liquid membranes.

Emulsion liquid membranes (ELMs) are based in the formation of a double emulsion, being the concept shown in Figure 7. These are a three phase system in which the liquid membrane, containing the organic phase (extractant and diluent) and a surfactant, forms relatively large droplets which encapsulated the smaller drops of the strip solution. The emulsion so formed is stable due to the convenient size of the drops and the content of the emulsifier or surfactant. Then, this emulsion is dispersed into the aqueous feed solution containing the metal (solute) to be extracted, and the process of pertraction or mass transfer takes place. Once the process is ended, the feed solution and the liquid membrane/strip solution emulsion are separated, and this last emulsion is break-up to recover the loaded strip phase, which is further processed, and the liquid membrane, which is back-recycled to form a new first emulsion with new strip solution.

The process is quick due to the high interfacial areas produced by the emulsion, making ELMs attractive for the treatment of low concentrations of the permeating species. On the other hand, problems encountered with the operation are related with emulsion stability and liquid membrane rupture.

In the case of emulsion stability, this is refereed to the stability of the liquid membrane/strip solution emulsion, which is not always break-up easily. Sometimes this can be improve by the use of bi-functional surfactants, acting as extractant as well as emulsifiers.

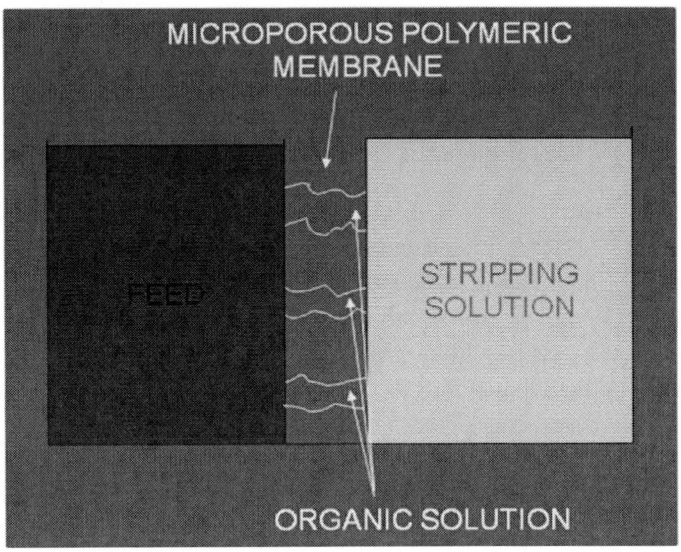

Figure 8. Schematic view of supported liquid membranes (flat-sheet configuration).

The problem of liquid membrane rupture is closely linked to membrane elasticity or rigidity, depending of the point of view. Sometimes this elasticity is a function of the temperature, thus, an increase in the temperature will help in the rupture of the emulsion. Also, long chain polymers can be added to the liquid membrane to decrease the rigidity of the membrane film.

Supported liquid membranes (SLMs) are based on the use of a porous solid membrane (polymer or ceramic) which support or holds the organic phase and separates the feed and stripping solutions (Figure 8). The pores of the thin solid membrane are completely filled with the organic solution or carrier phase and this impregnation produces relatively stable heterogeneous solid/liquid membranes). Often, this solid support is hydrophobic in its nature, which facilitated wetting by the organic solution and the reject of the aqueous phases.

Though flat-sheet supported liquid membranes are useful to obtain first data about the permeation of a given solute, for practical purposes, this configuration is not very effective and the hollow fiber and spiral modules are then used since they provide high surface area to volume ration and allow to work continuously and not only in batch conditions.

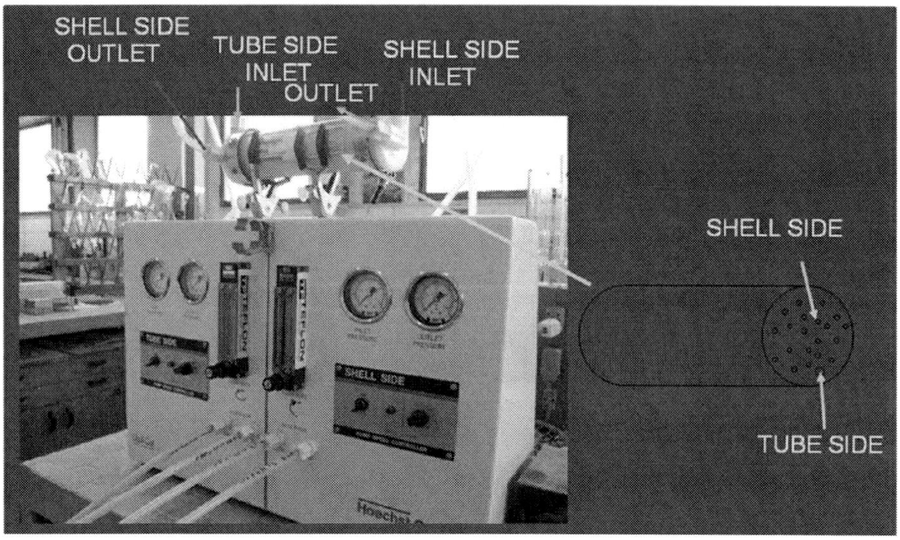

Figure 9. Hollow fiber module equipment.

In hollow fiber modules (Figure 9), the organic phase is confined, by capillarity, within the pores of the hollow fibers, and the aqueous feed and strip solutions circulate by each side of the fiber, thus, the solid membrane serves both as a support for the organic phase and as a uniform barrier between the aqueous solutions, which usually circulate within the inner (feed solution) and the outer (strip solution) of the fiber. This results in two aqueous/organic interfaces with well defined transfer areas.

Spiral wound modules consist of a sandwich of flat-sheet membranes alternated by spacers wrapped around a central axe.

From a practical point of view, one of the claimed advantages of supported liquid membranes is that the extraction, stripping and regeneration of the organic phase operations are combined and process simultaneously in a single step. Also, the small amount of organic phase used in operations makes feasible the use of expensive and more specific extractants. Lack of membrane stability under long-term operations is the major drawback associated with the use of SLMs because long-lived membranes are needed in order to minimize secondary wastes.

The instability of SLMs may be due either to the organic phase, the aqueous phases or the solid support itself.

Several attemps to improve the stability of these membranes include:

devices for the refilling, with fresh organic solution, of the solid support to compensate for the loss of the organic phase, sandwich type membranes, in which a small reservoir of the organic solution is placed between two or more supports,

gelation of SLMs, a gel is formed to avoid the flowing of the organic phase away from the gelled membrane,

polymer inclusion membranes, in which the solid support is created with the carrier phase included within,

other advanced technologies, though not essentially supported liquid membranes in themselves, used a solid microporous barrier to separate the aqueous phases from the organic solution. Included in this concept are: non-dispersive solvent extraction (NDSX), pseudo-emulsion hollow fiber strip dispersion (PEHFSD) and hollow fiber renewal liquid membrane (HFRLM) technologies.

In NDSX operation mode two hollow fibers modules, similar to those used in supported liquid membranes with hollow fiber modules, are needed to complete the operation; in the first module, the organic phase circulates by one side of the fibers, whereas the feed solution circulates by the other side of the fiber bundle, the loaded organic phase then passed to the second module where the stripping operation is carrying out. Then, the organic solution is recirculated to the first module.

By the use of dual function modules, the extraction and stripping of the solute are carried out simultaneously in a unique hollow fiber module containing one set of fibers for extraction and another set for stripping.

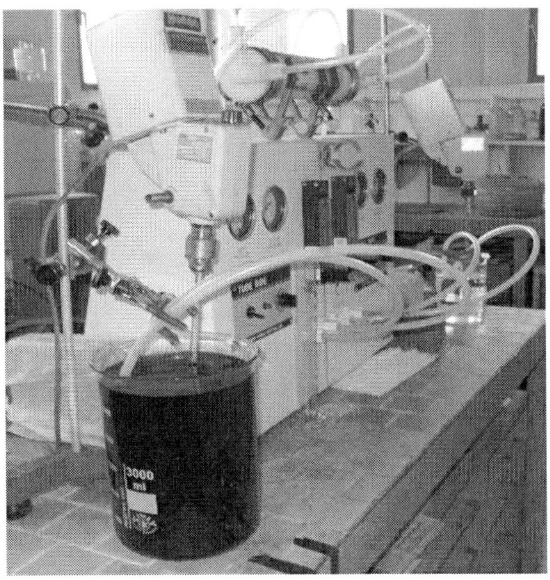

Figure 10. Pseudo-emulsion based hollow fiber with strip dispersion before operation.

PEHFSD processes comprise a unique membrane module for extraction and stripping, one stirred tank for feed phase homogenization and one stirred tank for preparation of a pseudo-emulsion (Figure 10). This pseudo-emulsion consists of an organic phase and a strip solution which is dispersed into the organic solution (Figure 11). Both phases pump the module by two gear pumps capable of variable flows. When the process begins (Figure 12), the organic phase of the pseudo-emulsion wet the microporous wall of the fiber because of its hydrophobic nature (Figure 13). When the operation ends (Figure 14), the characteristics of the pseudo-emulsion should be such that it should have clear and fast phase separation when mixing is stopped (Figure 15). The recovery of the extracted metal (solute) from pseudo-emulsion can be accomplished, pseudo-emulsion breaks down after the mixing of the organic and receiving phases stopped, and strip and organic phases separate (typically in very few minutes) automatically.

Figure 11. Detail of the organic/stripping phases reservoir tank before operation.

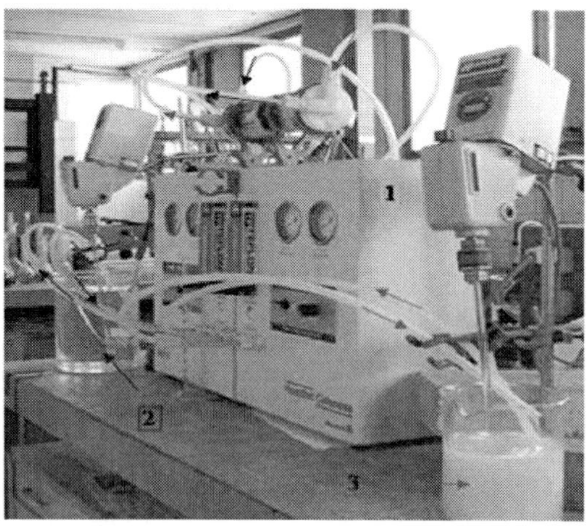

Figure 12. View of pseudo-emulsion based hollow fiber with strip dispersion under operation. 1: hollow fiber contactor, 2: feed; and 3: pseudo-emulsion solution.

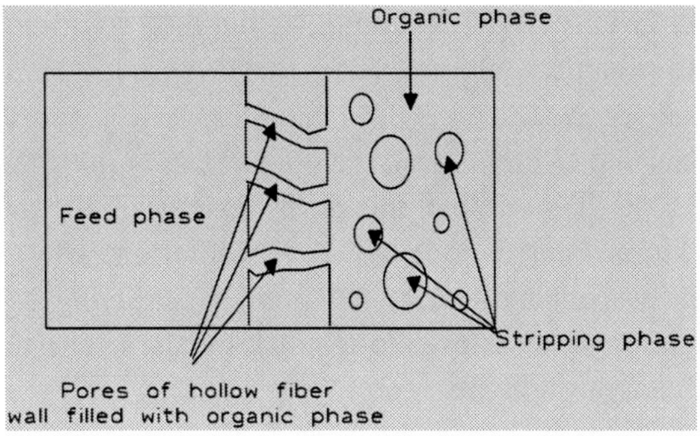

Figure 13. Enlarged view of pseudo-emulsion based hollow fiber with strip dispersion under operation and inside the module.

Figure 14. View of pseudo-emulsion based hollow fiber with strip dispersion at the end of the operation. Compare the aspect of the feed phase reservoir tank with that of Figure 10.

Figure 15. Detail of the organic/stripping phases reservoir tank after operation.

Typically, in NDSX and PEHFSD technologies the feed solution flows through the tube side of the fiber and the organic solution (NDSX) or the pseudo-emulsion phase (PEHFSD) flow through the shell side of the fiber.

In HFRLM operation mode the same device as above is used, but the organic phase is dispersed into the above and the pseudo-emulsion flows through the tube side of the fibers.

In the three above advanced membrane technologies, the interface between the aqueous/organic phases was maintained at the pore by applying a higher pressure to the aqueous phase than to the organic or pseudo-emulsion phases. The differential pressure was always kept below the breakthrough pressure.

Also, these technologies can work either in countercurrent or cocurrent flows of the respective phases, and in the recirculation or one through modes of the feed solution.

Chromium (VI) removal from liquid effluents using the various liquid membrane technologies described above drew great interest during the time period of 2004-2008, which is under consideration in this chapter. Several extractants had also been investigated as mobile carriers for this toxic element. Moreover, the use of a hydrid process (ion exchange plus NDSX) had been proposed as flow-sheet for the efficient chromium (VI) removal from an aqueous phase.

7. Adsorption and Ion-Exchange

The major difference between these technologies and LLE and LMs technologies are that in adsorption and ion-exchange, the active substance for mass transfer is contained in a solid matrix or is the solid matrix itself, and they can be used in non-clarified solutions. As it was pointed out before (Figure 2), these technologies are useful when the target species is in the ppm or less concentration range, at higher metal (solute) concentrations the size of the equipment should be such that make such technologies unviable.

As it can be seen from Figure 16, operationally both processes involve the scheme of sorption or loading, washing, elution and finally another washing, though this last step is not always needed. In columns, the adsorptive material packed the equipment, flowing the aqueous solution downstream (more often) or upstream.

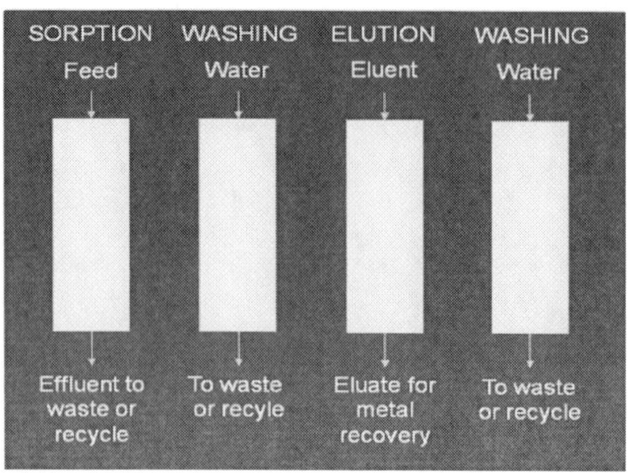

Figure 16. Sequence of operations in adsorption and ion-exchange processes.

A typical material for adsorption processes is activated carbon. They are prepared by heating (600º C), in absence of any gas, solid carbonaceous materials: coal, wood, nut shells, etc. The as-obtained material is further heated (400-800º C), in a gas environment, to render the final product. Most of them provided large surface area and high porosity and they can be wetted by water.

Table 4. Selected inorganic materials for Cr(VI) removal from aqueous solutions

Aluminium oxide	Fe@Fe2O3 nanowires (FCSNs)
Activated alumina	Magnetic nanoparticles
Activated clay	Manganese nodules residues
Bentonite derivative	Titanium oxide
CTAB-silica gelatine composite	Zero-valent iron

Other alternatives to the use of activated carbon to remove metals or solutes are: tailored bentonite clays, blast furnace sludges and rolling mill scales. Various other inorganic materials had also been investigated for chromium (VI) removal from different aqueous solutions (Table 4).

Table 5. Selected biomass materials for Cr(VI) removal from aqueous solutions

Agaricus bisporus	*Chlorella miniata*
Aspergillus flavus	*Cystoseira indica*
Bacillus sp.	*Fagus orientalis L.*
Borassus aethiopum	*Rhizopus arrhizus*
Cellulomonas sp. strain ES6	*Quercus ithaburensis*
Chalmydomonas reinhardtii	*Spirulina platensis/Chlorella vulgaris*

Biomaterials presented a higher selectivity degree that inorganic adsorbants and even ion exchange resins. Some of the biomaterials used to remove hazardous metals from liquid effluents are: chitosan, algaes, fungi, bacterial biomass, etc. To improve the effectiveness of biosorbants, they can be immobilised onto an inert framework. Table 5 showed some of biomass species used for chromium (VI) compound removal from different aqueous media.

Besides the above commented usefulness of adsorption techniques, sorbants are relatively cheap, not the case with activated carbon, as many of the sorbants are waste products themselves, and they are possible to use as decontaminant natural barriers.

The main drawback for such materials is their apparent lack of selectivity.

Table 6. Classification of synthetic resins

Type	Functional group	pH range of utility
Strong acid	$R\text{-}SO_3H$	0-14
Weak acid	R-COOH	4-14
Strong basic	$R\text{-}CH_2N^+(CH_3)_3X^-$	0-14
Weak basic	$R\text{-}CH_2N(CH_3)_2$	0-5

R is the inert matrix. X is the counter ion.

Ion-exchange resins can be classified as natural and synthetic. Among the first, zeolites (analite, chabazite), and fluoroapatites are included. Synthetic resins are produced by the reaction of styrene with divinyl benzene (which provides cross-linking capacity to the material). To this matrix a functional group is added. According to this functional group, four main types of resins are used (Table 6).

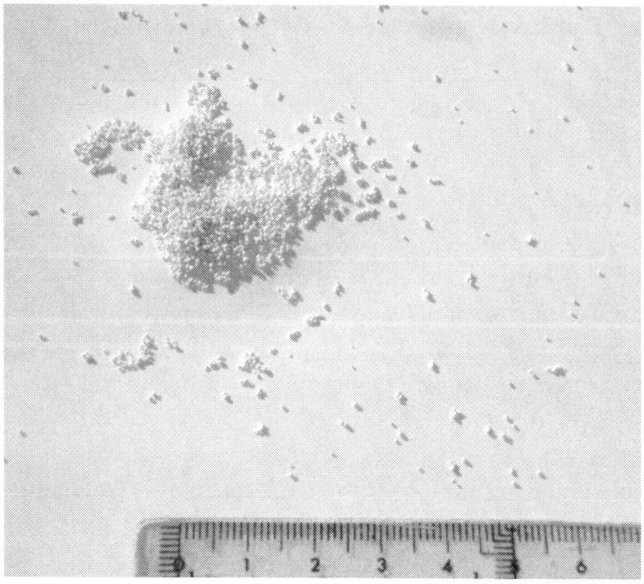

Figure 17. View of a typical ion-exchange resin.

Due to the characteristics of Cr(VI) compounds with strong and weak acid resins have no utility in the removal of such compounds from liquid effluents or waters.

These resins are usually produced as beads, generally in the less than 1 mm diameter range, a typical aspect of such resins is shown in Figure 17.

Similarly to the case of LLE extractants, ion-exchange resins are commonly known by the name given by their respective manufacturers rather than other terminology: Lewatit, Dowex, Amberlite, etc.

The mechanism of exchange with a given anion is similar to that showed in the case of amine extractants, i.e.:

$$R-CH_2N^+(CH_3)_3Cl^-_r + HCrO^-_{4\,aq} \Leftrightarrow R-CH_2N^+(CH_3)_3HCrO^-_{4\,r} + X^-_{aq} \quad (5)$$

where r and aq represents the resin and aqueous phases, respectively.

The main drawbacks for the use of resins are: their apparent lack of selectivity, they're somewhat difficult to eluate (strong basic resins) and are prone to swelling, physical breakdown, chemical degradation and solubility (osmotic shock).

Improvements to the use of these resins included:

improving their performance under extreme conditions,

improving their selectivity: coordination and chelating resins; solvent impregnation resins (SIRs); and solid extractants (Levextrel and TVEX resins).

Coordination and chelating ion exchange resins are polymers with covalently bound side chains containing donor atoms that are able to form complexes with selected metals or their complexes.

SIRs are prepared by adsorption of a liquid extractant on polymer supports.

Levextrel and TEVX resins are macroporous styrene-divinylbenzene copolymers containing a metal (solute) selective extractant entrapped within the polymer matrix.

The main difference between the SIRs and Levextrel and TEVEX resins, is that in SIRS the extractant impregnated the solid matrix, whereas in Levextrel and TEVX resins the extractant is added directly to the mixture of monomers during the bead polimerization process.

The adsorption techniques have received comparatively more attention in the literature, from 2004-2008, that the other separations technologies under consideration in this chapter. In particular, the possibility of using bisorption processes for Cr(VI) removal from various effluents (real and simulated) has been extensively investigated.

CONCLUSION

Three inter-related technologies capable of separating a Cr(VI)-bearing hazardous product streams from liquid effluents have been discussed.

The probable complexity of such liquid effluents makes exploring the usage of hybrid systems of interest: namely a combination of two of the above technologies, to, i.e. increase selectivity.

The target is the consecution of a *zero effluent* or even a *closed cycle process*, whereby all waste streams are recycled and the only output from the process is "at best" a saleable product.

And a last thought:

The waste from one can be the resource for another.

ACKNOWLEDGMENT

To the CSIC for support.

SELECTED LITERATURE

General

Alguacil, F. J., Villegas, M. A. Liquid membranas and the treatment of metal-bearing wastewaters. Revista de Metalurgia Madrid 38, 45-55 (2002).

Aguilar, M., Cortina, J. L. (Eds.) Solvent Extraction and Liquid Membranes (Fundamental and Applications in New Materials). CRC Press. Boca Raton (2008).

Gupta, C. K., Mukherjee, T. K. Hydrometallurgy in Extraction Processes, Vol. I. CRC Press. Boca Raton (1990).

Nriagu, J. O., Nieboer, E. (Eds.) Chromium in the Natural and Human Environments. Wiley Interscience. (1988).

Ritcey, G. M., Ashbrook, A. W. Solvent Extraction, Part I. Elsevier. Amsterdam (1984).

Trochimczuk, A. W., Kabay, N., Arda, M., Streat, M. Stabilization of solvent impregnated resins (SIRs) by coating with water soluble polymers and chemical crosslinking. Reactive and Functional Polymers 59, 1-7 (2004).

Sarkar, B. (Ed.) Heavy Metals in the Environment. Marcel Dekker. New York (2002).

Liquid-Liquid Extraction

Agrawal, A., Pal, C., Sahu, K.K. Extractive removal of chromium (VI) from industrial waste solution. Journal of Hazardous Materials 159, 458-464 (2008).

Beni, A., Karosi, R., Posta, J. Speciation of hexavalent chromium in waters by liquid-liquid extraction and GFAAS determination. Microchemical Journal 85, 103-108 (2007).

Bozkurt, S., Karakucuk, A., Sint, A., Yilmaz, M. Synthesis of two calyx[4]arene diamide derivatives for extraction of chromium (VI). Tetrahedron 61, 10443-10448 (2005).

Samaratunga, S., Nishimoto, J., Tabata, M. Separation of Cr(VI) from Cr(III) in $CaCl_2$ aqueous salt medium using hydrocarbons (as toluene, n-hexane, n-heptane or n-octane) mixed solvents of *ter*-butanol. Hydrometallurgy 89, 207-216 (2007).

Senol, A. Amine extraction of chromium (VI) from aqueous acidic solutions. Separation and Purification Technology 36, 63-75 (2004).

Venkateswaran, P., Palanivelu, K. Solvent extraction of hexavalent chromium with tetrabutyl ammonium bromide from aqueous solution. Separation and Purification Technology 40, 279-284 (2004).

Zhang, W., Liu, J., Ren, Z., Du, C., Ma, J. Solvent extraction of chromium (VI) with tri-n-butyl phosphate from aqueous acidic solutions. Journal of Chemical Engineering Data 52, 2220-2223 (2007).

Liquid Membranes

Alguacil, F. J., Alonso, M., Lopez, F., Lopez-Delgado, A. Uphill permeation of Cr(VI) using Hostarex A327 as ionophore by membrane-solvent extraction processing. Chemosphere 72, 684-689 (2008).

Alguacil, F.J., Lopez-Delgado, A., Alonso, M., Sastre. A.M. The phosphine oxides Cyanex 921 and Cyanex 923 as carriers for facilitated transport of chromium (VI)-chloride aqueous solutions. Chemosphere 57, 813-819 (2004).

Bringas, E., Fresnedo San Roman, M., Ortiz, I. Separation and recovery of anionic pollutants by the emulsion pertraction technology. Remediation of polluted groundwaters with Cr(VI). Industrial & Engineering Chemistry Research 45, 4295-4303 (2006).

Chiha, M., Samar, M. H., Hamdaoui, O. Extraction of Cr(VI) from sulphuric acid aqueous solutions by a liquid surfactant membrane (LSM). Desalination 194, 69-80 (2006).

Eliceche, A. M., Corvalan, S. M., Fresnedo San Roman, M., Ortiz, I. Minimum membrane area o fan emulsion pertraction process for Cr(VI) removal and recovery. Computers & Chemical Engineering 29, 1483-1490 (2005).

Galan, B., Calzada, M., Ortiz, I. Recycling of Cr(VI) by membrane solvent extraction: Long term performance with the mathematical model. Chemical Engineering Journal 124, 71-79 (2006).

Galan, B., Castañeda, D., Ortiz, I. Integration of ion exchange and non-dispersive solvent extraction processes for the separation and concentration of Cr(VI) from ground waters. Journal of Hazardous Materials 152, 795-804 (2008).

Kozlowski, C. A., Walkowiak, W. Applicability of liquid membrane in chromium (VI) transport with amines as ion carriers. Journal of Membrane Science 266, 143-150 (2005).

Kozlowski, C. A. Kinetics of chromium (VI) transport from mineral acids across cellulose triacetate (CTA) plasticized membranes immobilized by tri-n-octylamine. Industrial & Engineering Chemistry Research 46, 5420-5428 (2007).

Kulkarni, P. S., Kalyani, V., Mahajani, V. V. Removal of hexavalent chromium by membrane-based hybrid processes. Industrial & Engineering Chemistry Research 46, 8176-8182 (2007).

Kumbasar, R. A. Studies on extraction of chromium (VI) from acidic solutions containing various metal ions by emulsion liquid membrane using Alamine 336 as extractant. Journal of Membrane Science 325, 460-466 (2008).

Kumbasar, R. A., Selective separation of chromium (VI) from acidic solutions containing various metal ions through emulsion liquid membrane using trioctylamine as extractant. Separation and Purification Technology 64, 56-62 (2008).

Ren, Z. Q., Liu, J. T., Zhang, W. D., Du, C. S., Ma, J. N. Treatment of chromium (VI)-containing wastewater with hollow fiber renewal liquid membrane process. Electroplating Finishing 11, 20-23 (2006).

Venkateswaran, P., Palanivelu, K. Studies on recovery of hexavalent chromium from plating wastewater by supported liquid membrane using tri-n-butyl phosphate as carrier. Hydrometallurgy 78, 107-115 (2005).

Adsorption & Ion-Exchange

Acar, F. N., Malcok, E. The removal of chromium (VI) from aqueous solutions by *Fagus orientalis* L. Bioresource Technology 94, 13-15 (2004).

Ai, Z., Cheng, Y., Zhang, L., Qiu, J. Efficient removal of Cr(VI) from aqueous solution with Fe@Fe_2O_3 core-shell nanowires. Environmetal Science & Technology 42, 6955-6960 (2008).

Alvarez-Ayuso, E., Garcia-Sanchez, A., Querol, X. Adsorption of Cr(VI) from synthetic solutions and electroplating wastewaters on amorphous aluminium oxide. Journal of Hazardous Materials 142, 191-198 (2007).

Baral, S. S., Das, S. N., Rath, P. Hexavalent chromium removal from aqueous solutions by adsorption on treated sawdust. Biochemical Engineering Journal 31, 216-222 (2006).

Basha, S., Murthy, Z. V. P., Jha, B. Biosorption of hexavalent chromium by chemically modified seaweed, *Cystoseira indica*. Chemical Engineering Journal 137, 480-488 (2008).

Bhattacharya, A. K., Naiya, T. K., Mandal, S. N., Das, S. K. Adsorption, kinetics and equilibrium studies on removal of Cr(VI) from aqueous solutions using different low-cost adsorbents. Chemical Engineering Journal 137, 529-541 (2008).

Bishoni, N. R., Bajaj, M., Sharma, N., Gupta, A. Adsorption of Cr(VI) on activated rice husk carbon and activated alumina. Bioresource Technology 91, 305-307 (2004).

Chun, L., Hongzhang C., Zuohu, L. Adsorptive removal of Cr(VI) by Fe-modified steam exploded wheat straw. Process Biochemistry 39, 541-545 (2004).

Copello, G. J., Varela, F., Martinez Vivot, R., Diaz, L. E. Immobilized chitosan as biosorbent for the removal of Cd(II), Cr(III) and Cr(VI) from aqueous solutions. Bioresource Technology 99, 6538-6544 (2008).

Debnath, S., Ghosh, U. C. Kinetics, isotherm and thermodynamics for Cr(III) and Cr(VI) adsorption from aqueous solutions by crystalline hydrous titanium oxide. The Journal of Chemical Thermodynamics 40, 67-77 (2008).

Deepa, K. K., Sathishkumar, M., Binupriya, A. R., Murugesan, G. S., Swaminathan, K., Yun, S. E. Sorption of Cr(VI) from dilute solutions and wastewater by live and pretreated biomass of *Aspergillus flavus*. Chemosphere 62, 833-840 (2006).

Demiral, H., Demiral, I., Tumsek, F., Karabacakogiu, B. Adsorption of chromium (VI) from aqueous solution by activated carbon derived from olive bagasse and applicability of different adsorption models.

Elangovan, R., Philip, L., Chandraraj, K. Biosorption of hexavalent and trivalent chromium by palm flower (*Borassus aethiopum)*. Chemical Engineering Journal 141, 99-111 (2008).

Ertugay, N., Bayhan, Y. K. Biosorption of Cr(VI) from aqueous solutions by biomass of *Agaricus bisporus*. Journal of Hazardous Materials 154, 432-439 (2008).

Galan, B., Castañeda, D., Ortiz, I. Removal and recovery of Cr(VI) from polluted ground waters: A comparative study of ion-exchange technologies. Water Research 39, 4317-4324 (2005).

Garg, U. K., Kaur, M. P., Garg, V. K., Sud, D. Removal of hexavalent chromium from aqueous solution by agricultural waste biomass. Journal of Hazardous Materials 140, 60-68 (2007).

Gode, F., Moral, E. Column study on the adsorption of Cr(III) and Cr(VI) using Pumice, Yarikkaya brown coal, Chelex-100 and Lewatit MP 62. Bioresource Technology 99, 1081-1991 (2008).

Gode, F., Pehlivan, E. Removal of Cr(VI) from aqueous solution by two Lewatit-anion exchange resins. Journal of Hazardous Materials 119, 175-182 (2005).

Gokhale, S. V., Jyoti, K. K., Lele, S. S. Kinetic and equilibrium modeling of chromium (VI) biosorption on fresh and spent *Spirulina platensis/Chiorella vulgaris* biomass. Bioresource Technology 99, 3600-3608 (2008).

Han, X., Wong, Y. S., Wong, M. H., Tam, N. F. Y. Effects of anion species and concentration in the removal of Cr(VI) by a microalgal isolate, *Chiorella miniata*. Journal of Hazardous Materials 158, 615-620 (2008).

Hu, J., Lo, I. M. C., Chen, G. Comparative study of various magnetic nanoparticles for Cr(VI) removal. Separation and Purification Technology 56, 249-256 (2007).

Kabay, N., Solak, O., Arda, M., Topal, Ü., Yuksel, M., Trochimczuk, A., Streat, M. Packed column study of the sorption of hexavalent chromium by novel solvent impregnated resins containing Aliquat 336: Effect of chloride and sulfate ions. Reactive and Functional Polymers 64, 75-82 (2005).

Khezami, L., Capart, R., Removal of Cr(VI) from aqueous solution by activated carbons: Kinetic and equilibrium studies. Journal of Hazardous Materials 123, 223-231 (2005).

Kiran, B., Kaushik, A. Cyanobacterial biosorption of Cr(VI): Application of two parameter and Bohart-Adams models for batch and column studies. Chemical Engineering Journal 144, 391-399 (2008).

Kumar, R., Bishnoi, N. R., Bishnoi, G. K. Biosorption of chromium (VI) from aqueous solution and electroplating wastewater using fungal biomass. Chemical Engineering Journal 135, 202-208 (2008).

Leyva-Ramos, R., Jacobo-Azuara, A., Diaz-Flores, P. E., Guerrero Coronado, R. M., Mendoza-Barron, J., Berber-Mendoza, M. S. Adsorption of chromium (VI) from an aqueous solution on a surfactant-modified zeolite. Colloids and Surfaces A: Physicochemical and Engineering Aspects 333, 35-41 (2008).

Li, H., Liu, T., Li, Z., Deng, L. Low-cost supports used to immobilize fungi and reliable technique for removal hexavalent chromium in wastewater. Bioresource Technology 99, 2234-2242 (2008).

Malcok, E., Nuhoglu, Y. Determination of kinetic and equilibrium parameters of the batch adsorption of Cr(VI) onto waste acorn of *Quercus ithaburensis*. Chemical Engineering and Processing: Process intensification 46, 1020-1029 (2007).

Mallick, S., Dash, S. S., Parida, K. M. Adsorption of hexavalent chromium on manganese nodule leached residue obtained from NH_3-SO_2 leaching. Journal of Colloid and Interface Science 297, 419-425 (2006).

Maryuk, O., Pikus, S., Olszewska, E., Majdan, M., Skrzypek, H., Zieba, E. Benzyldimethyloctadecylammonium bentonite in chromates adsorption. Materials Letters 59, 2015-2017 (2005).

Memon, S., Tabakci, M., Roundhill, D. M., Yilmaz, M. A useful approach toward the synthesis and metal extraction with polymer appended thioalkyl calyx[4]arenes. Polymer 46,1553-1560 (2005).

Memon, S., Tabakci, M., Roundhill D. M., Yilmaz, M. Synthesis and evaluation of the Cr(VI) extraction ability of amino/nitrile calyx[4]arenes immobilized onto a polymeric backbone. Reactive and Functional Polymers 66, 1342-1349 (2006).

Misaelides, P., Zambouilis, D., Sarridis, Pr., Warchol, J., Godelitsas, A. Chromium (VI) uptake by polyhexamethylene-guanidine-modified natural zeolitic materials. Microporous and Mesoporous Materials 108, 162-167 (2008).

Mohan, D., Singh, D. M., Singh, V. K. Removal of hexavalent chromium from aqueous solution using low-cost activated carbon derived from agricultural waste materials and activated carbon fabric cloth. Industrial & Engineering Chemistry Research 44, 1027-1042 (2005).

Mor, S., Ravindra, K., Bishoni, N. R. Adsorption of chromium from aqueous solutions by activated alumina and activated charcoal. Bioresource Technology 98, 954-957 (2007).

Nakajima, A., Baba, Y. Mechanism of hexavalent chromium adsorption by persimmon tannin gel. Water Research 38, 2859-2864 (2004).

Nishihama, S., Nishimura, G., Hirai, T., Komasawa, I. Separation and recovery of Cr(VI) from simulated plating waste using microcapsules containing quaternary ammonium salt extractant and phosphoric acid extractant. Industrial and Engineering Chemistry Research 43, 751-757 (2004).

Nkhalambayausi-Chirwa, E. M., Wang, Y.-T. Modelling hexavalent chromium removal in a *Bacillus* sp. fixed-film bioreactor. Biotechnology and Bioengineering 87, 874-883 (2004).

Oh, Y. J., Song, H., Shin, W. S., Choi, S. J., Kim, Y.-H. Effect of amorphous silica and silica sand on removal of chromium (VI) by zero-valent iron. Chemosphere 66, 858-865 (2007).

Oliveira, D. Q. L., Gonçalves, M., Oliveira, L. C. A., Guilherme, L. R. G. Removal of As(V) and Cr(VI) from aqueous solutions using solid waste from leather industry. Journal of Hazardous Materials 151, 280-284 (2008).

Park, D., Yun, Y.-S., Kim, J. Y., Park, J. M. How to study Cr(VI) biosorption: Use of fermentation waste for detoxifying Cr(VI) in aqueous solution. Chemical Engineering Journal 136, 173-179 (2008).

Park, H.-Y., Tavlarides, L. L. Adsorption of chromium (VI) from aqueous solution using a imidazole functionalized adsorbent. Industrial and Engineering Chemistry Research 47, 3401-3409 (2008).

Pehlivan, E., Altun, T. Biosorption of chromium (VI) ion from aqueous solutions using walnut, hazelnut and almond shell. Journal of Hazardous Materials 155, 378-384 (2008).

Preetha, B., Viruthagiri, T. Batch and continuous biosorption of chromium (VI) by *Rhizopus arrhizus*. Separation and Purification Technology 57, 126-133 (2007).

Rajesh, N., Jalan, R. K., Hotwany, P. Solid phase extraction of chromium (VI) from aqueous solutions by adsorption of its diphenylcarbazide complex on an Amberlite XAD-4 resin column. Journal of Hazardous Materials 150, 723-727 (2008).

Ramnani, S. P., Sabharwal, S. Adsorption behaviour of Cr(VI) onto radiated crosslinked chitosan and its possible application for the treatment of wastewater containing Cr(VI). Reactive and Functional Polymers 66, 902-909 (2006).

Razmovski, R., Sciban, M. Biosprption of Cr(VI) and Cu(II) by waste tea fungal biomass. Ecological Engineering 34, 179-186 (2008).

Rivero, M. J., Primo, O., Ortiz, I. Modelling of Cr(VI) removal from polluted groundwaters by ion exchange. Journal of Chemical Technology & Biotechnology 79, 822-829 (2004).

Saha, B., Gill, R. J., Bailey, D. G., Kabay, N., Arda, M. Sorption of Cr(VI) from aqueous solution by Amberlite XAD-7 resin impregnated with Aliquat 336. Reactive and Functional Polymers 60, 223-244 (2004).

Sankararamakrishnan, N., Dixit, A., Iyengar, L., Sanghi, R. Removal of hexavalent chromium using a novel cross linked xanthated chitosan. Bioresource Technology 97, 2377-2382 (2007).

Silva, B., Figueiredo, H., Quintelas, C., Neves, I. C., Tavares, T. Zeolites as supports for the biorecovery of hexavalent and trivalent chromium. Microporous and Mesoporous Materials 116, 555-560 (2008).

Suksabye, P., Thiravetyan, P., Nakbanpote, W. Column study of chromium (VI) adsorption from electroplating industry by coconut coir pith. Journal of Hazardous Materials 160, 56-62 (2008).

Venditti, F., Ceglie, A., Palazzo, G., Colafemmina, G., Lopez, F. Removal of chromate from water by a new CTAB-silica gelatine composite. Journal of Colloid and Interface Science 310, 353-361 (2007).

Viamajala, S., Peyton, B. M., Gerlach, R., Sivaswamy, V., Apel, W. A., Petersen, J. N. Permeable reactive biobarriers for in situ Cr(VI) reduction: Bench scale tests using *Cellulomonas* sp. Strain ES6. Biotechnology and Bioengineering 101, 1150-1162 (2008).

Weng, C.-H., Sharma, Y. C., Chu, S.-H. Adsorption of Cr(VI) from aqueous solutions by spent activated clay. Journal of Hazardous Materials 155, 65-75 (2008).

Wu, Y., Ma, X., Feng, M., Liu, M. Behavior of chromium and arsenic on activate carbon. Journal of Hazardous Materials 159, 380-384 (2008).

Yakup Arica, M., Tuzun, I., Yalçin, E., Ince, O., Bayramoglu, G. Utilisation of native, heat and acid-treated microalgae *Chalmydomonas reinhardtii* preparations for biosorption of Cr(VI) ions. Process Biochemistry 40, 681-689 (2005).

In: Management of Hazardous Residues Containing Cr(VI) ISBN: 978-1-61470-637-3
Ed: Maria José Balart Murria

Chapter 7

THE RECOVERY OF CR (VI) IN MAGNETIC FE-MG OXIDE-COMPLEX FROM WASTEWATER

Wei Ma *

Dalian University of Technology, Department of Chemistry, 116023

ABSTRACT

At present, many methods have been used to treat waste water which contains Cr (VI). Magnetic particle adsorption has become a research hot spot and important subject in recent years because of its low cost and convenient operation by extra-magnetic separation. We use synthesized magnetic Fe-Mg oxide particles to recover chromate from Cr (VI)-containing wastewater. Series of magnetic ferric oxide with magnesium material were synthesized by chemical method. The crystal structure and Cr (VI) adsorption capacity were investigated, and the results indicated that the layer structure has the better Cr (VI) recovery capacity. Effects of pH, time, calcinated temperature, TDS, Cr (VI) initial concentration, ionic strength, and sorbent dosage on the Cr (VI) adsorption by Fe-Mg oxide were investigated. Results showed in optimal pH condition for Cr (VI) adsorption, the Cr (VI) removal efficiency was up to 90%. The adsorption process was very fast and finished within 40 minutes. The recovery rate decreased with the increase of ionic strength and sorbent dosage; decreasing with increased initial concentration. The common coexisting cation and anion in the solution have no effect on Cr (VI) adsorption capacity. Experimental data was analyzed by competed adsorption isotherms. Desorption, sorbent regeneration and recycle tests were also studied in 0.1M NaOH solution, which illustrated desorption processes completed with short cycle time and high efficiency. The inorganic magnetic Fe-Mg oxide can be used many times and is in favor of Cr (VI) recovery. The chemical elements analysis, XRD, SEM combined binding energy were used to characterize Fe-Mg oxide before and after adsorption process. The adsorption mechanism of Cr (VI) onto Fe-Mg oxide was proposed and the hydroxide group action was discussed in this study.

---

* Tel: +86-411-84706030 Fax: +86-411-84707416
Email: chmawv@yahoo.com

1. INTRODUCTION

Toxic heavy metal pollutants in water are severe environmental problems because they are not biodegradable. Chromium has been put on the top list of toxic pollutants by the USEPA [1] and the Ministry of Environmental protection of China. Removal of hazardous compounds from industrial effluents is one of the growing needs in the present day. Various techniques have been employed for treating Cr (VI) contaminated water, such as chemical redox followed by precipitation, ion exchange, reverse osmosis and adsorption/biosorption. But a major drawback related to ion exchange and reverse osmosis is their high operating costs; so they are not economically attractive. The precipitation process produces large amounts of sludge which causes second pollution. Adsorption has emerged as a cost-effective technique for removing metal ions from wastewater and has been widely studied during recent decades [1-7].

Among these methods, magnetic particles as adsorbents were paid huge attention to because they not only can be manipulated or recovered rapidly by an external magnetic field but also possess quite good performance owing to efficient specific adsorption capability. Magnetic adsorbents are composed of magnetic cores and polymeric shells. Compared to the traditional adsorbents, modification of polymeric shells, the surface functionality of magnetic nanoparticles can be tailored to various applications in the fields of high-density data storage, ferrofluids, magnetic resonance imaging (MRI), bioseparation, drug delivery, diagnosis, therapy, and immunoassays. It is also widely applied to the removal of ions from wastewater, such as the removal of dyes from their aqueous solutions by magbetic silica, the removal of copper by iron oxide-coated zeolite, and the recovery of amaranth by modified iron oxide magnetic nanoparticles [8-16]. However, it is less effective for the adsorption of polyvalent metal cations and not valid for the adsorption of anionic species [17]. Ferric with other metal binary oxide or hydroxide compounds have presented adsorbing capacity for polyvalent anions [18]. Recent research presets the mixed oxides as adsorbents for chromate [19-21]. Among them, synthetic and hydrous oxides such as iron (III) oxide, and iron (III) - aluminum (III) mixed oxide had been used for the anions fluoride ions removal studies [22]. Ferric oxide, highly abundant in the earth's crust, is one natural water purifier. We also presented a ferric-nickel oxide as a chromate adsorbent [23]. It possessed high anions binding capacity and good selected performance. Unfortunately, the drawback is that it may release heavy metal nickel ions from adsorbents if the pH of solution was very low, which have hindered the application in some conditions. So in this work, we attempted to develop a convenient method for the preparation of Ferric-Magnetic binary oxide or hydroxide for recovery chromate in acid condition. Furthermore, characterized by the XRD, SEM and FT-IR; the adsorption performance such as the effect of coexistence, pH effect, total dissolved substance (TDS) effect and including the isotherm and kinetic.

2. EXPERIMENT

2.1. Synthesis of Adsorbent

In the lab, ferrite-magnesium hybrid oxide was prepared by developing the coprecipitation method. $FeCl_3.6H_2O$ and $Mg(NO_3)_2.6H_2O$ were dissolved in 200ml of deionized water with mole rate of Fe and Mg 1:1 to 5:1,respectively. 2M NaOH was added into the above mixture under vigorous stirring and the composite precipitates were formed, when pH reached 8.0, the stirrer was turned off, then filtrated through a 0.45um filter film, the composite precipitates were rinsed with copious quantities of deionized water. After that the composite precipitates were calcined for 16h at 80℃, 230℃ and 440℃ respectively. The ferric-magnesium was milled and sieved into size fractions. The magnetic particles of diameter in <0.10mm were used in this work.

2.2. Material Characterizations

Chemical analysis was studied using Atomic Absorption Spectroscopy(AAS) (solaar 969, UNICAM), the synthesized material was dissolved in 1:1 H_2SO_4 and Fe^{3+} and Mg^{2+} concentration were analyzed. The synthesized material was examined by an X-ray diffractometer (XRD) (PW-1830 Philips) at ambient temperature. The instrument was equipped with a copper anode generating Cu Kα radiation(λ=1.5406Å). The elemental composition and chemical oxidation state of magnetic material and Cr-loaded magnetic material were investigated by an Electron Probe Microanalyzer (EPMA-1600 Shimadzu). The bonding analysis between Cr (VI) and magnetic material was conducted by the Fourier Transform-Infrared spectroscopy(FT-IR) (Avatar-360 Nicolet). The surface morphology was measured by a Scanning electron microscope (SEM) (JSM-5600LV). pH_{ZPC} was tested by titration method.

2.3. Chemicals

All chemicals were of grade and purchased from Tianjin chemical reagent company. Stock metal solutions were prepared using potassium dichromate and deionized water to produce a solution of 1000mg/L. This stock solution was diluted in accurate proportions to produce solutions of different initial concentrations.

2.4. Cr (VI) Concentration Analysis

The solutions were analyzed for hexavalent and total chromium. Hexavalent chromium (Cr (VI)) was detected by the diphenylcarbazide method: 1,5-diphenyl carbazide forms a pink complex in the presence of Cr (VI) ions in acidic solutions. The concentration of Cr (VI) was calculated from the absorbance at 542 nm using a UV-visible spectrophotometer. Total chromium (Cr(III) and Cr (VI)) was determined by analyzing Cr (VI) after $KMnO_4$ oxidation and makes Cr(III) into Cr (VI). The difference in concentration between total and hexavalent chromium was taken as the concentration of trivalent chromium.

2.5. Uptake Studies

To determine the time required for maximum uptake of chromium, 25 mL of a 10 mg/L, 20 mg/L, 50mg/L, mg/L, 200 mg/L Cr (VI) solution was taken in a 100 mL stoppered conical flask containing 0.05 g of adsorbent material, The contents of the flask were agitated in a rotary mechanical shaker at a fixation temperature and the chromium content was measured at the required time. The adsorbent was separated by centrifugation (4000 rpm) over 5 min. Chromium concentrations in the supernatant were determined by a UV-visible spectrophotometer, as stated above. The uptake of chromium(VI) by magnetic material (q) was calculated from the difference between the initial and final concentration as follows：

$$q = [(C_0\text{-}C_f)\times V]/M \qquad (1)$$

$$Rcr\ \%=(C_0\text{-}C_f)/C_0\times 100 \qquad (2)$$

where q is the chromium(VI) uptake (mg/g), C_o is the initial chromium(VI) concentration, C_f is the final chromium(VI) concentration (mg/L), M is the adsorbent dosage (g) and V is the solution volume (L), Rcr % is removal percentage of chromium(VI) .

3. Results and Discussion

3.1. Effect of Calcined Temperature on the Uptake of Chromate

The samples were calcined at the temperatures of 80 ℃, 230 ℃ and 440 ℃ were characterised by the SEM and XRD, results show in Figure 1 and 2. The adsorption experimental results present in Figure 3.

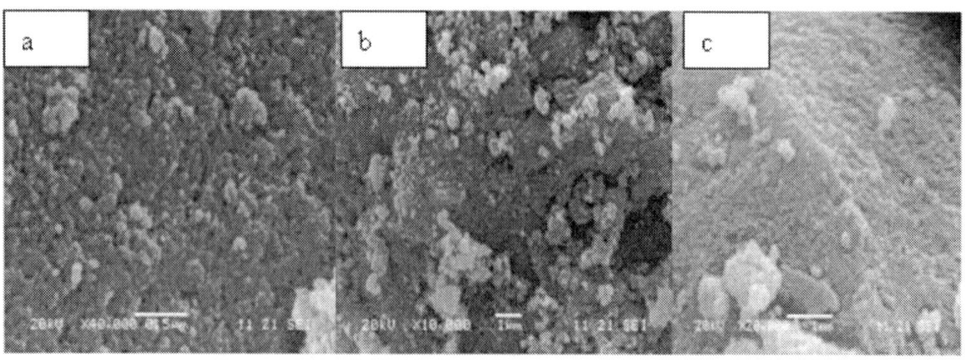

a -obtained at 80 ℃, b-obtained at230 ℃,c-obtained at 440 ℃

Figure 1. SEM of sample adsorbents.

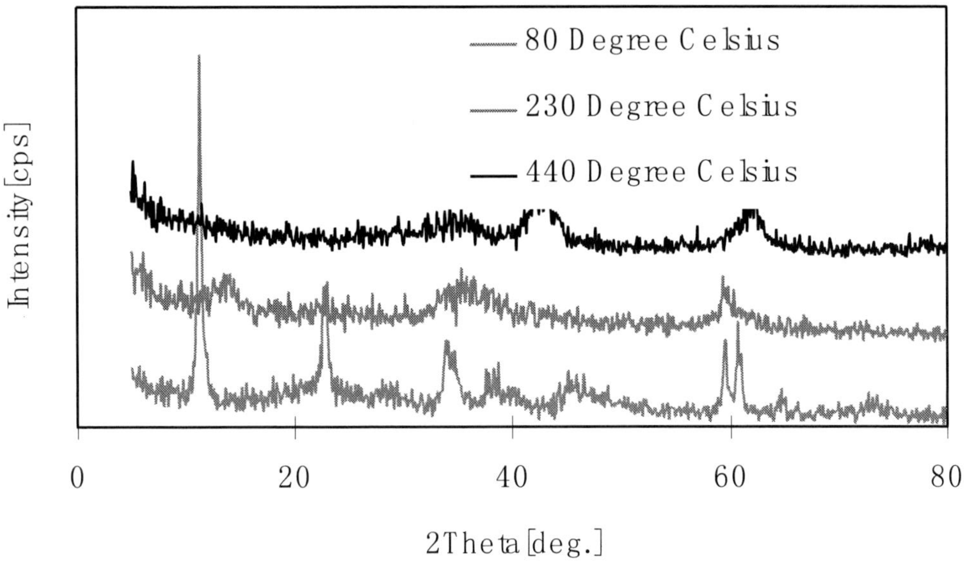

Figure 2. The XRD of the sample in the different calcined temperature.

The results of SEM and XRD indicate that the samples had a hydrotalcine structure about the size of 0.05 μ m and its structure was destroyed with increase in the calcined temperature and particle size was increased due to felted action.

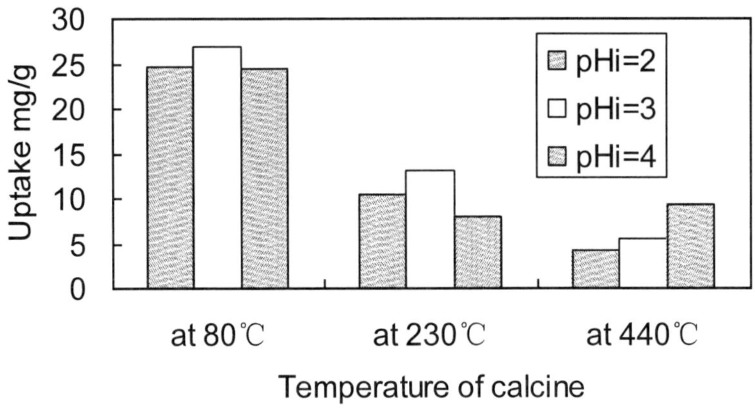

Figure 3. Effect of calcine temperature on the uptake in the initial pH range 2-4.

As can be seen from the Figure 3, the hydrotalcine structure sample presents better adsorption capacity of chromate in the pH range from 2-4 and the calcined temperature of 80 °C, so the sample was selected for use as the adsorbents for removal of chromate for following experimental tests.

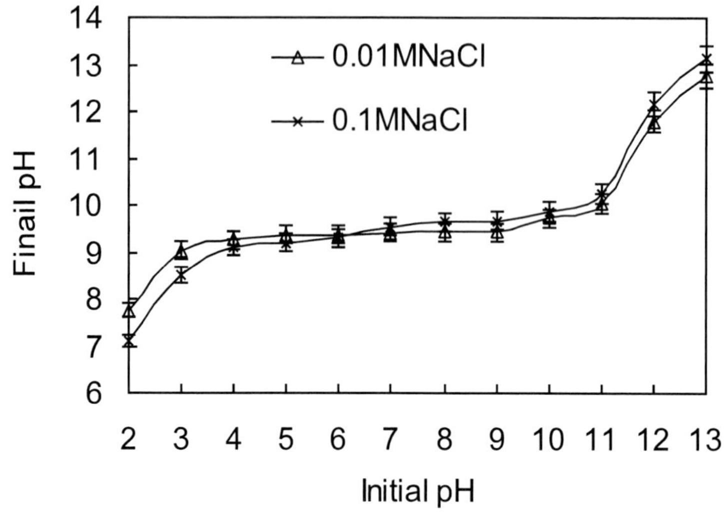

Figure 4. The pH_{zpc} results of the Ferric-Magnesium hybrid.

Figure 4 noted that the pH $_{zpc}$ of Fe-MgFe-Mg hybrid oxide with hydrotalcine structure is about 9.4. It means that there is positive charge on the surface of the sample at pH range 2-4 due to less pHzpc.

3.2. The Ph Effect on the Uptake of Cr onto the Layers of Mg/Fe Magnetic Oxide

The test process was carried out in the batch system with the layers of 0.05g Mg/Fe magnetic oxide adding to with 20 mg/L chromate solution to a shaker for 2 hours at a temperature of 25℃. The results were presented in Figure 5.

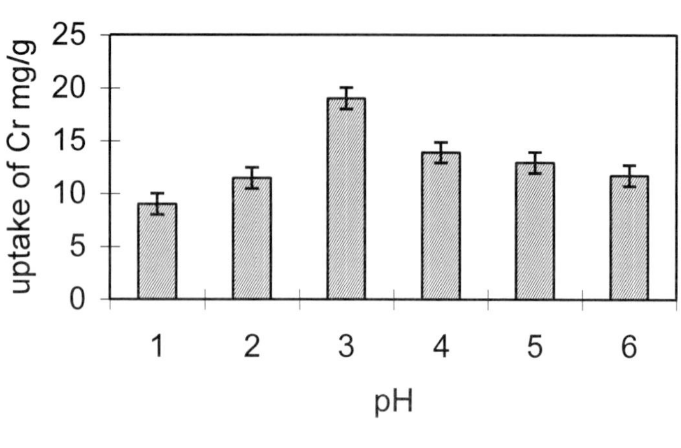

Figure 5. Effect of pH on the uptake of chromate.

As can be seen, the initial pH values is 3, so the best uptake value 19.07 mg/g was obtained, which indicated that following experimental tests of effect from concentration of chromate should be carried out at initial pH 3.

3.3. Effect of Total Dissolved Substance

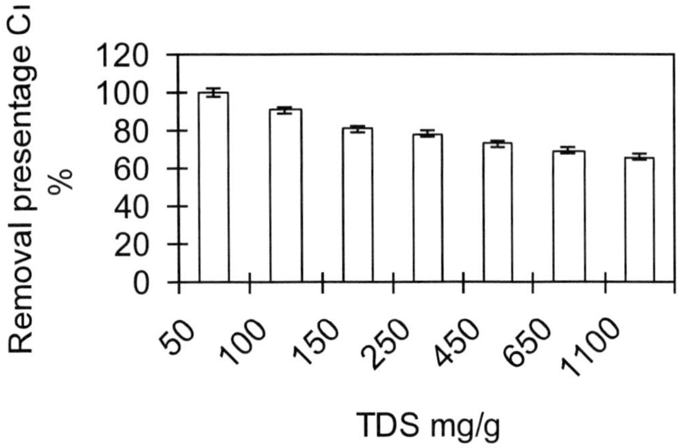

Figure 6. The Effect of Total Dissolved Substance (TDS).

Figure 6 is the result of increasing total dissolved substance using NaCl at pH 3 with other conditions the same as above 3.5. It shows that the adsorbent's capacity decreases with increasing TDS from 50 to 1100mg/L and the uptake capacity is reduced by about 20%. This trend means that the surface charge could be changed with the ionic strength increasing.

3.4. Effect on the Uptake From Other Ions

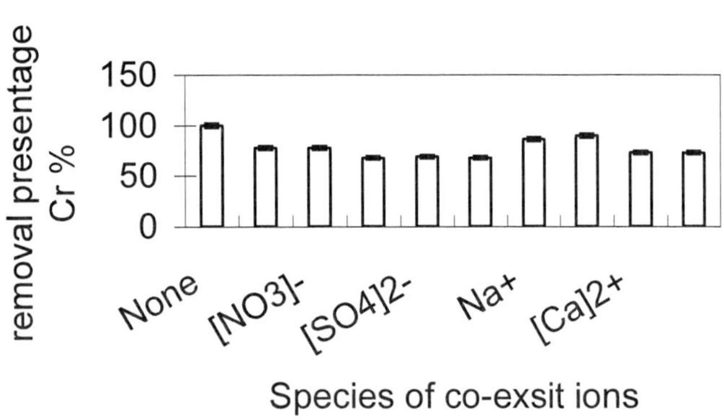

Figure 7. The effect of Co-existing species ions.

The ferric-magnesium hybrid oxide was added into 50ml solution of a single system of initial chromate concentration of 20 mg/L (named as none), and a binary system of mixed Cl^-, NO_3^-, CO_3^{2-}, SO_4^{2-}, PO_4^{3-},Na^+, NH^+, Ca^{2+} and Cu^{2+}, respectively. The concentration of co-exist ions are the same as chromium 20 mg/L at pH 3. The results present in the Figure 7 with any date were average value by 3 times. As can be seen, the biding site of adsorbents were occupied by others ions at first that may reduce removal rate of chromate. The bar value indicated that polyvalent ions have evidence affected by the removal of chromate. For instance, the removal percentage of chromium is 75% only at coexist PO_4^{3-}. Monovalent cation such as Na^+ and NH_4^+ has small effect on the adsorption capacity. The sequence of effect from coexisting ions is: PO_4^{3-}> CO_3^{2-}, SO_4^{2-}, Ca^{2+} and Cu^{2+}> Cl^-, NO_3^-> Na^+, NH^+.

3.5. Effect on the Contacted Time and Kinetic Behaviors

The effect of contact time and kinetic behaviors on the chromium adsorption onto the ferric-magnesium hybrid oxide was examined at pH 3 and an initial chromate concentration 20 mg/L.

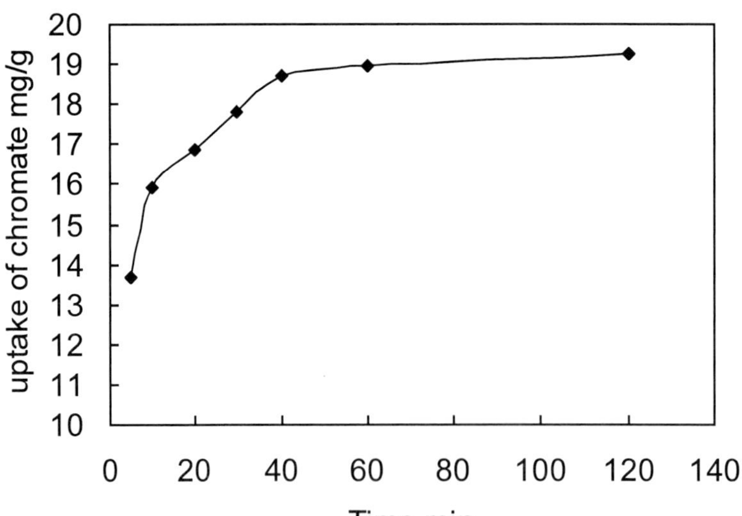

Figure 8. Effect of contacting time.

The results present in Figure 8. From the plot can be seen the uptake of chromate by Fe-Mg hybrid oxide increases with time increasing and attains a maximum value at about 40 min., thereafter, it almost does not further increase.

Several kinetic models have been used to evaluate the adsorption behaviors. The pseudo-first order equation, the pseudo-second order equation, and the intraparticle diffusion model were employed with the equations [24-26].

$$1/q_e = (k_1/q_1)(1/t) + 1/q_1 \quad (3)$$

$$t/q_e = (1/k_2 q_2^2) + t/q_2 \quad (4)$$

$$q_e = kpt^{1/2} + C \quad (5)$$

where, qt is the amount of chromate adsorbed (mg/g) at time t, q_1 is the maximum adsorption capacity (mg/g) for pseudo first-order adsorption, k_1 is the pseudo-first order rate constant for the chromate adsorption process (min^{-1}), q_2 is the maximum adsorption capacity for the pseudo-second order adsorption (g/mg min), k_2 is the pseudo second order rate constant (min^{-1}), kp is the intraparticle diffusion rate constant (mg/g $min^{1/2}$), and C is the intercept.

By plotting 1/qe is versus 1/t according to pseudo-first order equation, a straight line with $1/q_e = 0.10931/t + 0.0517$ was obtained in Figure 9-a. From the linear correlation analysis it is estimated that the values of the correlation coefficient ($R^2{}_1$) for the pseudo-first-order model is 0.978 and its q_1 is 19.34 mg/g. Figure 9-b is obtained by plotting t/q_e versus t, in which the linear equation is $t/q_e = 0.0508t + 0.1316$ and correlation coefficient R_2^2 is near 1 and q_2 is 19.69 mg/g. It implies the applicability of the pseudo-second order model for adsorption pf chromium (VI) onto the Fe-Mg hybrid oxide. Figure 9-c presents the trend of qt versus $t^{1/2}$ for intra-particle diffusion model given in Equation 5. Using the first 5 points, e.g the results within 40 min fit to Equation 5, the line is given by $q_e = 1.2061t^{1/2} + 11.23$ with correlation coefficient R_3^2 is 0.9896, it clearly points out that intra particle diffusion is the rate-controlling step which 40min of adsorption process and other parameters are listed in Table 1.

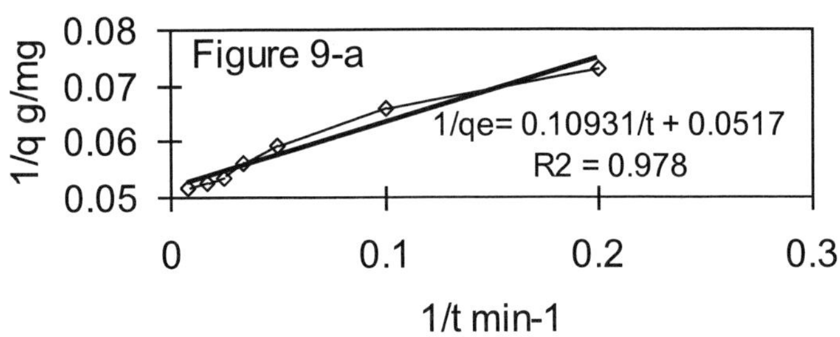

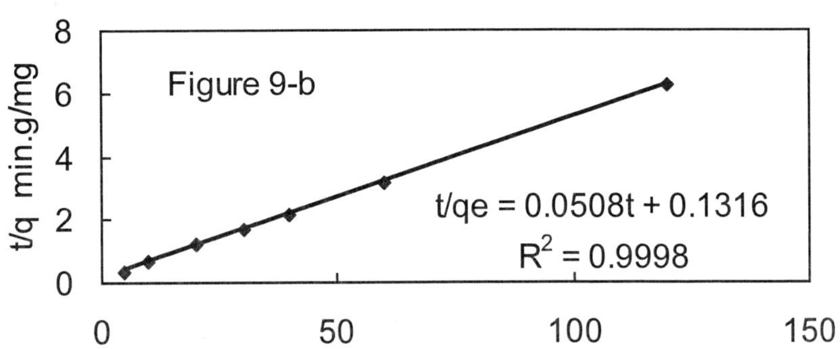

Figure 9 (Continued)

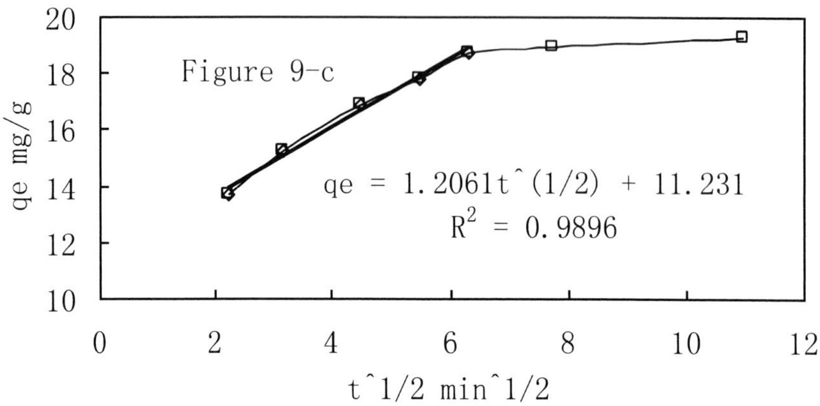

a-pseudo-first model, b-pseudo-second model, c-inter particle diffusion model.

Figure 9. The fit results of kinetic models.

Table 1. The kinetic parameters of pseudo-first order, pseudo second order and intra-particle diffusion model

Model	Rate constant	Adsorption constant(mg/g)	Correlation coefficient R^2
Pseudo-first order	k_1 (min^{-1}): 2.114	q_1:19.34	0.978
Pseudo-second order	k_2(mg/mg min):0.3409	$q_{2:}$ 19.69	0.9998
Intra-particle diffusion	k_p(mg/g $min^{1/2}$):1.2601	C: 11.23	0.9896

The above results show that the pseudo-second order model is suitable to describe kinetic behaviors of chromium (VI) adsorption onto Fe-Mg hybrid oxide; the max capacity is close to the experimental date. The intra-particle diffusion is a control step for this process at initial stages and may be control for a chemical reaction.

3.6. Effect of Chromate Concentration and Adsorption Equilibrium Isotherm

The results of adsorption of chromium onto Fe-Mg hybrid oxide at initial concentration of CrO_4^- range from 5mg/L to 200mg/L (e.g. 0.05 mmol/L to 2.00 mmol/L) at initial pH is 1, 3, 5, 7, respectively, and are present in Figure 10. The finial pH and equilibrium concentration of chromium are recorded and pointed out in the 3D picture in Figure 11. In order to conveniently account for competitive behaviors of hydroxyl, the unit of chromate replaced mg/g with mmol/L in Figure 10 and 11. Figure 10 presents that the optimal pH is 3 and also clarifies the relation between capacities and the chromium ion concentration which shows that the adsorption capacity increased with the increasing of the initial metal ion concentration. This is most probably due to increase in the initial ion concentrations which provided more chances for the adsorbents.

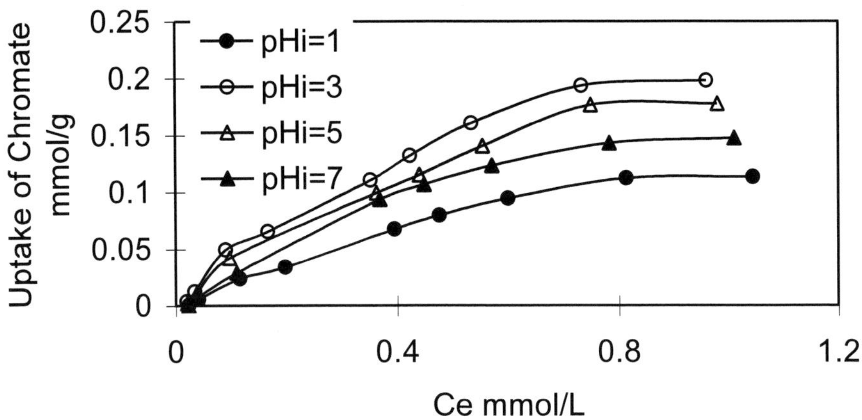

Figure 10. The plots of uptake of chromate versus equilibrium concentration at pH range 1 to 7.

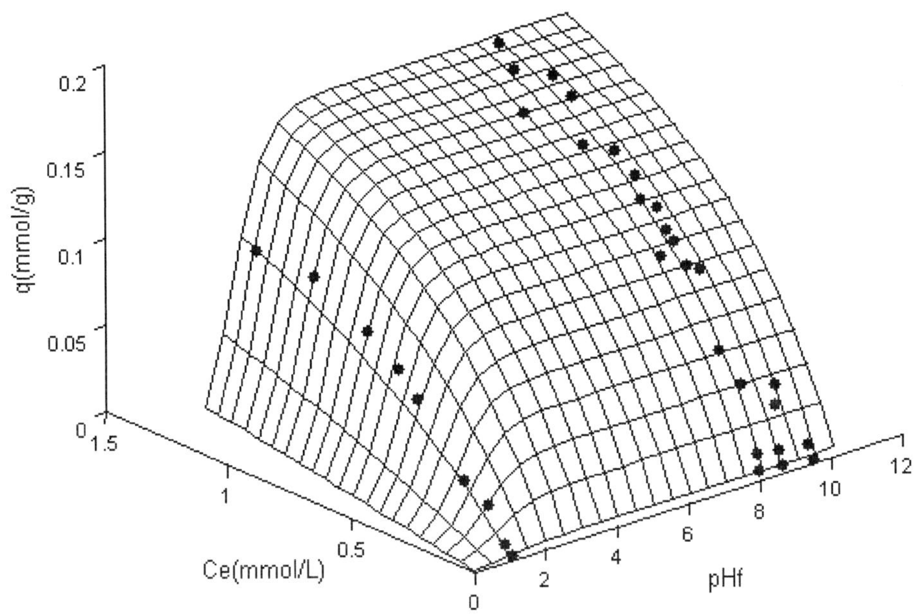

Figure 11. The 3D graph and OH-Anions competitive model.

With regards to adsorption isotherms such as the general isotherm, Langmuir and Freundlich meant that the pH is kept in contact number during the adsorption processes [26]. As can be seen from Figure 11, the finial pHs were transfer of initial pHs values.

We have been developing competitive adsorption isotherms in order to take into account the adsorption of H^+ with other competitive cations[27-28], such as the following in Equation 6. In this study, they replaced cation and hydrogen with anions $HCrO_4^-$ and hydroxyl ions, respectively. Then the anions –hydroxyl competitive model is given by Equation 7, (named the OH-Anions competitive model).

$$q(M) = \frac{q_{\max}\{(1/K_m)C_e[M]\}}{1+((1/K_m)Ce[M])+((1/K_h)C_e[H])} \qquad (6)$$

$$q(MO) = \frac{q_{\max}\{(1/K_m)C_e[MO]\}}{1+((1/K_m)Ce[MO])+((1/K_{Oh})C_e[HO])} \qquad (7)$$

where C_e is equilibrium metal ion concentration (mmol/L), Kh or Koh is equilibrium constant for the binding of proton (mmol/L), K_m is equilibrium constant for the binding of metal ions (mmol/L), q_{max} is the maximum amount of adsorption (mmol/g), q is amount of metal bound to the Fe-Mg oxide complex (mmol/g).

Among this, According Figure 12, pH range 1-7 the major species is $[HCrO_4^-]$, there assuming that sorption equilibrium can be expressed as

$$\text{Fe-Mg-O(OH)}^- + [HCrO_4]^- \leftrightarrow \text{Fe-Mg-O}-HCrO_4 + OH \qquad (8)$$

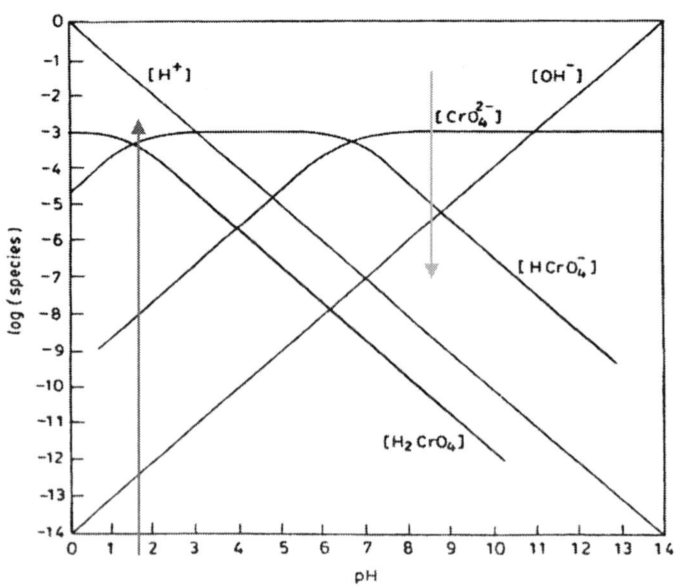

Adopted from 30.

Figure 12. Logarithmic concentration diagram for 10^{-3} M $H_2Cr_2O_4$ solution.

The 3D plots are calculated and drawn by Equation 7 with computer software MatLab and presented in Figure 13. The model parameters are indicated in Table 2.

Table 2. The parameters of OH-anions competitive model

Rate constant	Adsorption constant(mg/g)
Adsorption constant	
qmax(mmol/g)	
k_{oh} (L/mmol)	0.300
	0.652
K_M (L/mmol)	64.086
Correlation coefficient R^2	0.952

The results in Figure 13 and the correlation coefficient (R^2) is 0.952, it notes that the OH-Anions Competitive model describeds the ligand exchange process of chromate onto Fe-Mg hybrid oxide very well.

3.7. Regeneration and Recovery of Chromium

Regenerations are conducted by 0.1M NaOH solution in the colum set for recovery high chromate concentration solution for 2 times. And dried Fe-Mg hybrid oxide is used to 4 consecutive cycles of adsorption-desorption-washing-drying processes. The results show that the Chromate can be recovered by $NaCrO_4$ and Fe-Mg hybrid oxide kept adsorption capacity well (date is not shown here).

CONCLUSION

The following conclusions can be drawn from the current study.

Synthetic Fe-Mg hybrid oxide with hydrotalcine layers structure is suitable for removal and recovery of hexavalent chromium at acid solution. The influence from co-existing ions follows the order of polyvalent > monovalent such as PO_4^{3-}> SO_4^{2-}>Cl^- ~ NO_3^-.The regeneration using the solution of NaOH is suitable and higher concentration of Na_2CrO_4 solution is obtained in the regeneration process.

The pseudo-second order model is suitable to describe kinetic behaviors of chromium (VI) and the intra-particle diffusion is a control step at initial stages then may be control by chemical reaction.

The OH-Anions Competitive model can be used to well describe the process of adsorption performance of chromate onto Fe-Mg hybrid oxide and is the major mechanism is ligand exchange.

References

[1] Baral, A., Engelken, R.D., Chromium-based regulations and greening in metal finishing industries in the USA. *Environ, Sci. Police*, 5 (2002) 121–133.

[2] Park, D., Yun, Y.-S., Park, J.M.,. Reduction of hexavalent chromium with the brown seaweed Ecklonia biomass, *Envir. Sci. Technol.* 38 (2004),4860–4864.

[3] Aksu, Z., Açikel,Ű.Kabasakal,E.,Tezer,S.,Equilibrium modeling of individual and simultaneous biosorption of chromium(VI)and nickel(II)onto dried activated sludge, *Wat. Res.* 36 (2002) 3063-3073.

[4] Altundogan, H.S., Cr (VI) removal from aqueous solution by iron (III) hydroxide-loaded sugar beet pulp, *Process Biochemistry*, 40 (2005) 1443–1452.

[5] Álvarez-Ayuso,E., Garc´ıa-S´anchez, A. Querol, X. Adsorption of Cr (VI) from synthetic solutions and electroplating wastewaters on amorphous aluminium oxide, *J. Hazar. Mater.* B122 (2006) 231-245.

[6] Oguz ,E., Adsorption characteristics and the kinetics of the Cr (VI) on the Thuja oriantalis, Colloids and Surfaces A: *Physicochem. Eng. Aspects*, 252 (2005) 121–128.

[7] Bayrak, Y., Yesiloglu,Y. Gecge, U.,Adsorption behavior of Cr (VI) on activated hazelnut shell ash and activated bentonite, *Microporous and Mesoporous Mater.* 91 (2006) 107–110.

[8] Asem A. Atia, Ahmed M. Donia,Waheeba A. Al-Amrani, Adsorption/desorption behavior of acid orange 10 on magnetic silica modified with amine groups, *Chemical Engineering Journal*, 150 (2009) 55–62.

[9] Zargar B., Parham H., Hatamie A., Fast removal and recovery of amaranth by modified iron oxide magnetic nanoparticles, *Chemosphere*, xxx (2009) xxx–xxx.

[10] Liao, M., Wu, K., Chen, D., 2003. Fast removal of basic dyes by a novel magnetic nano-adsorbent. *Chem. Lett.* 32, 488–489.

[11] Mak, S., Chen, D., 2004. Fast adsorption of methylene blue on polyacrylic acid bound iron oxide magnetic nanoparticles. *Dye Pigment*, 61, 93–98.

[12] Mayo, J.T., Yavuz, C., Yean, S., Cong, L., Shipley, H., Yu, W., Falkner, J., Kan, A., Tomson, M., Colvin, V.L., 2007. The effect of nanocrystalline magnetite size on arsenic removal. *Sci. Tech. Adv. Mater.* 8, 71–75.

[13] Runping H, Lina Z, Xin Z, Yanfang X, Feng X, Yinli L, Yu W, Characterization and properties of iron oxide-coated zeolite as adsorbent for removal of copper(II) from solution in fixed bed column, *Chemical Engineering Journal*, 149 (2009) 123–131

[14] Uheida, A., Salazar-Alvarez, G., Bjkman, E., Yu, Z., Muhammed, M., 2006. Fe3O4 and c-Fe2O3 nanoparticles for the adsorption of Co2+ from aqueous solution. *J. Colloid Interf. Sci.* 298, 501–507.

[15] Mak S.Y., Chen D.H., Binding and sulfonation of poly(acrylic acid) on iron oxide nanoparticles: a novel, magnetic, strong acid cation nano-adsorbent, *Macromol. Rapid Commun.* 26 (2005) 1567–1571.

[16] Hu, J., Lo, I.M.C., Chen, G., Fast removal and recovery of Cr (VI) using surface-Modified Jacobsite (MnFe2O4) Nanoparticle. *Langmuir*, 21 (2005) 11173-11179.

[17] Hu, J., Chen, G., Lo, I.M.C., Removal and recovery of Cr (VI) from wastewater by magnetite nanoparticles, *Wat.Res.* 39 (2005) 4528–4536.

[18] Martinez-Gallegos, S., Pfeiffer, H. Lima , E. Espinosa, M. Bosch, P. Bulbulian, S. Cr (VI) immobilization in mixed (Mg,Al) oxides, *Microporous and Mesoporous Mater.* 94 (2006) 234–242.

[19] Gaosheng Z., Huijuan L., Ruiping L., Jiuhui Q. , Removal of phosphate from water by a Fe–Mn binary oxide adsorbent, *Journal of Colloid and Interface Science*, xxx (2009) xxx–xxx.

[20] Linsen W., Gang Y, Ren W.., Wei M.; Selective adsorption and separation of chromium(VI) on the magnetic iron-nickel oxide from waste nickel liquid. *Journal of Hazardous Materials*, 164(2009) 1159-1163.

[21] Krishna B., Kaushik G., Uday Chand G., Adsorption of fluoride by hydrous iron(III)–tin(IV) bimetal mixed oxide from the aqueous solutions, *Chemical Engineering Journal*, 149 (2009) 196–206

[22] Kannan, N.; Sundaram, M.M. (2001) Kinetics and mechanism of removal of methylene blue by adsorption on various carbons—a comparative study. *Dye. Pigment.*, 51: 25.

[23] Ho, Y.S.; Wase, D.A.J.; Forster, C.F. (1996) Kinetic Studies of Competitive Heavy Metal Adsorption by Sphagnum Moss Peat. *Environ. Technol.*, 17: 71.

[24] Weber J., W.J.; Morriss, J.C. (1963) Kinetics of adsorption on carbon from solution. *J.Sanitary Eng. Div. Am. Soc. Civ. Eng.*, 89: 31.

[25] Werner S., Chemistry of the solid-water interface, processes at the Mineral-Water and Partic-Water Interface in Natural System.A Wiley-interscience Publication.New York, (1992).

[26] Wei M. John M. T. Development of multimetal binding model and application to binarymetal biosorption onto peat biomass, *Water Res.* 37 (2003) 3967-3977.

[27] Wei M. John M. T.. Determination and modelling of effects of pH on peat biosorption of chromium, copper and cadmium, *Biochem. Eng. J.* 18 (2004) 33-40.

In: Management of Hazardous Residues Containing Cr(VI) ISBN: 978-1-61470-637-3
Ed: Maria José Balart Murria

Chapter 8

FACTORS AFFECTING THE BIOLOGICAL REMOVAL OF HEXAVALENT CHROMIUM USING ACTIVATED SLUDGES

***Edgardo M. Contreras*[1,2*], *Ana M. Ferro Orozco*[1] and *Noemí E. Zaritzky*[1,2]**

[1]Centro de Investigación y Desarrollo en Criotecnología de Alimentos (CIDCA) – CONICET CCT La Plata - Fac. de Cs. Exactas – UNLP

[2]Fac. de Ingeniería - UNLP. Argentina

ABSTRACT

For many years, conventional Cr(VI) removal was achieved by chemical reduction, ion exchange or adsorption. Because of the high costs associated with physical-chemical processes to remove Cr(VI), the search for new technologies has focused attention on the biotransformation of Cr(VI). Several researchers demonstrated that most of bacteria present in activated sludges are capable of reducing Cr(VI) to Cr(III). For this reason, removal of Cr(VI) using activated sludge is a promising technique. This chapter addresses the effect of important factors concerning the bioreduction of Cr(VI) by activated sludge in batch reactors, such as, biomass concentration, the presence, type, and concentration of carbonaceous substrates, the nitrogen source concentration, and Cr(VI) concentration. In addition, a mathematical model that describes the Cr(VI) bioreduction kinetics is also discussed.

The reduction of Cr(VI) to Cr(III) in activated sludge batch reactors is a catalytic mediated process associated with the presence of biomass. Although the biomass itself could serve as the electron donor for reducing Cr(VI) to Cr(III), an external electron donor (e.g. carbonaceous substrate) increased the rate and extent of Cr(VI) reduction. High Cr(VI) reduction rates were obtained with fermentable substrates such as cheese whey, lactose, and glucose. On the contrary, with non-fermentable substrates such as citrate and acetate, Cr(VI) removal rates were slower. The low fraction of electrons of the

* Email: edgardo.contreras@ing.unlp.edu.ar

carbonaceous substrate that reduce Cr(VI) to Cr(III) suggests that Cr(VI) reduction in activated sludge batch reactors is mainly via a cometabolic pathway.

Besides the carbonaceous substrate, the presence of a nitrogen source is also necessary to obtain high Cr(VI) removal rates. In addition, when one or both of carbonaceous or nitrogen source is depleted during the growth phase, Cr(VI) removal continues but at a lower rate. Thus, the bioreduction of Cr(VI) is the result of a fast Cr(VI) reduction process, associated with the biomass growth, and a slow one that does not depend on the presence of both a carbonaceous substrate and a nitrogen source. The second process may be important under certain conditions, such as in continuous cultures in which substrates concentrations are normally low.

In this chapter, the effect of the initial Cr(VI) concentration on the activated sludge growth kinetics was studied. As the Cr(VI) concentration increased, the Lag phase also increased; in addition, a threshold inhibitory concentration of Cr(VI) within the range of 100 to 300 mg/L was observed. The biomass growth yield and the observed specific growth rate decreased as the Cr(VI) concentration increased; conversely, the substrate oxidation coefficient increased with higher Cr(VI) concentrations. These effects were attributed to the additional energy needed by microorganisms to maintain an active metabolism due to the presence of Cr(VI).

1. INTRODUCTION

Chromium is a heavy metal often used in several industrial processes; it is widely used in metal finishing industries, petroleum refining, leather tanning, iron and steel industries, inorganic chemical production, textile manufacturing and pulp production (Katz and Salem, 1994). For this reason, chromium is often detected in sewage plants that combine industrial and municipal wastewater for treatment (Patterson, 1985). In contaminated habitats, this heavy metal may accumulate in microorganisms, aquatic flora and fauna, which in turn, may enter into the human food chain and result in health problems.

Chromium is usually found in the environment in oxidation states (III) and (VI). Each of the above mentioned oxidation states have different biological and chemical properties. Typical hexavalent chromium compounds are the highly soluble chromates (CrO_4^{2-}, $HCrO_4^-$) and the dichromate ($Cr_2O_7^{2-}$); these compounds are toxic and carcinogenic for a variety of organisms due to their strong oxidizing capability (U.S. EPA, 1998a). Trivalent chromium can form both anionic ($Cr(OH)_4^-$, $CrCl_6^{3-}$) or cationic ($Cr(H_2O)_6^{3+}$, $Cr(OH)^{2+}$, $Cr(OH)_2^+$) compounds. These compounds are considered to be non-labile, inert species in the environment and essential for mammals in trace amounts (U.S. EPA, 1998b). Therefore, reducing Cr(VI) to Cr(III) is beneficial for eliminating the toxicity of Cr(VI) to the environment.

For many years, conventional methods for removing Cr(VI) include chemical reduction to Cr(III) followed by a precipitation step under alkaline conditions, ion exchange, or adsorption using activated carbon (Patterson, 1985). All these methods have their own disadvantages; alkali precipitation produces large quantities of chemical sludge, whereas ion exchange and adsorption are generally costly and less specific for Cr(VI) removal in the presence of other ions. For this reason, the search for new technologies has focused attention on the biotransformation of metals by using micro-organisms. In contrast to the conventional methods, biodetoxification is cost-effective (Li et al., 2007). A great number of bacterial genera were described as capable of reducing Cr(VI) to Cr(III) including *Escherichia* (Wang and Shen, 1997), *Pseudomonas* (Wang and Xiao, 1995), *Bacillus* (Philip et al., 1998), *Enterobacter* (Rege et al., 1997), *Desulfovibrio* (Lovley and Phillips, 1994), *Shewanella* (Myers et al., 2000), *Serratia* (Mondaca et al., 2002), *Rhodobacter* (Nepple et al., 2002) and *Arthrobacter* (Asatiani et al., 2004). However, the main disadvantage of using pure cultures to remove Cr(VI) compounds is related to the use of sterile conditions to prevent external microbial contamination. For this reason, removal of Cr(VI) using mixed cultures, such as activated sludge, is a promising technique.

In general, heavy metals negatively affect the biological treatment due to their toxicity for micro-organisms. However, literature data concerning the toxic effect of Cr(VI) on the activated sludge process (e.g. substrate removal, respiration activity, or bacterial growth) is contradictory. Stasinakis et al. (2002) reported that Cr(VI) concentrations higher than 10 mg/L inhibited the growth of non-acclimatized activated sludge. Chua (1998) found that the effect of 0.05 mg/L of Cr(VI) on the performance of sequencing batch reactors was a function of the sludge age. Yetis et al. (1999) reported the stimulation of an acclimated sludge in the presence of 25 mgCr(VI)/L. Vankova et al. (1999) found that the Cr(VI) concentration that inhibited 50% the respiration rate ranged between 40 to 90 mg/L. However, Madoni et al. (1999) reported that 100 mgCr(VI)/L only reduced the activated sludge respiration rate by 21.5%.

This chapter addresses the effect of important factors concerning the bioreduction of hexavalent chromium by activated sludge in batch systems, such as biomass concentration, the presence, type, and concentration of carbonaceous substrates, the nitrogen source concentration, and Cr(VI) concentration in order to optimize the design and operation of a biological Cr(VI) reduction process. In addition, a mathematical model that describes the Cr(VI) bioreduction kinetics is also discussed.

2. Experimental Procedures

2.1. Activated Sludges

Activated sludges used in all the experiments showed in the present chapter were harvested from an aerobic laboratory-scale (4.5 L) activated sludge reactor with partial biomass recycle. The reactor was fed with a synthetic wastewater of the following composition: dehydrated cheese whey 1.5 g, $(NH_4)_2SO_4$ 0.94 g, and $NaHCO_3$ 1.03 g, all dissolved in 1 L of tap water. Soluble chemical oxygen demand (COD_S) of the synthetic wastewater was 1500 mg L^{-1}. The hydraulic retention time was 2 days; the sludge age was maintained at 45 days by daily wasting of mixed liquor directly from the reactor. During the experiments the temperature of the reactor was 20 ± 2 °C. Under steady-state conditions dissolved oxygen (DO) concentration was above 4 mg L^{-1}, pH was 7.5 ± 0.4, COD_S of the effluent ranged between 30 and 80 mg L^{-1}, and total suspended solid (TSS) concentration ranged from 3000 to 4000 mgTSS L^{-1}.

2.2. Cr(VI) Removal Experiments

Hexavalent chromium removal experiments were performed in aerated vessels (0.5 L) at constant temperature (20 ± 2 °C) and initial pH = 7.0 ± 0.1; this pH was selected because it is the optimum value for the metabolic activity of most of the micro-organisms that are present in a typical activated sludge. pH was measured continuously using a polymer body pH probe (Broadley-James Corp.) connected to an on/off pH controller (Masstek, Argentina). Small air pumps were employed to aerate and to agitate the vessels. Inocula for Cr(VI) removal assays were obtained from the activated sludge reactor; biomass samples were washed three times with phosphate buffer (KH_2PO_4 2 g L^{-1}, K_2HPO_4 0.5 g L^{-1}, pH = 7) before performing the assays. All Cr(VI) removal experiments were performed in a mineral basal medium added with the tested carbonaceous substrates (Ferro Orozco, 2009). The medium composition was the following: carbonaceous substrate 5 gCOD L^{-1}, ammonium sulfate 212 mgN L^{-1}, and micronutrient solutions M1 and M2 (1 mL L^{-1}). The composition of M1 was (expressed as g/100ml): $FeSO_4.7H_2O$ 1.5, $ZnSO_4.7H_2O$ 0.5, $MnSO_4.H_2O$ 0.3, $CuSO_4.5H_2O$ 0.075, $CoCl_2.6H_2O$ 0.015, and citric acid 0.6. M2 solution contained the following (g/100ml): $(NH_4)_6Mo_7O_{24}.4H_2O$ 0.05, BO_3H_3 0.01, KI 0.01. A Cr(VI) stock solution (10 gCr(VI) L^{-1}) was prepared using analytical grade $K_2Cr_2O_7$; in all the experiments an appropriate volume of this solution was added to obtain an initial Cr(VI) concentration ranged from 10 to 300 mgCr(VI) L^{-1}. At predetermined intervals samples were taken to determine biomass concentration (X) as total suspended solids (TSS). An aliquot was centrifuged at 13000 rpm for 5min (Eppendorf centrifuge 5415C, Hamburg, Germany) in order to remove the biomass, and the supernatant was analyzed. Then, depending on the experiment, the following determinations were performed: soluble Cr(VI), total soluble Cr, chemical oxygen demand (COD), total ammoniacal nitrogen (TAN), and respiration rate of the biomass (oxygen uptake rate, OUR).

From this data, Cr(VI) removal rates were obtained from the slope of Cr(VI) as a function of time; then, from the ratio between the Cr(VI) removal rate and biomass concentration, the specific Cr(VI) removal rate (q_{Cr}) was obtained. In addition, Cr(VI) reduction yield (the amount of Cr(VI) removed per substrate consumed, $Y_{Cr/S}$) was calculated from the slope of Cr(VI) as a function of the carbonaceous substrate concentration. The oxygen consumed (OC) by the biomass during the experiment was calculated as the integral of the OUR as a function of time. The biomass growth yield ($Y_{X/S}$) and substrate oxidation coefficient ($Y_{O/S}$) were obtained from the slope of X and OC as a function of S concentration, respectively. The observed specific growth rate (μ_{obs}) was calculated from the slope of the linear part of the plot of Ln(X) as a function of time. Besides, carbonaceous substrate and total ammoniacal nitrogen were measured to verify the presence of both substrates during the determination of Cr(VI) removal rates or μ_{obs}.

2.3. Analytical Techniques

Hexavalent chromium concentration was determined by the diphenylcarbazide method (APHA, 1989).

Total soluble chromium concentration was determined as follows. Samples containing Cr(VI) and Cr(III) were treated with ammonium persulphate and potassium permanganate to oxidize all Cr(III) to Cr(VI); then, total Cr(VI) was determined using the diphenylcarbazide method. Soluble Cr(III) concentration was estimated as the difference between total soluble Cr and Cr(VI). Precipitated and/or adsorbed (non-soluble) Cr was calculated as the difference between the initial total Cr and the total soluble Cr concentration.

Chemical oxygen demand (COD), and total ammoniacal nitrogen (TAN) were determined using commercial reagents (Hach Company, Loveland, CO).

Biomass concentration was determined as total suspended solids (TSS); 8-mL of the culture were poured into pre-weighted centrifuge tubes, centrifuged and washed twice with distilled water, and placed at 105 ^{o}C for 24h; TSS was calculated as the difference between final (dry sample + tube) and initial (tube alone) weights divided by the sample volume (8-mL).

Oxygen uptake rates (OUR) were measured using a closed respirometer (Ros, 1993).

3. Chromium (VI) Removal Potential of Activated Sludges

3.1. Cr(VI) Removal in Activated Sludge Batch Reactors is a Catalytic Mediated Process Associated to the Presence of Biomass

Because of the high costs associated to physical-chemical processes to remove Cr(VI), the search for new technologies has focused attention on the biotransformation of metals by using micro-organisms. Several researchers demonstrated that a great number of bacterial genera are capable of reducing Cr(VI) to Cr(III) (Wang and Shen, 1997; Wang and Xiao, 1995; Philip et al., 1998; Rege et al., 1997, Lovley and Phillips, 1994; Myers et al., 2000; Mondaca et al., 2002; Nepple et al., 2002; Asatiani et al., 2004). The biomass obtained from activated sludge systems should also have the potential to reduce Cr(VI) because most of these genera are present in biological wastewater treatments.

In order to determine if the micro-organisms present in activated sludges were capable of reducing hexavalent chromium, batch Cr(VI) removal assays in the presence of lactose (5g/L) were performed (Ferro Orozco et al., 2007a). Two Cr(VI) removal systems were studied: abiotic experiments (without the addition of biomass), and Cr(VI) removal experiments with an initial biomass concentration of 2.3 gTSS/L. Figure 1 shows the removal of Cr(VI) in abiotic systems was negligible. Conversely, for the system with biomass, Cr(VI) concentration decreased within the first 67 hs. When the biomass was removed by centrifugation (indicated by the arrow), no subsequent change in Cr(VI) concentration was observed. Besides, as the initial biomass concentration increased, the removal rate of Cr(VI) also increased (Figure 2). Because neither detectable adsorption of chromium species, reduction of Cr(VI), or oxidation of Cr(III) on the vessel walls were produced, these experiments demonstrated that the removal of Cr(VI) in activated sludge batch reactors is a catalytic mediated process associated to the presence of the biomass (Ferro Orozco et al., 2007a,b).

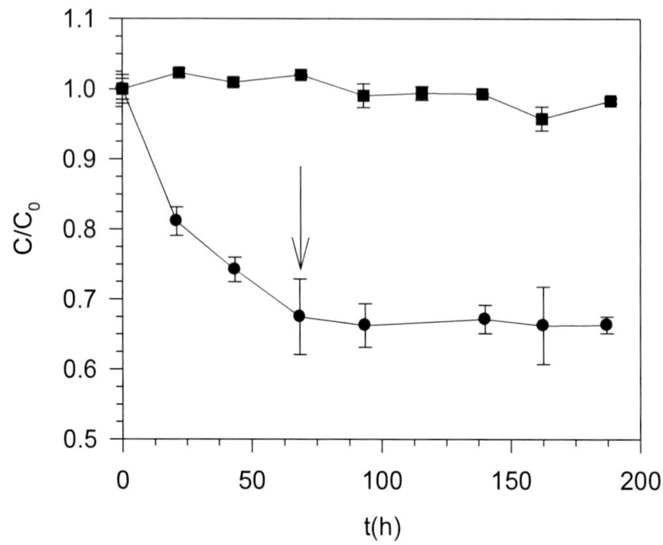

Figure 1. Removal of Cr(VI) by activated sludge (circles) as a function of time in the presence of lactose (5 g/L) as electron donor. The initial biomass concentration was 2.3 gTSS/L. The arrow indicates the time when biomass was removed by centrifugation. Results represent mean values ± one standard deviation from two determinations. Squares: abiotic control.

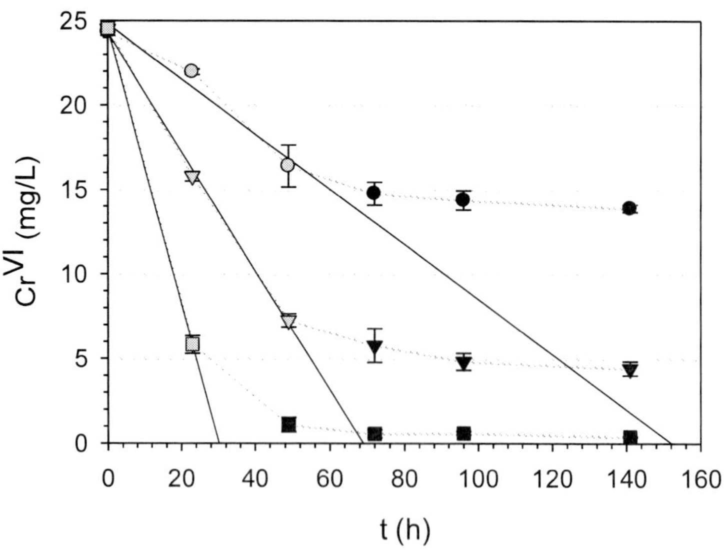

Figure 2. Effect of the initial biomass concentration on the removal of Cr(VI): (●) 0.7 gTSS/L, (▼) 1.7 gTSS/L, and (■) 3.5 gTSS/L. Grey symbols indicate the data used to calculate the Cr(VI) removal rate. Bars indicate one standard deviation.

3.2. Effect of Electron Donors (Organic Substrates) on the Rate of Cr(VI) Removal by Activated Sludge

One of the main mechanisms proposed in the literature for Cr(VI) removal by pure or mixed cultures under aerobic conditions is the reduction of Cr(VI) to Cr(III) (Imai and Gloyna, 1990; Wang and Shen, 1997; Stasinakis et al., 2004). This Cr(VI) reductase activity depends on the presence of a suitable electron donor (Bopp and Ehrlich, 1988; Philip et al., 1998; Shen and Wang, 1994; 1995; Wang and Shen, 1997; Wang and Xiao, 1995). Although the biomass itself could serve as the electron donor, this may lead to a decrease in the biomass concentration, and therefore, in the Cr(VI) reductase activity. For this reason, an external electron donor is preferred.

Batch assays were performed in order to compare Cr(VI) removal by activated sludge alone, and by activated sludge with an external electron donor; in this case, lactose (5 g/L) was used. Figure 3 shows that in both cases the Cr(VI) concentration decrease with time; however, a higher Cr(VI) removal rate was obtained when lactose was present in the culture medium. Therefore, although the biomass itself could serve as the electron donor for reducing Cr(VI) to Cr(III), an external electron donor, such as lactose, increased the rate and extent of Cr(VI) removal.

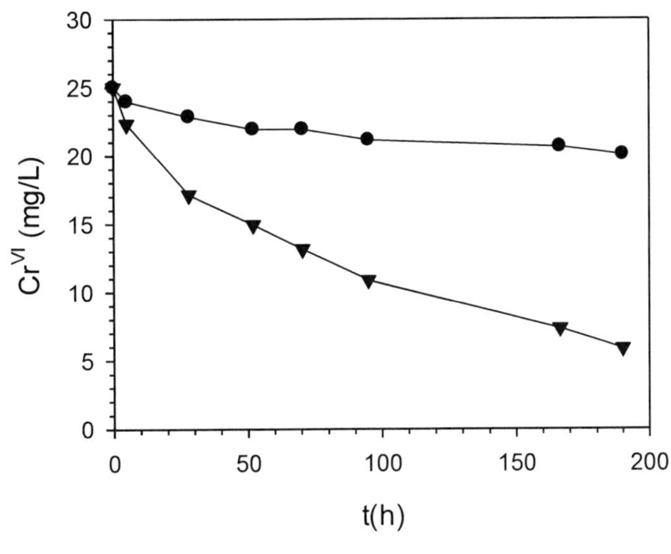

Figure 3. Cr(VI) removal by activated sludge alone (●), and activated sludge with addition lactose (5 g/L) (▼). In both cases the biomass concentration was 2.6 gTSS/L, and pH = 7.0.

Table 1 shows the distribution of the following chromium fractions observed in the assayed systems: soluble Cr(VI), soluble Cr(III) and non-soluble Cr; the last one includes adsorbed and/or precipitated Cr(III) and Cr(VI) species. The occurrence of Cr(III) is a direct evidence that the main mechanism for removal of Cr(VI) by activated sludge is through its reduction to Cr(III). With regard to activated sludge alone, the formation of soluble Cr(III) was negligible; on the contrary, in the systems with lactose, soluble Cr(III) represented 10 to 100% of the added Cr (Ferro Orozco et al., 2007b; Ferro Orozco, 2009). It must be pointed out that solubility of Cr(III) at pH 7 is rather low, about 10^{-5} M, due to the formation of solid

$Cr(OH)_3$ (Stumm and Morgan, 1995). However, soluble Cr(III) concentrations found in most of the tested chromium removal systems were higher than the solubility of Cr(III) at pH 7. These results can be explained considering that trivalent chromium can form a wide variety of coordination complexes; once formed, these complexes are quite stable in aqueous solutions. Nitrogen containing compounds such as NH_3, amino acids and proteins are potential ligands for the formation of coordination complexes with Cr(III) (Katz and Salem, 1994).

Table 1. Chromium fractions obtained in Cr(VI) removal assays by activated sludge alone, and activated sludge with lactose as external electron donor. In all cases the contact time was 190 hs. Values were expressed as percentage of the total Cr concentration

	Activated sludge alone			Activated sludge with electron donor (lactose)		
Initial Cr(VI) (mgCr/L)	Soluble Cr(VI) (%)	Soluble Cr(III) (%)	Prec+Ads* Cr (%)	Soluble Cr(VI) (%)	Soluble Cr(III) (%)	Prec+Ads* Cr (%)
10	95	0	5	0	100	0
25	80	0	20	24	16	60
50	89	1	10	35	14	51
100	91	0	9	66	10	24

*precipited and adsorbed

3.3. Thermodynamic Analysis of Cr(VI) Reduction by Organic Compounds Commonly Found in Wastewaters

Although hexavalent chromium is a known strong oxidant even under the neutral pH conditions usually found in aquatic environments (U.S. EPA, 1998a), Figure 1 shows that at neutral pH, lactose cannot reduce Cr(VI) to Cr(III) during the first 200 hs. Equilibria of hexavalent chromium compounds are pH and concentration dependent (Katz and Salem, 1994). Calculations using the software CHEAQS (Verweij, 2005) demonstrate that chromate ion (CrO_4^{2-}) represents more than 98% of the total Cr(VI) species for pH higher than 8. Chromate monoacid ($HCrO_4^-$), and dichromate ($Cr_2O_7^{2-}$) ions are the main species when the pH is lower than 5. In addition, as the total Cr(VI) concentration increases, the fraction of $HCrO_4^-$ decreases due to the formation of dichromate. In all cases, chromic acid (H_2CrO_4) fraction is negligible. Within the normal pH range of most biological wastewater treatment systems (pH about 5 to 9) and for Cr(VI) concentrations usually found in wastewaters (usually less than 2 mM), $Cr_2O_7^{2-}$ ion account for less than 3% of the total Cr(VI); thus, $HCrO_4^-$ and CrO_4^{2-} are the dominant species (Figure 4).

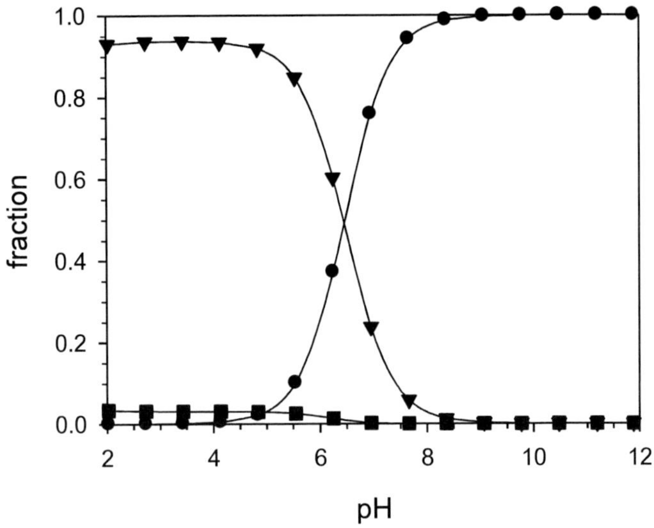

Figure 4. Effect of pH on the distribution of the following Cr(VI) species: (●) CrO_4^{2-}, (▼) $HCrO_4^-$, and (■) $Cr_2O_7^{2-}$. The total Cr(VI) concentration is 1 mM.

Table 2 shows reduction half-reactions and free energies for some organic compounds usually found in wastewaters; in addition, the reduction half-reaction corresponding to CrO_4^{2-} is also included. Based on the data shown in Table 2, the change of free energy at pH 7.0 and 25 °C ($\Delta G^o(W)$) corresponding to the reaction between acetic acid (electron donor) and chromate ion (electron acceptor) is about -83.3 kJ per mole of electrons transferred. Thus, from a thermodynamic point of view, Cr(VI) is capable of oxidizing acetic acid and most of the organic compounds commonly found in wastewaters.

Table 2. Reduction half-reactions and free energies for some organic compounds and chromate ion (CrO_4^{2-})

Half-reaction	$\Delta G^o(W)$ kJ/mole	Reference
Domestic wastewater		
$9/50\ CO_2 + 1/50\ HCO_3^- + 1/50\ NH_4^+ + H^+ + e^- \rightarrow 1/50\ C_{10}H_{19}O_3N + 9/25\ H_2O$	31.8	(1)
Carbohydrate		
$1/4\ CO_2 + H^+ + e^- \rightarrow 1/4\ CH_2O + 1/4\ H_2O$	41.8	(1)
Grease		
$4/23\ CO_2 + H^+ + e^- \rightarrow 1/46\ C_8H_{16}O + 15/46\ H_2O$	27.6	(1)
Acetic acid		
$1/4\ CO_2 + H^+ + e^- \rightarrow 1/8\ CH_3COOH + 1/4\ H_2O$	27.6	(1)
Chromate		
$1/3\ CrO_4^{2-} + 4/3\ H_2O + e^- \rightarrow 1/3\ Cr(OH)_3(s) + 5/3\ OH^-$	-55.7	(2)

$\Delta G^o(W)$ values refer to pH 7.0 and 25 °C; other species are assumed to be at unit activities.

(1) Adapted from Orhon and Artan (1994); (2) Calculated from Schriver et al. (1994)

Results shown in Figure 5 demonstrate that at neutral pH and 20°C, abiotic (in the absence of biomass) Cr(VI) removal by simple sugars (glucose, lactose), organic acids (acetate, citrate), or by a complex organic substrate such as cheese whey was negligible (Ferro Orozco, 2009). The stability of Cr(VI) to its reduction by the tested organic compounds occurs because electron-transfer reactions are frequently slow; thus, in the absence of a catalyst, such as the activated sludge biomass, redox systems are far from equilibrium (Stumm and Morgan, 1996). For this reason, the presence of cells that act as catalyst and a suitable electron donor are both necessary to achieve the reduction of Cr(VI).

3.4. Effect of the Type of Carbonaceous Substrate on the Cr(VI) Removal by Activated Sludge

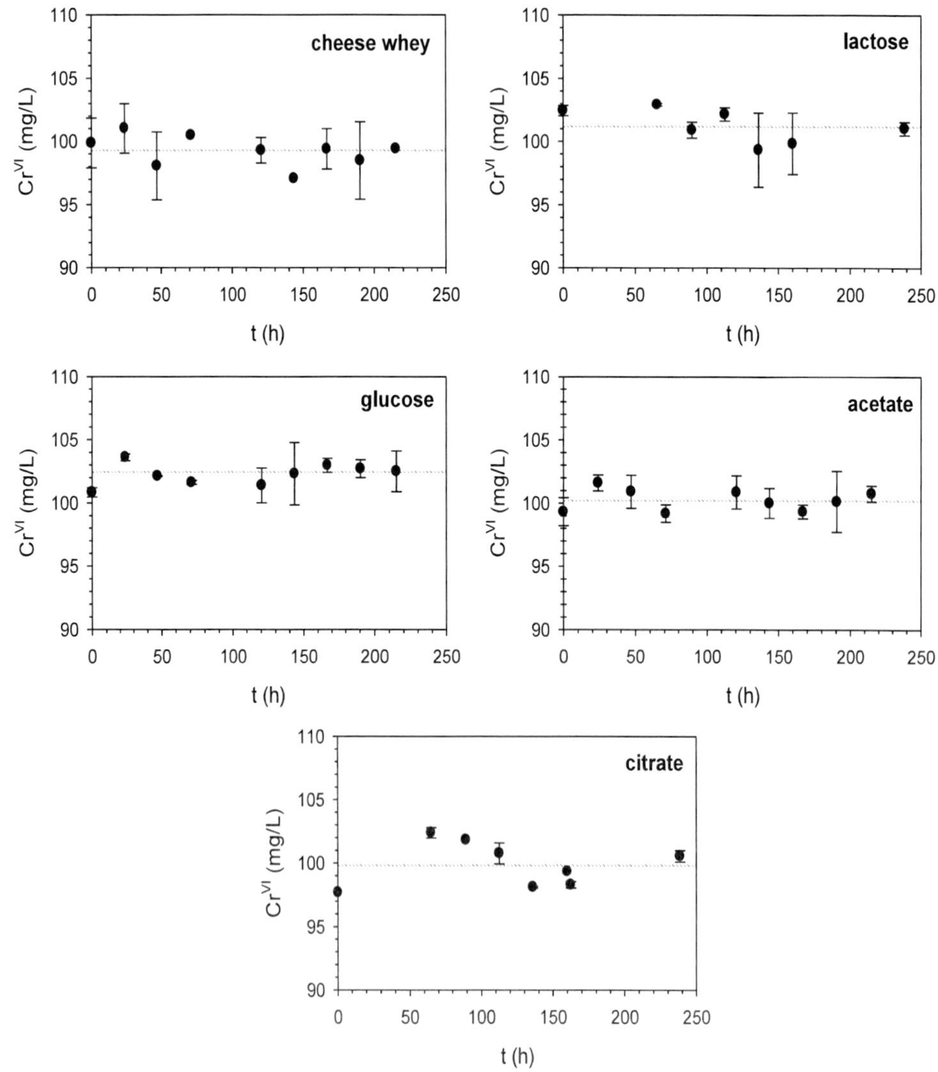

Figure 5. Abiotic (without biomass) Cr(VI) removal experiments performed at neutral pH and 20 °C with different organic substrates. Dotted lines indicate the overall mean value for Cr(VI).

In the previous sections it was demonstrated that chromate reductase activity of the biomass depends on the presence of a carbonaceous substrate that acts as an electron donor. For this reason, the effect of the type of carbonaceous substrate on the Cr(VI) removal by activated sludge was analyzed. Tested substrates were cheese whey, lactose, glucose, citrate and acetate. Cheese whey was selected because it is a residue from dairy industries; glucose and lactose were tested because these sugars are main components of the cheese whey (Atkinson and Mavituna, 1991; Ghaly et al., 2003). Citrate and acetate were chosen because they are non-fermentable compounds.

Table 3 shows that the specific Cr(VI) removal rate (q_{Cr}) depended on the type of the tested carbonaceous substrate; the following order of preference for substrates by activated sludge related to q_{Cr} were observed: cheese whey ≈ lactose > glucose > citrate > acetate. In general, high q_{Cr} values were obtained with fermentable substrates, such as cheese whey, lactose, and glucose. Conversely, with non-fermentable substrates, such as citrate and acetate, relatively low specific Cr(VI) removal rates were obtained. The lowest q_{Cr} value was obtained when Cr(VI) removal experiments were performed without the addition of a carbonaceous substrate, confirming the need of an external electron donor to enhance the Cr(VI) removal rate. Besides, considering the requirement of adding a carbonaceous substrate for the biological removal of Cr(VI), cheese whey could be an adequate and inexpensive choice, because it is constituted from residue from dairy industries (Ferro Orozco et al., 2007a; Ferro Orozco, 2009).

The amount of Cr(VI) removed per carbonaceous substrate consumed ($Y_{Cr/S}$) was similar for cheese whey, lactose, and glucose; $Y_{Cr/S}$ values corresponding to acetate and citrate were slightly lower that those obtained with fermentable substrates (Ferro Orozco et al., 2007a; Ferro Orozco, 2009). Values corresponding to $Y_{Cr/S}$ shown in Table 3 fall within the range reported by other authors. Bae et al. (2000) reported $Y_{Cr/S}$ values ranged between 1.5 and 2.9 mgCr(VI) $gCOD^{-1}$ for *E. coli* ATCC 33456 grown in glycerol. Caravelli and Zaritzky (2009) reported $Y_{Cr/S}$ = 1.0 mgCr(VI) $gCOD^{-1}$ corresponding to the removal of Cr(VI) by *Sphaerotilus natans* in continuous culture with citrate as carbon and energy source.

Table 3. Effect of the type of carbonaceous substrate on the specific Cr(VI) removal rate (q_{Cr}), on the amount of Cr(VI) removed per substrate consumed ($Y_{Cr/S}$), and on the fraction of carbonaceous substrate electrons that reduce Cr(VI) to Cr(III) (F_{e^-/e^-})

	q_{Cr} (mgCr(VI) $gTSS^{-1}$ h^{-1})	$Y_{Cr/S}$ (mgCr(VI) $gCOD^{-1}$)	F_{e^-/e^-} * (%)
Cheese whey	0.124 ± 0.002	4.2 ± 0.3	0.19 ± 0.01
Lactose	0.112 ± 0.004	4.4 ± 0.4	0.20 ± 0.02
Glucose	0.084 ± 0.002	3.7 ± 0.4	0.17 ± 0.02
Citrate	0.060 ± 0.016	1.5 ± 0.2	0.07 ± 0.01
Acetate	0.044 ± 0.003	3.1 ± 0.2	0.14 ± 0.01
Without substrate	0.025 ± 0.011	n.a.	n.a.

* F_{e^-/e^-} was calculated by means of eq.(3); n.a.: not applicable

Considering that the source of electrons used to reduce Cr(VI) to Cr(III) was the carbonaceous substrate, it is interesting to determine the electron transfer efficiency from the substrate to Cr(VI). The fraction of electrons of the carbonaceous substrate that reduce Cr(VI) to Cr(III) can be calculated by means of an electron balance (Orhon and Artan, 1994). Considering the following reduction half-reactions,

$$4e^- + 4H^+ + O_2 \rightarrow 2H_2O \tag{1}$$

$$3e^- + Cr^{6+} \rightarrow Cr^{3+} \tag{2}$$

reaction (1) indicates that if 32 gCOD substrate (one mole of O_2) are consumed, then 4 moles of electrons must be transferred to biomass, molecular oxygen, and Cr(VI) for maintaining the electron balance. From reaction (2), 3 moles of electrons are necessary to reduce 52 gCr(VI) to Cr(III). Considering these equivalences, the fraction of the electrons of the carbonaceous substrate that reduce Cr(VI) to Cr(III) (F_{e^-/e^-}) can be calculated from the coefficient $Y_{Cr/S}$ as follows:

$$F_{e^-/e^-} = Y_{Cr/S} \frac{32\,gCOD}{4\,e^-} \frac{3\,e^-}{52\,gCr} \tag{3}$$

where $Y_{Cr/S}$ is expressed in units gCr $gCOD^{-1}$. Table 3 shows that $Y_{Cr/S}$ corresponding to lactose as electron donor (the highest value) was $4.4*10^{-3}$ gCr $gCOD^{-1}$. Based on eq.(3), in the most favorable case only 0.2% of the transferable electrons of the substrate are actually used to reduce Cr(VI) to Cr(III), the other 99.8% was used for biomass synthesis and energy production by reducing molecular oxygen (Ferro Orozco et al., 2007a; Ferro Orozco, 2009). A similar conclusion can be derived from $Y_{Cr/S}$ values reported by other authors (Bae et al., 2000; Caravelli and Zaritzky, 2009).

The low fraction of electrons of the carbonaceous substrate that reduce Cr(VI) to Cr(III) suggests that Cr(VI) reduction in activated sludge batch reactors is mainly via a cometabolic pathway. Many hazardous substances are transformed into less harmful compounds by cometabolism. The term cometabolism can be defined as the transformation of a nongrowth (also called secondary) substrate by growing cells in the presence of a growth (or primary) substrate (Alvarez-Cohen and Speitel Jr., 2001). The cometabolic reactions yield no carbon or energy benefits to the transforming cells. Therefore, a growth substrate must be available to grow new cells, provide an energy source, and induce production of cometabolic enzymes. Cometabolism results from the lack of specificity of enzymes and cofactors; for this reason, the electron transfer efficiency from the primary substrate to the secondary one is usually rather low (Alvarez-Cohen and Speitel Jr., 2001).

3.5. Effect of Hexavalent Chromium Concentration on Cr(VI) Removal Efficiency (R_E) by Activated Sludge

In general, heavy metals negatively affect the biological treatment due to their toxicity for the microorganisms. With regard to hexavalent chromium, Stasinakis et al. (2002) reported that Cr(VI) concentrations higher than 10 mg/L inhibited the growth of non-acclimatized activated sludge. Vankova et al. (1999) found that the Cr(VI) concentration that inhibited 50% the respiration rate ranged between 40 to 90 mg/L. Madoni et al. (1999) reported that 100 mgCr(VI)/L only reduced activated sludge respiration rate by 21.5%. These reports indicate that the initial Cr(VI) concentration could negatively affects the Cr(VI) removal efficiencies (R_E) by activated sludge.

In order to compare Cr(VI) removal efficiencies (R_E) obtained by using activated sludge alone, and by activated sludge with an external electron donor (lactose 5 g/L), batch assays were performed with different initial Cr(VI) concentrations, ranged from 10 to 100 mgCr/L (Ferro Orozco et al., 2007b). Figure 6 shows that Cr(VI) removal efficiencies (R_E) corresponding to systems without an external electron donor (lactose) were lower than 20% (Figure 6a) for the tested Cr(VI) concentrations; on the other hand, R_E values in the systems containing lactose ranged between increased from 35 to 100%. However, for Cr(VI) initial concentrations higher than 10 mg/L the capacity to reduce Cr(VI) of the activated sludge (Figure 6b) decreased as the initial Cr(VI) concentration increased, reflecting the loss of metabolic activity of the activated sludge due to the toxicity of Cr(VI). R_E final values at 190 hs. were 77, 65 and 34 % for the initial Cr(VI) concentrations of 25, 50, and 100 mg/L respectively (Figure 6b). Recently, studies demonstrated a protective effect of activated carbon on micro-organisms in the presence of hexavalent chromium (Ferro Orozco et al., 2004; 2007b; 2008). Thus, the addition of powdered activated carbon could be a suitable method to reduce the loss of metabolic activity of activated sludge.

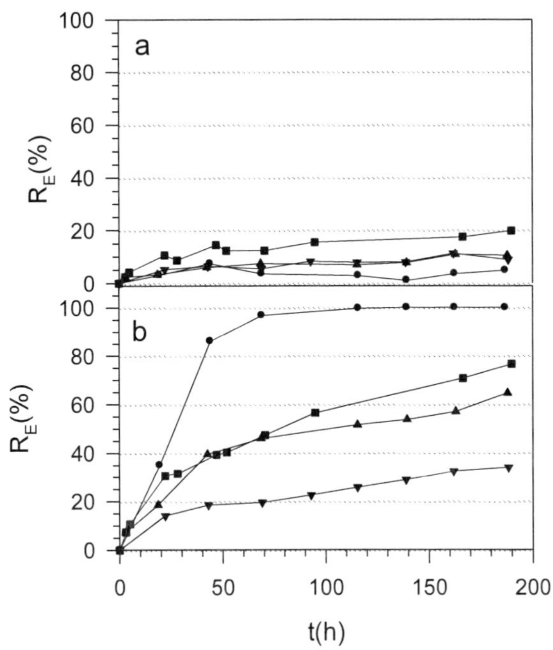

Figure 6. Effect of Cr(VI) concentration on Cr(VI) removal efficiencies (R_E) as a function of time by (a) activated sludge, and (b) activated sludge with lactose (5 g/L). Initial Cr(VI) concentrations: (●) 10 mg/L, (■) 25 mg/L, (▲) 50 mg/L, and (▼) 100 mg/L. Biomass concentration: 2.6 ± 0.4 gTSS/L.

3.6. A Mathematical Model for Cr(VI) Removal by Activated Sludge.

Among the mathematical models available in the literature concerning the biological removal of Cr(VI), one of the most used is the model developed by Wang and co-workers (Shen and Wang, 1994; Wang and Xiao, 1995; Wang and Shen, 1995, 1997; Nkhalambayausi-Chirwa and Wang, 2004). In this model, it is assumed that the rate and extent of biological Cr(VI) reduction in batch cultures depends on the initial biomass concentration, regardless of the subsequent growth. Therefore, the new cells generated in the presence of hexavalent chromium would not have the Cr(VI) reduction capacity due to the mutagenic effects of Cr(VI). According to this model, the biological Cr(VI) reduction rate in batch systems can be expressed as a Monod-type equation with respect to the Cr(VI) concentration:

$$\frac{dCr}{dt} = -\frac{q_{Crm}\, Cr}{K_C + Cr} X \tag{4}$$

where Cr and X are the instantaneous Cr(VI) and the active biomass concentrations respectively, q_{Crm} is the maximum specific Cr(VI) reduction rate, and K_C is the half-saturation constant. In this model, the variable X refers not to the total biomass concentration (e.g. TSS) but to the active biomass concentration, that is, biomass with the ability to reduce Cr(VI) or chromate reductase activity.

The toxic effect of Cr(VI) against the active biomass is taken into account assuming that the active biomass concentration decreases proportionally to the Cr(VI) reduction rate:

$$\frac{dX}{dt} = -\frac{1}{R_C}\frac{q_{Crm}\, Cr}{K_C + Cr} X \tag{5}$$

where R_C is the maximum specific Cr(VI) reduction capacity of the active biomass (Wang and Shen, 1997). The introduction of the coefficient R_C is the key point of this model; combining eqs.(4) and (5), and integrating, the following expression was obtained:

$$(Cr_0 - Cr) = R_C\,(X_0 - X) \tag{6}$$

where Cr_0 and X_0 are the initial Cr(VI) and active biomass concentrations respectively. Equation (6) shows that the amount of Cr(VI) removed, (Cr_0 – Cr), is proportional to the decrease of the active biomass, (X_0 – X), being R_C the proportionality constant. The maximum amount of Cr(VI) that could be removed (ΔCr_{max}) is reached when the biomass have used all its Cr(VI) reduction capacity during the Cr(VI) reduction process, that is, when the final active biomass concentration (X_F) is zero:

$$\Delta Cr_{max} = R_C\, X_0 \tag{7}$$

Thus, eq. (7) shows that for this model, the maximum amount of Cr(VI) that can be removed in batch assays only depends on the initial biomass concentration. Conversely, eq.(6) can be used to determine the active biomass concentration at the end of a Cr(VI) removal batch test (X_F); in this case, the minimum value for X_F that could be achieved is reached when the final Cr concentration is zero:

$$X_F = X_0 - Cr_0/R_C \quad (8)$$

From eq.(8), it is clear that the higher R_C, the higher X_F. Considering the limit case $R_C \to \infty$ ($1/R_C \to 0$), then $X_F \to X_0$, and no toxic effect of Cr(VI) on the biomass could be observed; therefore, the inverse of R_C can be interpreted as a measure of the resistance of the tested biomass to Cr(VI) toxicity.

Combining eqs.(4) and (6) and integrating, the following equation can be obtained:

$$t = \frac{K_C}{q_{Crm}\left(\frac{Cr_0}{R_C} - X_0\right)} \ln\left[\frac{Cr\,X_0}{Cr_0\left(X_0 - \frac{Cr_0 - Cr}{R_C}\right)}\right] + \frac{R_C}{q_{Crm}} \ln\left[\frac{X_0}{X_0 - \frac{Cr_0 - Cr}{R_C}}\right] \quad (9)$$

Hexavalent chromium removal assays were performed to evaluate the model coefficients q_{Crm}, K_C, and R_C by means of eq. (9) (Ferro Orozco et al., 2008; Ferro Orozco, 2009). Figure 7 shows the data of Cr(VI) removal by activated sludge as a function of time for different initial Cr(VI) concentrations (10 to 100 mgCr/L). In these experiments lactose (5 gr/L) was added as electron donor and high initial biomass concentration (X_0) was used (2600 ± 400 gTSS/L). Equation (9) was fitted to the experimental data of Cr(VI) concentrations as a function of time in order to obtain the parameters q_{Crm}, K_C, and R_C; Figure 7 shows that a satisfactory agreement between the proposed model and the experimental data was obtained for all the tested initial Cr(VI) concentrations.

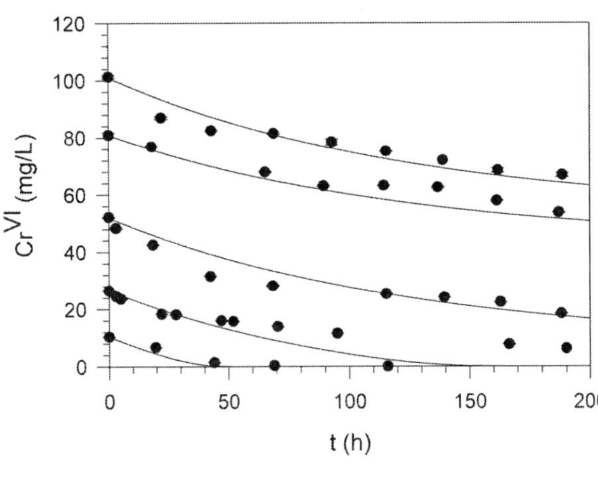

Figure 7. Cr(VI) removal by activated sludge in batch reactors corresponding to different initial Cr(VI) concentrations. In all cases lactose (5 gr/L) was added was added as the electron donor and the initial biomass concentration (X_0) was 2600 ± 400 mgTSS/L. Lines indicate Cr(VI) predictions by the model (eqs.(4) and (5)) using the coefficients shown in Table 4.

Obtained coefficients were within the range of those reported by other authors (Table 4). Schmieman et al. (1998) studied the Cr(VI) removal kinetics of enriched mixed cultures from soil. Because the enrichment medium contained 10 mgCr(VI)/L, those authors obtained mixed cultures acclimated to Cr(VI). For this reason, maximum specific Cr(VI) reduction rates (q_{Crm}) reported by Schmieman et al. (1998) were higher than the corresponding to non-acclimated activated sludges. In general, reported q_{Crm} values for pure cultures are higher than those corresponding to mixed cultures. In addition, half-saturation constants (K_C) corresponding to both types of mixed cultures (soil samples and activated sludge) are lower than for pure cultures. Table 4 also shows that R_c values corresponding to activated sludge is higher than for pure cultures; according to eq.(8), mixed cultures are more resistant to Cr(VI) toxicity than pure cultures. Finally, it must be considered that the use of pure cultures to remove Cr(VI) on the industrial scale implies high costs associated to maintain sterile conditions. For this reason, mixed cultures, such as activated sludge, could be a suitable option to remove Cr(VI).

Wang's model has several restrictions that must be considered before it is applied. Equation (5) referred to the active biomass only includes an inactivation term; thus, active biomass growth was neglected. However, this approximation is correct only in the cases when the initial biomass concentration (X_0) is much higher than the initial growth limiting substrate concentration (S_0), or if the biomass growth yield ($Y_{X/S}$) is close to zero due to the presence of an inhibitor, such as Cr(VI), for example. Another drawback of this model is that Cr(VI) reduction rate does not depend on the presence of a carbonaceous substrate (S). However, in the previous sections it was shown that the depletion of S determines a strongly decrease on the Cr(VI) removal rate. Thus, the model could be applied only if the carbonaceous substrate is present in excess within the studied time interval. In this sense, further research is needed in order to develop a more general biological Cr(VI) removal model that takes into account all these effects.

Table 4. Comparison between kinetic coefficients of the model using (eqs.(4) and (5)) available in the literature

Type of culture	q_{Crmax} mgCr gTSS^{-1} h^{-1}	K_{Cr} mgCr/L	R_C mgCr/gTSS	Reference
Bacillus sp.[(a,b)]	0.32	5.4	6.4	(c)
D. vulgaris ATCC 29579[(a,b)]	1.98	19.2	--	(c)
E. coli ATCC 33456[(a,b)]	1.5	8.6	22.1	(c)
P. fluorescens LB300 [(a,b)]	0.26	5.6	9.7	(c)
E. cloacae HO1 [(a,b)]	1.29 ± 0.26	1.56 ± 0.57	--	(d)
Surface soil mixed culture[(a,b)]	0.90 ± 0.41	1.48 ± 0.11	--	(d)
Subsurface soil mixed culture[(a,b)]	3.01 ± 0.79	0.39 ± 0.51	--	(d)
River sediment mixed culture[(a,b)]	1.19 ± 0.16	0.19 ± 0.17	--	(d)
Activated sludge[(b)]	0.19 ± 0.03	1.1 ± 0.4	23.3 ± 1.9	(e)

(a) Assuming that the dry weight of one cell corresponds to 3 10^{-13} grams (Atkinson and Mavituna, 1991); (b) assuming a cell volatile fraction of 0.95, and standard biomass composition ($C_5H_7O_2N$); (c) Wang and Shen, 1997; (d) Schmieman et al., 1998; (e) Ferro Orozco et al., 2008.

3.7. Relationship between the Cr(VI) Reduction Process and Microbial Growth

In previous sections it was discussed that when the carbonaceous substrate is completely consumed, the reduction rate of Cr(VI) decreases almost to a stop. These results suggest a relationship between the Cr(VI) reduction process and the microbial growth. Nitrogen is one of the main constitutents of the biomass; several studies showed that the nitrogen content for most activated sludges is a constant. Therefore, if the initial carbonaceous substrate concentration (S_0) is maintained at a constant, by controlling the initial nitrogen source concentration (N_0), the amount of biomass produced at the end of a batch assay can also be controlled.

Hexavalent chromium removal batch assays similar to those described in the previous section were performed; however, in these experiments the added nitrogen source (ammonium sulfate) concentration ranged from 0 to 250 mgN L^{-1}; in all cases carbonaceous substrate concentration (lactose) was 5 g L^{-1}, biomass concentration was 700 mgTSS L^{-1}, and initial Cr(VI) concentration was 25 mg/L (Ferro Orozco, 2009). Figure 8 shows examples of batch Cr(VI) removal assays performed in the presence of two different initial nitrogen source concentrations (N_0). In both cases, three phases in the growth curve can be clearly observed. The first stage corresponds to the Lag phase in which the biomass is acclimated to the new environmental conditions. After that, removal of hexavalent chromium, consumptions of the carbon and nitrogen sources, and biomass growth begin. Maximum Cr(VI) removal rates are obtained during this exponential growth phase, where no carbon or nitrogen limitation occurs. Finally, biomass growth and Cr(VI) removal end when one of the substrates (carbon or nitrogen source) is exhausted (Ferro Orozco, 2009).

Figure 8 shows that when the initial total ammoniacal nitrogen (TAN) was 210 mgN L^{-1}, soluble COD (lactose) was depleted but TAN values were about 25 mgN L^{-1} at t = 120 hs.; therefore, this case corresponded to a carbon limited batch assay. In addition, at this time Cr(VI) removal rate decreased drastically and TAN concentration increased due to the biomass decay. When the initial TAN concentration was 60 mgN L^{-1}, nitrogen source was depleted at t = 120 hs.; although the carbonaceous substrate concentration was around 3000 mgCOD L^{-1}, the biomass growth stopped. Hence, in this case, the biomass growth was limited by the nitrogen source. Because lactose was the carbon but also the energy source, biomass concentration remained almost constant; however, Cr(VI) removal rate decreased drastically due to the depletion of the nitrogen source. These results confirmed that the presence of both a carbonaceous substrate and a nitrogen source are necessary to obtain high Cr(VI) removal rates. When one of both substrates is depleted during the growth phase, Cr(VI) removal continues but at a lower rate. Therefore, the bioreduction of Cr(VI) is the result of a fast Cr(VI) removal process, associated with the biomass growth, and a slow one that does not depend on the presence of both a carbonaceous substrate and a nitrogen source. It must be pointed out that the second process may be important under certain conditions, such as in continuous cultures in which substrates concentrations are normally low (Ferro Orozco, 2009).

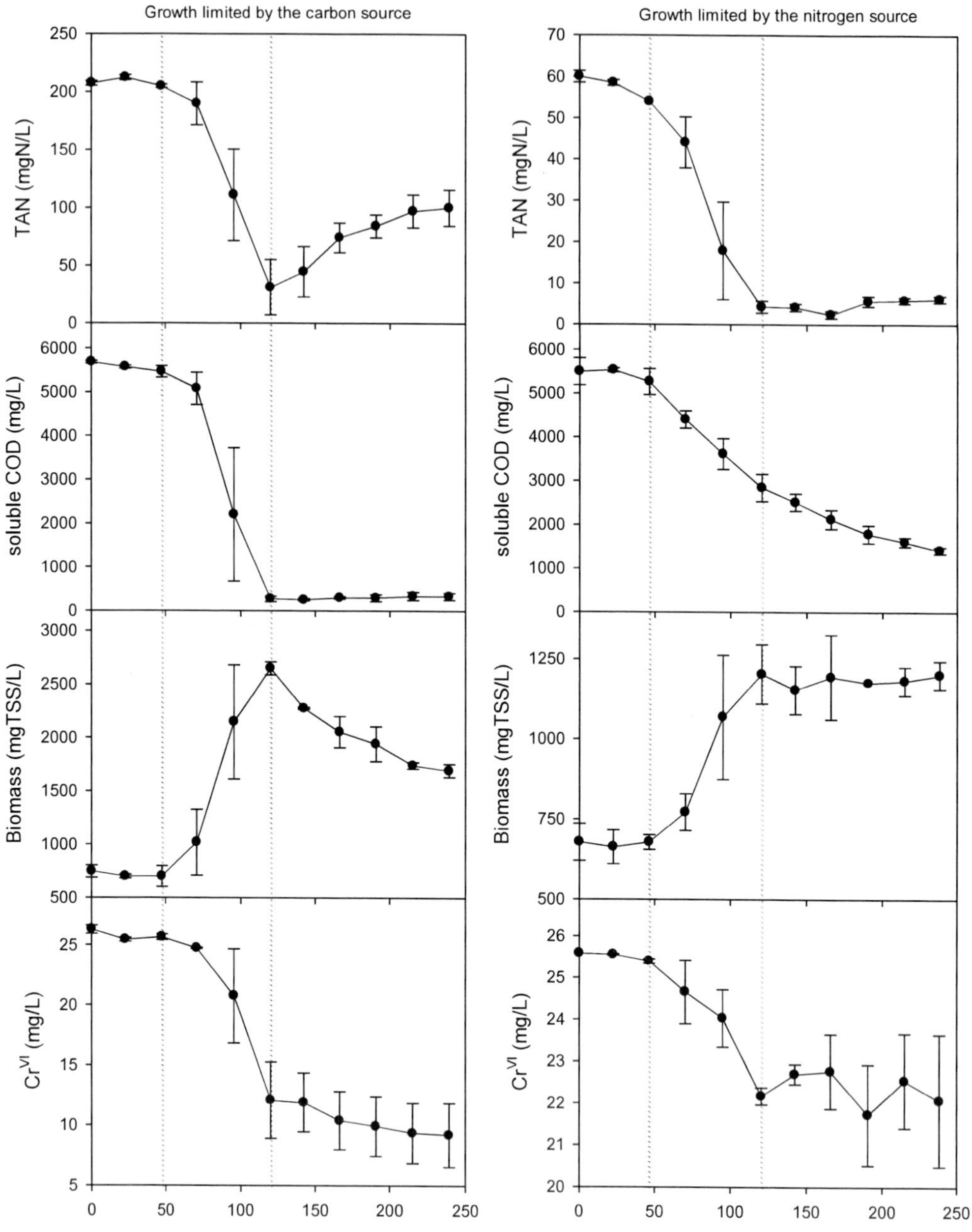

Figure 8. Hexavalent chromium removal by activated sludge performed under carbon source limiting conditions (left panel) or nitrogen source limiting conditions (right panel). In both cases, the time between dashed lines corresponds to the exponential growth phase.

The effect of the initial nitrogen source concentration (N_0) on the Cr(VI) removal is shown in Figure 9. Cr(VI) removal increased as the initial nitrogen source concentration (N_0) increased from 10%, corresponding to assays with no nitrogen addition, to about 60% for N_0 = 130 mgN/L. In these cases, biomass growth was limited by the nitrogen source. For N_0 values higher than 130 mgN/L, Cr(VI) removals were constant because the carbonaceous substrate was completely consumed before the nitrogen source; thus, in these cases, growth and Cr(VI) removal were limited by the carbonaceous source (Ferro Orozco, 2009).

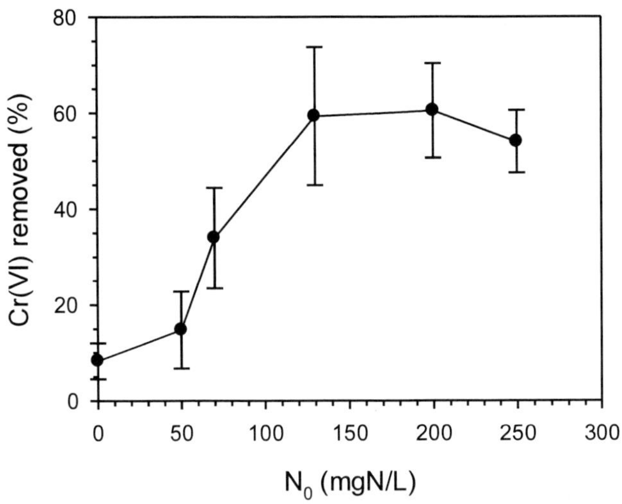

Figure 9. Effect of the initial nitrogen source concentrations (N_0) on Cr(VI) removal. In all cases, initial conditions were the following: lactose = 5 g L^{-1}, biomass = 700 mgTSS L^{-1}, Cr(VI) = 25 mg/L.

Depending on the ratio between initial concentration of the nitrogen source and the carbonaceous substrate (N_0:S_0), one of both compounds may limit the biomass growth. The stoichiometric N_0:S_0 ratio, $(N_0:S_0)_{St}$, can be defined as the initial N_0:S_0 ratio at which both N and S are depleted at the same time; in this case both N and S are limiting substrates simultaneously. If $N_0:S_0 > (N_0:S_0)_{St}$ then N is in excess with respect to the stoichiometric requirements and S is the growth limiting substrate. Conversely, when $N_0:S_0 < (N_0:S_0)_{St}$ the carbonaceous substrate is in excess with respect to the stoichiometric requirements and N limits the biomass growth. If assimilation (incorporation of the nitrogen source into biomass) is the only nitrogen consumption process (e.g. if nitrification can be neglected), the stoichiometric N_0:S_0 ratio can be calculated as the product between the nitrogen content of the biomass (i_B), and the biomass growth yield ($Y_{X/S}$). According to several authors (Irvine and Bryers, 1985; Atkinson and Mavituna, 1991; Orhon and Artan, 1994), the most accepted value corresponding to i_B for activated sludge is 110 mgN $gTSS^{-1}$; however, $Y_{X/S}$ depends on many factors, such as wastewater characteristics, culture conditions, and the presence of toxic compounds (e.g. Cr(VI)). For these reasons, it is advisable to determine $Y_{X/S}$ for each particular case to get a better estimation of the stoichiometric N_0:S_0.

Knowledge of the stoichiometric N_0:S_0 ratio allows estimating amendments of the wastewater composition to optimize the removal of Cr(VI) by activated sludge and therefore, to aid process design and operation for the treatment of Cr(VI) containing wastewaters using batch reactors. Previous studies (Ferro Orozco, 2009) show that biomass growth yield ($Y_{X/S}$) corresponding to activated sludge growing on lactose in the presence of 25 mgCr(VI)/L is about 0.3 gTSS/gCOD; based on this value for $Y_{X/S}$ and considering i_B = 110 mgN/gTSS, the stoichiometric N_0:S_0 ratio calculated as the product between $Y_{X/S}$ and i_B results 33 mgN/gCOD. Considering this ratio and taking into account that the initial lactose concentration (S_0) was 5 g/L in Cr(VI) removal experiments shown in Figure 9, the initial nitrogen concentration necessary to accomplish the stoichiometric requirement is 165 mgN/L. In this case, if N_0 > 165 mgN/L then the nitrogen source is in excess with respect to the carbonaceous substrate; therefore, increasing N_0 above the stoichiometric value would not have an effect on biomass growth or on the Cr(VI) removal, in agreement with results shown in Figure 9.

3.8. Effect of the Initial Cr(VI) Concentration on the Activated Sludge Growth Kinetics

As was previously discussed, the initial biomass concentration (X_0), the presence, type and initial concentration of a carbonaceous substrate (S_0), the initial nitrogen source concentration (N_0), and the ratio N_0:S_0 are important factors that affect the removal of Cr(VI) by activated sludge. However, because hexavalent chromium is a known toxicant, the effect of the initial Cr(VI) concentration on activated sludge growth kinetics is another important feature to take into account in order to optimize design and operation of a biological Cr(VI) reduction process (Philip et al., 1998). In this section, the effect of the initial Cr(VI) concentration on the activated sludge growth kinetics is discussed.

The growth of activated sludge in aerobic batch reactors were studied. Experiments were performed at constant temperature (20°C) and pH = 7.0; inocula were obtained from the activated sludge reactor; initial biomass concentration was 700 ± 50 mgTSS/L. The culture medium contained cheese whey (5 gCOD/L) as carbonaceous source, and $(NH_4)_2SO_4$ (1g/L) as nitrogen source, diluted in a phosphate buffer. Appropriate amounts of a Cr(VI) stock solution were added to obtain 10, 25, 50, 100 and 300 mgCr(VI)/L. Control experiments, without the addition of Cr(VI), were also performed (Ferro Orozco, 2009).

Figure 10 shows that in the assay containing 300 mgCr(VI)/L, no microbial growth was observed, indicating that the threshold inhibitory concentration of Cr(VI) may be within the range of 100 to 300 mg/L. Regarding the experiments with hexavalent chromium concentration lower than 100 mgCr(VI)/L, results showed that the increase in Cr(VI) concentration caused longer lag phases. This effect of Cr(VI) on the lag phase was reported by other authors (Bae et al., 2000; Gikas and Romanos, 2006; Li et al., 2007). Vankova et al. (1999) attributed this phenomenon either to the gradual acclimation of the activated sludge, to the reduction of Cr(VI) concentration due to the adsorption on biomass flocs or due to the reaction with other substances present in the culture medium. Hexavalent chromium compounds have been shown to induce frame-shift mutations and base-pair substitutions both at G-C and A-T base pairs in bacterial cells (Bae et al., 2000). It has been proposed that bacterial SOS functions can repair the DNA damage produced by Cr(VI). An increase in the lag phase by Cr(VI) addition may be due to an increased time for adaptation of DNA repair during exposure to hexavalent chromium in the medium (Bae et al., 2000).

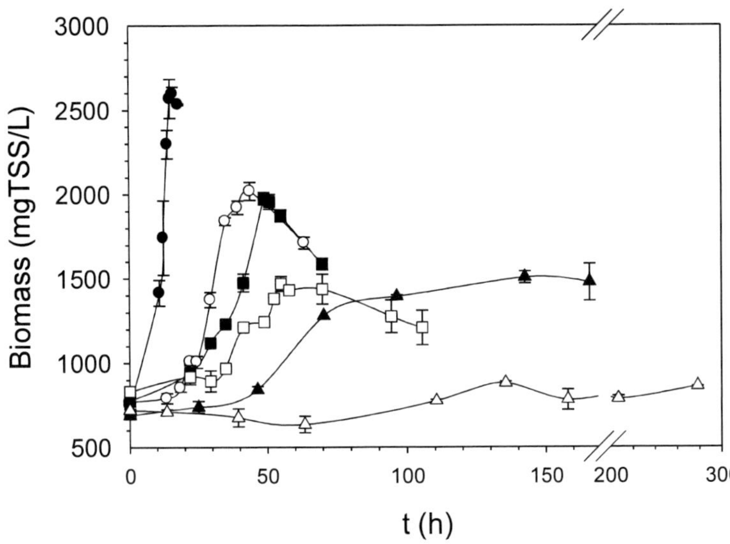

Figure 10. Effect of the initial Cr(VI) concentration on the biomass growth. Tested Cr(VI) concentrations were (●) 0, (O) 10, (■) 25, (□) 50, (▲) 100, and (△) 300 mgCr/L.

After the lag phase, biomass concentration (X) and OUR increased, and S concentration decreased as a function of time. From this data, biomass growth yield ($Y_{X/S}$), substrate oxidation coefficient ($Y_{O/S}$) and the observed specific growth rate (μ_{obs}) were calculated for each experiment. Table 5 shows that, compared to the control experiment (without Cr(VI) addition), both coefficients $Y_{X/S}$ and μ_{obs} decreased as the Cr(VI) concentration increased. On the other hand, $Y_{O/S}$ values increased with higher Cr(VI) concentrations. The effect of Cr(VI) on $Y_{X/S}$ and $Y_{O/S}$ can be attributed to the additional energy needed by microorganisms to keep an active metabolism due to higher stress conditions of the environment. However, the toxic effect of Cr(VI) on the activated sludge growth kinetics can be reduced by the use of powdered activated carbon (PAC), for example. The addition PAC into the aeration basin of activated sludge reactors has been proposed as a suitable technique to protect activated sludge against toxic wastewaters (Orshansky and Narkis, 1997; Costa and Márquez, 1998).

Table 5. Effect of Cr(VI) on the biomass growth yield ($Y_{X/S}$), the substrate oxidation coefficient ($Y_{O/S}$), and the observed specific growth rate (μ_{obs})

Initial Cr(VI) concentration (mg/L)	$Y_{X/S}$ (gTSS gCOD^{-1})	$Y_{O/S}$ (gO_2 gCOD^{-1})	μ_{obs} (h^{-1})
0	0.473 ± 0.010	0.385 ± 0.013	0.154 ± 0.008
10	0.378 ± 0.090	0.508 ± 0.116	0.047 ± 0.002
25	0.351 ± 0.083	0.544 ± 0.109	0.027 ± 0.005
50	0.255 ± 0.075	0.668 ± 0.098	0.016 ± 0.003
100	0.173 ± 0.038	0.775 ± 0.050	0.010 ± 0.001
300	n.a.	n.a.	n.a.

With regard to wastewaters containing hexavalent chromium, recent studies demonstrated that the addition of 4 gPAC/L exerts a protective effect on activated sludge against Cr(VI) toxicity (Ferro Orozco et al., 2004, 2007b, 2008; Ferro Orozco, 2009).

Conclusion

This chapter addresses the effect of important factors concerning the bioreduction of hexavalent chromium by activated sludge in batch systems. Throughout this chapter it was demonstrated that the reduction of Cr(VI) to Cr(III) is a catalytic mediated process associated with the presence of biomass.

Although the biomass itself could serve as the electron donor for reducing Cr(VI) to Cr(III), an external electron donor (e.g. carbonaceous substrate) increased the rate and extent of Cr(VI) reduction. High Cr(VI) reduction rates were obtained with fermentable substrates such as cheese whey, lactose, and glucose. On the contrary, with non-fermentable substrates such as citrate and acetate, Cr(VI) removal rates were slower. The low fraction of electrons of the carbonaceous substrate that reduce Cr(VI) to Cr(III) suggests that Cr(VI) reduction in activated sludge batch reactors is mainly via a cometabolic pathway.

Besides the carbonaceous substrate, the presence of a nitrogen source is also necessary to obtain high Cr(VI) removal rates. In addition, when one or both of carbonaceous or nitrogen source is depleted during the growth phase, Cr(VI) removal continues but at a lower rate. Thus, the bioreduction of Cr(VI) is the result of a fast Cr(VI) reduction process, associated to the biomass growth, and a slow one that does not depend on the presence of both a carbonaceous substrate and a nitrogen source. The second process may be important under certain conditions, such as in continuous cultures in which substrates concentrations are normally low.

With regard to the effect of the initial Cr(VI) concentration on the activated sludge growth kinetics, a threshold inhibitory concentration of Cr(VI) within the range of 100 to 300 mg/L was observed. In addition, the biomass growth yield and the observed specific growth rate decreased as the Cr(VI) concentration increased; conversely, the substrate oxidation coefficient increased with higher Cr(VI) concentrations. These effects were attributed to the additional energy needed by microorganisms to maintain an active metabolism due to the presence of Cr(VI).

The findings presented in this chapter increase our understanding of the fundamental processes concerning the bioreduction of hexavalent chromium by activated sludge in batch systems; thus, they are useful in helping to design cost-effective, high-performance Cr(VI) removal bioprocesses.

References

Alvarez-Cohen L., Speitel Jr. G.E. (2001) Kinetics of aerobic cometabolism of chlorinated solvents. Biodegradation 12:105-126.

APHA (1989) Standard Methods for the Examination of Water and Wastewater, 17th ed. American Public Health Association.

Asatiani N.V., Abuladze M.K., Kartvelishvili T.M., Bakradze N.G., Sapojnikova N.A., Tsibakhashvili N. Ya., Tabatadze L.V., Lejava L.V., Asanishvili L.L., Holman H-Y.

(2004). Effect of chromium (VI) action on *Arthrobacter oxydans. Current Microbiology.* 49, 321-326.

Atkinson B., Mavituna F. (1991). *Biochemical Engineering and Biotechnology Handbook.* Stockton Press, New York, USA.

Bae W.C., Kang T.G., Kang I.K., Won Y.J., Jeong B.C. (2000). Reduction of hexavalent chromium by *Escherichia coli* ATCC 33456 in batch and continuous cultures. *The Journal of Microbiology.* 38, 36-39. The Microbiological Society of Korea.

Bopp L.H., Ehrlich H.L. (1988). Chromate resistance and reduction in *Pseudomonas fluorescens* strain LB300. *Arch. Microbiol.* 150, 426-431.

Caravelli A.H., Zaritzky N.E. (2009). About the performance of *Sphaerotilus natans* to reduce hexavalent chromium in batch and continuous reactors. *Journal of Hazardous Materials* (in press). doi:10.1016/j.jhazmat.2009.03.018

Chua, H. (1998). Effects of trace chromium on organic adsorption capacity and organic removal in activated sludge. *The Science of the Total Environment.* 214, 239-245.

Ferro Orozco A.M. (2009) *Tratamiento biológico de efluentes con utilización simultánea de carbón activado para la eliminación de cromo hexavalente de aguas residuales*. Ph.D. Thesis, Universidad Nacional de La Plata, Argentina.

Ferro Orozco A.M., Contreras E.M., Zaritzky N.E. (2008) Modelling Cr(VI) removal by a combined carbon-activated sludge system. *Journal of Hazardous Materials* 150(1), 46-52.

Ferro Orozco A.M., Contreras E.M., Bertola N.C., Zaritzky N.E. (2004) Chromium (VI) removal using powdered activated carbon in activated sludge reactors. *Proceedings of the Fifth Latin American Biodegradation and Biodeterioration Symposium (LABS5)*, Campeche, México. pp:106-109.

Ferro Orozco A.M., Contreras E.M., Bertola N.C., Zaritzky N.E. (2007a) Efecto de la concentración inicial de biomasa y de distintos sustratos en la reducción de Cr(VI) por barros activados. *Ingeniería Sanitaria y Ambiental* 91, 69-74.

Ferro Orozco A.M., Contreras E.M., Bertola N.C., Zaritzky N.E. (2007b) Hexavalent chromium removal using aerobic activated sludge batch systems added with powdered activated carbon. *Water SA* 33(2), 239-244.

Ghaly A.E., Tango M.S.A., Adams M.A. (2003) Enhanced lactic acid production from cheese whey whith nutrient supplement addition. *Agricultural Engineering International: the CIGR Journal of Scientific Research and Development.* Manuscript FP 02 009. May, 2003

Gikas P., Romanos P. (2006). Effects of tri-valent (Cr(III)) and hexa-valent (Cr(VI)) chromium on the growth of activated sludge. *J. Hazard. Mater.* B133, 212-217.

Irvine R.L., Bryers J.D. (1985) Stoichiometry and kinetics of waste treatment. In: *Comprehensive Biotechnology.* Vol. 4. (C.W. Robinson, J.A. Howell, eds.). Pergamon, New York. Cap. 41, pp. 757-772.

Katz S.A., Salem H. (1994) *The Biological and Environmental Chemistry of Chromium.* VCH Publishers, Inc. New York, USA.

Li Y., Low G.K.-C., Scott J.A., Amal R. (2007). Microbial reduction of hexavalent chromium by landfill leachate. *J. Hazard. Mater.* 142, 153-159.

Lovley D.R., Phillips E.J.P. (1994). Reduction of chromate by *Desulfovibrio vulgaris* and its c3 Cytochrome. *Appl. Environ. Microbiol.* 60 726-728.

Madoni P., Davoli D., Guglielmi L. (1999). Response of SOUR and AUR to heavy metal contamination in activated sludge. *Water Res*. 33, 2459-2464.

Mondaca M.A, Campos V, Moraga R., Zaror C.A. (2002). Chromate reduction in *Serratia marcescens* isolated from tannery effluent and potential application for bioremediation of chromate pollution. *The Scientific World J.* 2, 972-977.

Myers C.R., Carstens B.P., Antholine W.E., Myers M.J. (2000). Chromium(VI) reductase activity is associated with the cytoplasmic membrane of anaerobically grown *Shewanella putrefasciens* MR-1. *J. Appl. Microbiol.* 88, 96-106.

Nepple B.B, Kessi J., Bachofen R. (2002). Chromate reduction by *Rhodobacter sphaeroides. J. Ind. Microb. Biotechnol.* 25, 198-203.

Nkhalambayausi-Chirwa and Wang (2004) Modeling hexavalent chromium removal in a *Bacillus* sp. fixed-film bioreactor. Biotechnol. Bioeng. 87, 874-883.

Orhon D., Artan N. (1994) *Modelling of activated sludge systems.* Technomic Publishing Company, Inc. Lancaster, USA.

Patterson J.W. (1985). Industrial Wastewater Treatment Technology. Butterworth publishers. Stoneham, MA. pp. 53-393.

Philip L., Iyengar L., Venkobachar C. (1998). Cr(VI) reduction by *Bacillus coagulans* isolated from contaminated soils. *J. Environ. Eng*. 124, 1165-1170.

Rege M.A., Petersen J.N., Johnstone D.L., Turick C.E., Yonge D.R., Apel W.A. (1997). Bacterial reduction of hexavalent chromium by *Enterobacter cloacae* strain HO1 grown on sucrose. *Biotechnol. Lett.* 19, 691-694.

Ros M. (1993). *Respirometry of Activated Sludge*. Technomic Publishing Co., Inc. Basel, Switzerland.

Schmieman E.A., Yonge D.R., Rege M.A., Petersen J.N., Turick C.E., Johnstone D.L., Apel W.A. (1998). Comparative kinetics of bacterial reduction of chromium. *Journal of Environmental Engineering*. 124, 449-455.

Schriver D.F., Atkins P., Langford C.H. (Eds.), *Inorganic Chemistry* (2nd ed.); Freeman: New York, 1994.

Shen H., Wang Y.T. (1994). Biological reduction of chromuim by *E. coli. Journal of Environmental Engineering.* 120, 560-572.

Shen H., Wang Y.T. (1995). Characterization of enzymatic reduction of hexavalent chromium by *Escherichia coli* ATCC 33456. *Appl. Environ. Microbiol.* 59, 3771-3777.

Stasinakis A.S., Mamais D., Thomaidis N.S., Lekkas T.D. (2002). Effect of chromium (VI) on bacterial kinetics of heterotrophic biomass of activated sludge. *Wat. Res.* 36, 3342-3350.

Stasinakis A.S., Thomaidis N.S., Mamais D., Lekkas T.D. (2004). Investigation of Cr(VI) reduction in continuous-flow activated sludge system. *Chemosphere*. 57, 1069-1077.

Stumm, W.; Morgan, J.J. *Aquatic Chemistry* (3rd ed.), Wiley, New York, 1996.

U.S. EPA. (1998a). Toxicological Review of Hexavalent Chromium (CAS No. 18540-29-9) U.S. Environmental Protection Agency. Washington DC.

U.S. EPA. (1998b). Toxicological Review of Trivalent Chromium (CAS No. 16065-83-1) U.S. Environmental Protection Agency. Washington DC.

Vankova S., Kupec J., Hoffman J. (1999). Toxicity of chromium to activated sludge. *Ecotoxicol. Environ. Safety*. 42, 16-21.

Verweij, W. (2005) "Chemical Equilibria in Aquatic Systems – CHEAQS" – PC calculating programm (http://home.tiscali.nl/cheaqs/index.html)

Wang Y.T., Shen H. (1995). Bacterial reduction of hexavalent chromium. *Journal of Industrial Microbiology*. 14, 159-163.

Wang, Y., Shen, H. (1997). Modelling Cr(VI) reduction by pure bacterial cultures. *Wat. Res.* 31, 727-732.

Wang Y., Xiao C. (1995). Factors affecting hexavalent chromium reduction in pure cultures of bacteria. *Wat. Res.* 29, 2467-2474.

Yetis U., Demirer G.N., Gokcay C.F. (1999). Effect of chromium (VI) on the biomass yield of activated sludge. *Enzyme and Microbial Technology*. 25, 48-54.

In: Management of Hazardous Residues Containing Cr(VI) ISBN: 978-1-61470-637-3
Ed: Maria José Balart Murria

Chapter 9

CHROMIUM (VI) GALVANIC BATH: CHEMICAL TREATMENTS AND POSSIBLE RECYCLING WAYS OF THE OBTAINED SLUDGES

M. Cannio, L. Barbieri and F. Bondioli*

Department of Materials and Environmental Engineering, University of Modena and Reggio Emilia - Via Vignolese, 905, Italy

ABSTRACT

Galvanic processes, that increase corrosion resistance properties and improve wear qualities of metal materials, are based on metal plating baths and rinsewaters. They generate effluents with a metal concentration varying with the installed process. Traditional systems reduce the toxicity of wastewater by aggregating disposed aqueous solutions of various concentrations, followed by treatment with chemicals to coagulate, flocculate, and settle out solid wastes for off-site disposal. In this paper the efficiency of different precipitating agents was evaluated in order to obtain a Cr (III) sludge.

There are several treatment routes proposed in the literature to avoid dumping, or in other words, recycle the wastes. Recent increase in the cost of landfill disposal and decrease in number of disposal sites have led to consideration of alternative routes and treatments. For this reason, the screening results of scientific activity conducted on a laboratory scale on the possibility to both inertize and valorize the obtained galvanic sludge in the traditional ceramic filed (glasses, bricks, tiles, pigments and glazes) were reported with particular emphasis on the possible problems and the possible ways to solve them.

* Email: Federica.Bondioli@unimore.it

1. THE GALVANIC PROCESS

Electrolytic hard chromium plating is an industrially widespread technology for the production of tough wear-resistant coatings (Figure 1). In the last years, much effort was devoted to the identification of possible alternative routes to the use of high chromic acid solutions, since it has been demonstrated that Cr^{6+} compounds are toxic and carcinogenic. Functional chromium coatings consist of a thick layer of chromium (from 1.3 to 300 μm) to provide a surface with functional properties such as corrosion resistance, hardness and wear resistance [1-5]. Decorative chromium coatings consist of a thin layer of chromium (0.003 to 0.5 μm) to provide a bright surface with wear and tarnish resistance when plated over a nickel layer. The electrodeposition processes used to deposit both hard and decorative chromium coatings are still based on the operating conditions set by Fink's patent in 1926 [6]: i) deposition bath made up of an aqueous solution containing CrO_3 and H_2SO_4 with a weight ratio 100:1 (normally 250 and 2.5 $g \cdot dm^{-3}$, respectively); ii) working temperature about 50°C; iii) current densities from 0.200 to 0.400 $A \cdot cm^{-2}$. The electrochemical deposition can be performed at constant current but, due to the difficulty in the measurement of actual area of complex shaped pieces, the voltage control is preferred. The current efficiency seldom reaches 18 %, as most of the current supplied to the cell is dissipated by the hydrogen reduction reaction.

Figure 1. System of tank used in industry for the hard chrome deposition.

The use of CrO_3 compounds is strongly rebutted because they are toxic, carcinogenic and can cause permanent genetic alterations; moreover its oxidising nature makes CrO_3 dangerous to the environment. [7,8]. A way to overcome the problem could be deposit chromium coatings from solutions of trivalent chromium salts, which do not show the same hazard to human health and for the environment. Unfortunately, coatings obtained using Cr(III) solutions are not quite satisfactory, because in aqueous solution the Cr(III) ions form highly stable aqueous complexes. To minimize this problem an appropriate ligand capable to form a Cr(III) complex with optimum chemical and electrochemical reactivity can be used [2], but the state of the art of Cr(III) plating still presents several drawbacks such as low current efficiency and solution instability. Another method for limiting the heavy environmental impact of highly concentrated solutions as well as reducing the high disposal costs of processing fluids could be the use of more diluted solutions, as reported in a previous research paper [8]. The significant drawback is the decrease of the solution conductivity, which can be overcome by an increase of the deposition time [8].

2. The Galvanic Sludge

The electrodeposition process produces a high quantity of refusals and products of discard of workmanship:

1) washing waters (galvanic processes generate effluents with a metal concentration depending on the employed process);
2) materials and equipment (aluminium protective ribbons, varnishes and plastics);
3) lead and iron dust on the fund of the tank;
4) barium carbonate residual ($BaCO_3$ is used to correct the sulphates content);
5) the exhausted bath;
6) rectification and pre-treatment moods composed by heavy metals such as iron, cast iron, copper, chromium.

Traditional systems reduce the toxicity of wastewater by adding Cr^{6+} reducing agents, followed by treatment with chemicals to coagulate, flocculate, and settle out solid wastes for off-site disposal [9]. In the Italian factory that kindly furnished the sludge used in the work reported in this study, the galvanic solution is treated with a two-stage process. The first stage reduces hexavalent chromium to the chemically unstable $Cr_2(SO_4)_3$ by using chemical reducing agents such as sulfur dioxide (SO_2), sodium bisulfate ($NaHSO_3$) or sodium/calcium meta-bisulfite ($Na_2S_2O_5$).

$$4H_2CrO_4 + 3CaS_2O_5 \longrightarrow 2Cr_2(SO_4)_3 + 4H_2O + 3CaO \quad (1)$$

pH= between 2 and 3

In the trivalent state, chromium freely bonds to hydroxide and the final result is a non-toxic precipitate, chromium hydroxide $Cr(OH)_3$, which can be easily separated and disposed in landfills:

$$Cr_2(SO_4)_3 + 3\,Ca(OH)_2 \longrightarrow 2Cr(OH)_3 + 3CaSO_4 \quad (2)$$

pH = 8 or higher to obtain the hydroxides precipitation.

3. Optimization on a Laboratory Scale of the Removal Process of Chromium From Galvanic Residual Water by Chemical Precipitation

The purpose of this study is to improve the technical process for removing trivalent chromium as Cr hydroxide from galvanic wastewater via alkaline precipitation. The industrial effluent presents in a clear and transparent dark green color, with pH equal to 3,88 to 17°C. The chemical composition of these industrial waters is reported in Table 1.

The residual Cr^{6+} content of this solution, determined by digital photometer NANOCOLOR® 500 D, after the industrial reduction process with $Ca(OH)_2$ was <2ppm, reaching the national discharge limit. Therefore, four alkaline precipitating agents were evaluated on the laboratory scale in order to optimize the reduction process and the chemical composition of the final sludge: NaOH, NH_4OH, $Ca(OH)_2$ and $Mg(OH)_2$. After some preliminary tests, calcium and magnesium hydroxides were refused since their solubility product equilibrium constants in water are very low (K_{ps} $Ca(OH)_2$=$5.5 \cdot 10^{-6}$ and K_{ps} $Mg(OH)_2$=$1.2 \cdot 10^{-11}$, respectively) even if the $Ca(OH)_2$ is today still the more used industrially. Experiments were conducted to evaluate the influence of the alkaline agent, pH and precipitation time on the nature of the final product. The selected alkaline reagents, NaOH and NH_4OH, both 1 M, were added to a known volume of effluents (100 or 200 mL), furnished by the Italian galvanic industry (Modena, Italy), stirred at a constant rate of 800 rpm by means of a magnetic stirrer; during the addition of the reagent, pH and temperature were monitored. Following the precipitation reaction step, the obtained gel was aged for 12 hours, then it was centrifuged at 4000 rpm for 40 minutes. The Cr content determined by inductively coupled plasma optical emission spectrometry (ICP, Varian®, model Liberty 200™), in the treated wastewater after the Cr (III) precipitation as a function of precipitation pH is reported for both precipitant agents in Figure 2 and Table 2.

Table 1 - Chemical composition of the industrial wastewater determined by ICP

Element	Concentration (ppm)
Si	57,0
Al	27,6
Ca	511,3
Mg	68,8
Na	7655
K	7,9
Fe	159,0
Ti	1,7
Ni	6,5
Cu	233,9
Mn	2,9
Co	3,2
Cd	1,4
Cr	3557

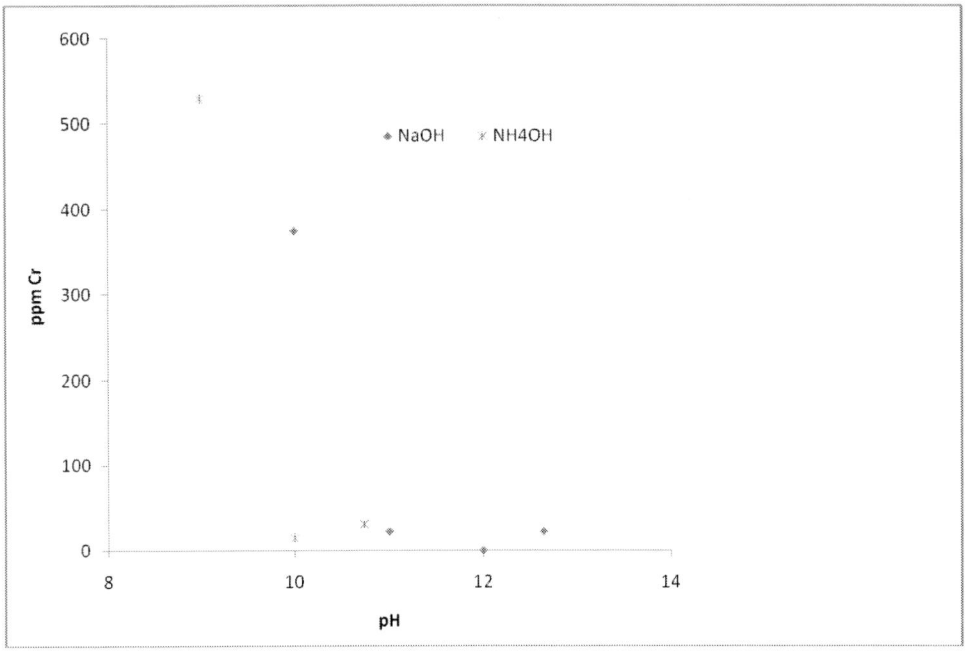

Figure 2. pH influence on the Cr (III) precipitation. Alkaline agent: NaOH 1 M and NH_4OH 1 M.

Table 2. Results obtained through the addition of NaOH and NH_4OH 1 M to the industrial wastewater

	pH	Characteristic of sludge	Residual liquor characteristic	Cr concentration in the residual liquor by ICP (ppm)
NaOH	10	Scarce precipitation of a dark green gel	Green, clear, transparent	375,1
NaOH	11	Good precipitation of a dark green gel	Green very clear, clear, transparent	22,28
NaOH	12	Good precipitation of a dark green gel	Colourless, clear, transparent	0,107
NaOH	12.64	Good precipitation of a dark green gel	Green very clear, clear, transparent	22,7
NH_4OH	9	Scarce precipitation of a dark green gel	Green, clear, transparent	529,3
NH_4OH	10	Good precipitation of a dark green gel	Quite colourless, clear, transparent	15,67
NH_4OH	10.73	Good precipitation of a dark green gel	Blue, clear, transparent.	31,27

The results underlined that the maximum amount of $Cr(OH)_3$ recovered was mainly dependent on pH and alkali type. From the analysis of the obtained results it can be concluded that the process of Cr(III) precipitation from the industrial wastewaters indicates a removal efficiency of 100% and 99% respectively using NaOH at pH 12 and NH_4OH at pH 10. Furthermore, a chemical analysis by ICP was performed on the obtained sludge after the fusion with lithium metaborate ($LiBO_2$) in Pt crucible and dissolution by nitric (HNO_3) [10-11]. In particular, Table 3 shows that the investigated process leads to sludge with a higher content of Cr_2O_3 than that industrially obtained indicating that the efficiency of the industrial process could be seriously optimized substituting the $Ca(OH)_2$ precipitant agent with NaOH or NH_4OH.

Table 3. Chemical composition of dried sludges determined by ICP

Oxides	Content % (sludge from NaOH)	Content % (sludge from NH_4OH)	Content % (average industrial galvanic sludge)
SiO_2	0,31	0,46	0,39
Al_2O_3	0,26	0,34	0,28
CaO	3,18	3,71	42,50
MgO	0,54	0,56	0,58
Na_2O	21,3	6,21	3,11
K_2O	0,28	0,96	0,05
Fe_2O_3	1,08	1,32	0,72
NiO	0,014	0,064	0,02
CuO	1,52	1,25	0,68
MnO	0,011	0,032	0,02
Cr_2O_3	26,26	32,50	15,91
CoO	0,013	0,040	0,02
CdO	0,004	0,016	0,01
L.O.I.	45,23	52,50	30

4. Inertisation and Valorisation of the Obtained Sludge

Galvanic sludges produced by the described physico-chemical treatment are potentially eco-toxic residues, because of the high mobility of metals, such as chromium, nickel, copper present in the wastes. In the specific case studied in this work, Cr(III) could be partially oxidized to the more toxic and more soluble Cr(VI), giving off very hazardous sludge. For these reasons disposal of industrial galvanic sludge are subjected to detailed regulations that require specific landfill sites. Moreover, recent increase in the cost of land disposal and decreases in the number of disposal sites has led to a consideration of alternatives routes and treatments. Several inertization/stabilization processes of wastes from different industrial sediments (i.e. leather waste, galvanic sludges…) are reported in literature, e.g. vitrification and devitrification processes, incorporation into ceramic matrices such as bricks, tiles and stoneware, etc… [12-17].

In general, the addition of industrial wastes to raw material or different industrial matrix requires a carefully study of the overall effects. In particular, specific requirements involve:

1) characterization of raw material;

2) waste characterization: for a correct valorization and recycling of industrial wastes it is necessary to take into account not only the kind and the amount of the waste, but also its chemical composition;
3) investigation concerning the nature of obtained phases and final microstructure;
4) leaching characteristics of the final products in different media;
5) analysis about the technological properties of the obtained products.

Vitrification is a promising technological option for the treatment of waste. The high temperature involved in the process (1200-1500°C) leads to the destruction of organic pollutants and the glassy network can incorporate hazardous heavy metals in a stable assemblage. The recycling and reuse of waste containing vitrifying, melting and stabilizing oxides into glass and glass ceramic materials seems to be a promising solution for the reduction of residue volumes. On the other hand, if a waste does not contain the characteristic elements of the glassy network, but it is rich in transition oxides, it can be introduced in a glassy matrix in order to confer color and to promote the crystallization process. This method was tested for several kinds of industrial waste [12-17], but there are few works describing the stabilization/solidification of chromium galvanic sludge in glass [11, 18]. In the section "Inertisation of galvanic sludge: the vitrification process" it is reported that the study aimed to determine the potential inertization of galvanic waste rich in chromium and calcium by its incorporation as additive to glass cullet in the percentages of 0.12, 0.25, 0.5 and 1 Cr_2O_3 wt%. In order to investigate the effects on $Cr(OH)_3$ stabilization, the glasses were characterized in terms of structure, chemical durability, leaching test and crystallisation tendency.

Bricks and tiles are heterogeneous materials based on a mix of clay raw materials with a wide-ranging overall composition. For this reason, such materials can tolerate the presence of different types of wastes even in significant amounts. Chemical fixation of species might involve macro or micro-encapsulation, according to the nature of components involved and to the operational conditions. Ceramics are prepared from malleable, earthly materials, such as clay that becomes rigid at high temperatures. Besides clays, which confer cohesion and plasticity, and feldspar, as melting agents, ceramic mixture also include inert material, i.e. quartz, SiO_2, that provide structural support, necessary to retain shape during drying and firing. The firing step is determinant in the final immobilization degree, since major combining reactions between clay and waste particles occur upon heating and phenomena of sintering also promote the physical consolidation of the material by reaching a stable suitable microstructure. The chemical nature of phases and the final microstructure of the material strongly affects its stability and solubility degree. In the literature there are several examples related to the inertization of different kinds of industrial waste [12-17], but there are few works describing the stabilization of chromium galvanic sludge in clay [19]. In this work, it is reported the study aiming to determine the potential inertization of galvanic waste rich in chromium and calcium by its incorporation in two different clay-based matrices. In order to investigate the effects on $Cr(OH)_3$ stabilization, the galvanic waste was added in different amounts into two different industrial batch compositions used for the manufacturing of ceramic bricks for building application. In particular, two thermal treatments were used for the firing of the investigated bricks: a slow treatment working at 950°C for 24 hours and a fast firing cycle using an isotherm at 1150°C for 2 hours.

Finally, the sludge valorisation can be obtained by achievement of thermal and chemical stable crystal structures, able to colour ceramic material. Ceramic pigments were synthesized by solid state reactions and the colour developed in a suitable ceramic glaze was investigated in comparison with the colour developed by the commercial pigments prepared from pure Cr_2O_3 [10]. Since few natural deposits of chromium exist, a stable ceramic pigment obtained from an economical source is of great interest for the ceramic industry from an economic point of view. This opportunity was already tested for production of chromium-containing pigments from chromium recovered from leather waste [20], but there are not so many previous works known to the authors related to the use of Cr sludge from the galvanic industry as secondary raw material for the production of ceramic pigments. Therefore, the purpose of this research is to investigate the possible recycling of Cr sludge from galvanic industry as raw material in the production of ceramic pigments, in particular $CaCr_{0.04}Sn_{0.97}SiO_5$ and $Ca_3Cr_2(SiO_4)_3$.

5. Physical-Chemical Characterization of the Galvanic Sludge

The galvanic sludge used in the experiments, obtained from a chromium electroplating process of an Italian factory and from the optimization process described in the above section, contained Cr as the main heavy metal. The sludge, previously dried at 110°C for 24 h, was mainly characterized by inductively coupled plasma optical emission spectrometry (ICP, Table 3), X-ray diffraction (XRD), scanning electron microscopy (SEM) and thermal analysis (TG-DTA). A complete description of the characterization of chromium galvanic wastes is presented elsewhere [11]. Briefly, the sludge is basically constituted by hydroxides, carbonates ($CaCO_3$) and sulphates ($CaSO_3 \cdot 0.5H_2O$). The average grain size distribution of agglomerates is narrow: all selected samples show maximum particle size of about 65μm after milling, indicating a suitable fineness for the mixing with ceramic materials. The moisture level is over 30 wt%. Table 3 gives the average chemical composition of the sludge and shows that the loss of ignition (LOI) is close to 30% for the industrial sludge and in the range 45-55% for the mud obtained using NaOH and NH_4OH as precipitating agents.

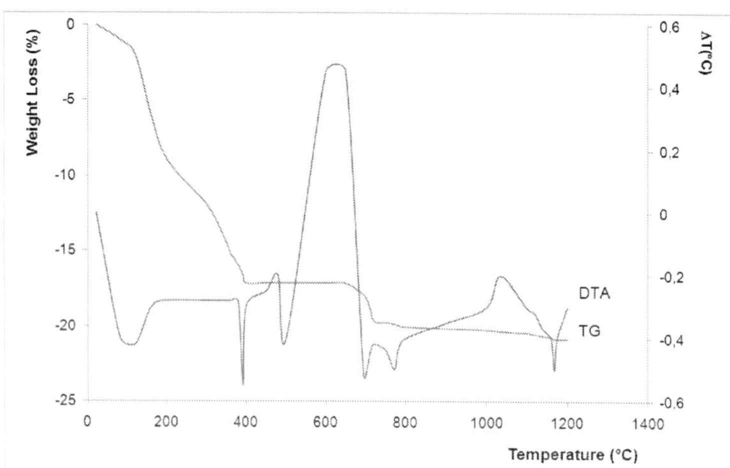

Figure 3. DTA and TG curves of as-received industrial galvanic sludge (↑exo).

Figure 3 shows TG and DTA curves of the as-received industrial galvanic sludge. The DTA curve shows three exothermic peaks at around 480, 700 and 1035°C, not correlated to weight loss, probably due to crystallization process.

Endothermic peaks appeared at around 393 (release of $CaSO_3 \cdot 0.5H_2O$ crystallization water), 695 ($CaCO_3$ decomposition), 771 and 1168°C. These peaks, not correlated to weight loss, are probably due to melting reactions. To better understand the thermal behavior of the galvanic sludge, X-ray diffraction was performed on samples previously submitted to different isothermal treatments from 400 to 1200°C with 1h of soaking time. The XRD data show that $CaSO_4$, Na_2SO_4, $CaCrO_4$ (yet at 400°C) and $CaCO_3$ are the main constituents below 800°C. The elementary analyses (Table 4) performed on the dried and thermally treated samples shows an absence of carbon for thermal treatment higher than 800°C, confirming the $CaCO_3$ decomposition. The presence of sulfur until 1200°C confirms the presence of $CaSO_4$ and Na_2SO_4 as crystalline phases. Leaching tests on non treated and thermally treated galvanic sludge were carried out by extracting at a liquid to soil ratio of 10:1 in aqueous and in sulphuric acid media for 24h. The concentrations of total Cr and Cr^{6+} in the leachates were determined using ICP and UV-Vis spectrophotometer (with absorbance recorded at 540nm) respectively. The results indicate a high mobility of Cr (95.8%) in H_2SO_4 medium and the presence of Cr^{6+} for specimens thermal treated from 400 to 1200°C confirming the XRD results.

Table 4. Elementary analysis of galvanic sludge dried at 105°C and thermally treated at different temperatures

Temperature (°C)	C %	S %
105	1.76	10.99
700	0.64	9.59
800	0.00	9.50
1000	0.00	6.53
1200	0.00	6.39

6. Inertisation of Galvanic Sludge: The Vitrification Process

This section reports the results of a study performed for characterizing the properties of glasses obtained by vitrification of chromium(III) galvanic sludges (Cr_2O_3 15.91 wt%, Table 3) added to a glass cullet, as deeply explain elsewhere [11]. In particular, the aim of this section is to evaluate environmental risk assessment of these coloured glass. Batch compositions were prepared by mixing this sludge with glass containers cullet; the container composition is rich in SiO_2 (69.89 wt %), Na_2O (12.32 wt %) and CaO (11.03 wt %), while the sludge have a high amount of CaO and Cr_2O_3 (Table 3). Galvanic sludge was added as additive to glass cullet in the percentages of 0.5, 1.5, 3, 6 wt% corresponding to Cr_2O_3 content of 0.12 (V1 glass), 0.25 (V2 glass), 0.5 (V3 glass) and 0.96 (V4 glass) wt% (Table 5). The melting step was carried out at 1450 °C in an electrical furnace for 2 h followed by quenching in water or on graphite mould (Figure 4).

The comparison between chemical composition from theoretical calculations and the ICP chemical analysis shows both a very low crucible refractory contamination (low increase in the Al_2O_3 amount [21]) and no significant variation of the Cr_2O_3 percentage confirming an effective immobilization of this oxide within the glassy network (Figure 5).

Table 5. Chemical composition (wt%) from theoretical calculations and from ICP chemical analysis of V1 (Cr_2O_3 0.12 wt%), V2(Cr_2O_3 0.25 wt%), V3(Cr_2O_3 0.5 wt%), and V4 (Cr_2O_3 0.96 wt%) glasses (± 0.01)

Oxide	V1		V2		V3		V4	
	Theor. (wt%)	Exp. ICP (wt%)	Theor. (wt%)	Exp. ICP (wt%)	Theor. (wt%)	Exp. ICP (wt%)	Theor. (wt%)	Exp. ICP (wt%)
SiO_2	69.57	68.00	69.21	68.50	68.52	68.00	67.18	67.15
Al_2O_3	2.64	6.20	2.62	7.10	2.60	7.80	2.55	8.80
CaO	11.28	10.57	11.56	11.50	12.10	10.74	13.50	11.42
MgO	2.24	1.92	2.24	2.22	2.23	2.20	2.20	2.11
Na_2O	12.27	11.61	12.24	12.30	12.20	10.15	12.10	10.01
K_2O	1.48	1.51	1.47	1.42	1.46	1.36	1.45	1.39
Fe_2O_3	0.12	0.19	0.12	0.13	0.13	0.90	0.15	0.15
BaO	0.13	0.12	0.13	0.11	0.13	0.10	0.13	0.11
ZrO	0.13	0.13	0.14	0.13	0.14	0.01	0.13	0.02
Cr_2O_3	0.12	0.11	0.25	0.22	0.49	0.45	0.96	0.93

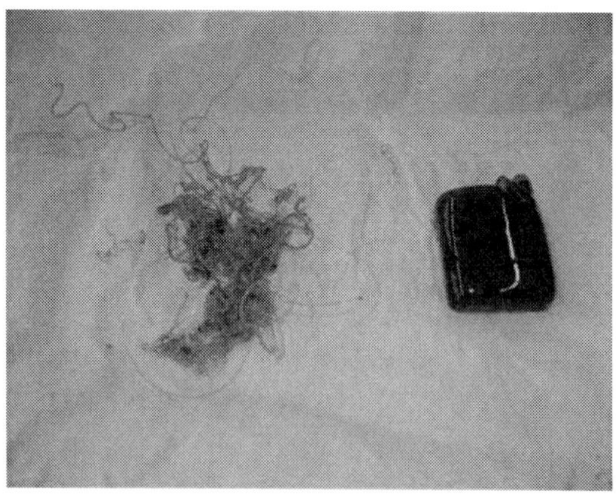

Figure 4. Glass sample obtained by quenching a) in water and b) on graphite mould (V4).

To assess the chemical characteristics of the glasses, durability tests were performed on representative specimens (Table 6) following ISO/R 719 for durability in water, ISO/695 for durability in alkali and DIN 12116 for durability in acid.

Figure 5. Particular of refractory corrosion due to the gas (clorine/and or sulfates) solid reaction.

The water chemical durability of the Cr obtained glass remains constant with respect to container glass up to 3% in waste showing a medium resistance (class 3). Moreover a decrease of pH and conductivity is observed with the increasing waste percentage with respect to glass cullet, corresponding to a lower amount of ions in the eluates, in particular for high concentration of Cr_2O_3, due to the network former role of Cr^{6+}. The examined glasses are only slightly attacked in basic (class 2) and acid medium as confirmed by the low weight loss per surface unit (class 3) (Table 7).

Table 6. Durability results of the prepared glasses

Samples Cr_2O_3 content (wt%)	Class in H_2O	pH*	Conductivity* [µs/cm]	Class in alkali	Class in acid
0	3	9.3	110	2	3
0.12	3	8.7			
0.25	3	7.9		2	3
0.5	3	7.6	93.60	2	3
0.96	3	7.4	84.8	2	3

*in aqueous environment

Finally, release procedure following the compliance leaching test, UNI 10802 and the EPA 1312, were performed on Cr-based glasses for evaluating the leaching of heavy metals. The UNI 10802 (Table 7 reports as example the leaching values obtained for V4 glass) and the EPA tests conducted on chromium glasses indicated that all the glasses were inert and the heavy metals were immobilized; in particular for all the glasses the release of chromium was lower than 0.05 ppm, according to Italian regulation DM 186/06 All. 3A, confirming the capacity of the glassy matrix to fix this type of waste.

Table 7. Metal releases in the leaching solution UNI10802, pH and conductivity

Concentration (ppm)	Sample V4
Cr	0,04
Cd	<0.01
Ni	<0.01
Cu	0.02
Pb	0.02
Al	0.05
As	0.003
Co	< 0.003
Fe	0.015
Mn	0.003
Zn	0.04
Na	1.56
K	0.514
Si	<0.005
Ca	<0.01
Mg	0.021
Ba	0.005
Conductivity [µS/cm]	**37.6**
pH	8.9

7. Inertisation of Galvanic Sludge: Encapsulation in Industrial Bricks

Two industrial bricks-making formulations A and B were used as inertization matrices, and their chemical compositions are reported in Table 8.

Table 8. Chemical composition of brick-making formulations determined by ICP

Composition	Oxide content (wt%)										
	SiO_2	Al_2O_3	CaO	MgO	Na_2O	K_2O	Fe_2O_3	TiO_2	NiO	MnO	LOI
A	53.04	12.94	11.73	3.46	1.29	2.35	6.29	0.76	/	0.15	7.99
B	56.9	19.2	/	2.08	2.11	4.1	8.8	0.81	0.5	/	4.8

The bricks are obtained by pressing in moulds at suitable plasticity the raw materials mixture. In particular, chromium waste were added to the base brick formulations in the following percentages of 0.5 and 2.5 wt% respectively. Two different thermal cycles were employed: a slow firing (950°C for 24 hours in a tunnel kiln) for the compositions based on the batch termed A, and a fast thermal treatment (1150°C for 2 hours) for the bodies prepared taking the batch termed B as base formulation (Figure 6).

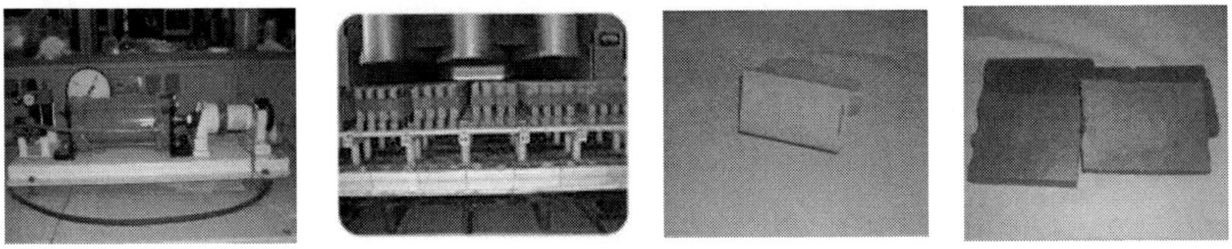

Figure 6. Scheme for the production of building materials containing galvanic waste.

Table 9 shows that the addition of chromium sludge into the bricks obtained by slow thermal treatment (A base composition) does not modify the technological properties up to a great extent. A slight increase of water absorption has been observed although it could be considered into the industrial tolerances. The galvanic waste introduced into the ceramic bodies modify the surface aspects of the bricks. It is thought that the increase of sulphate amount could lead to higher salt migration towards the surface (efflorescence).

XRD patterns of fired samples were analysed to investigate the presence of new phases. For the bricks obtained by the slow thermal cycle treatment the results show the occurrence of Cr(VI) containing phases, e.g tarapacaite (K_2CrO_4), in all the compositions and the increase of these phases with increasing galvanic sludge content in the bricks. Figure 7 reports the XRD pattern of the bricks obtained following the slow thermal treatment, using a composition as based matrix with an addition of 2.5% wt of galvanic sludge. The presence of Cr(VI) was confirmed by the leaching test in aqueous media carried out following DIN38414-S4 standard.

Table 9. Physical characteristics of base and sludge-doped formulations fired in slow thermal cycle

Added waste (wt%)	Shrinkage (%)	Absorption (instantaneous)	Absorption (after 24 h)	Flexural strength (MPa)	$CaCO_3$ (wt%)	Efflorescence
0	5.23	10.50	15.48	11.74	19.3	Slight
0.5	4.9	11.05	15.89	11.91	18.6	Strong
2.5	5.32	11.23	17.07	11.37	19.3	Very strong

The values obtained ranging from 400 to 700 mg/kg of Cr (VI) are completely out of the legal tolerance. This high level of Cr(VI) precludes the commercial use of the bricks obtained by the slow thermal cycle treatment.

Table 10. Physical characteristics of base and sludge-doped formulations fired in fast thermal cycle

Added waste (wt%)	Shrinkage (%)	Absorption (instantaneous)	Absorption (after 24 h)	Flexural strength (MPa)	$CaCO_3$ (wt%)	Efflorescence
0	3.4	2.01	2.8	/	0.8	Absent
0.5	3.8	1.92	2.99	/	0.12	Absent
2.5	3.08	3.33	5.21	/	0.3	Absent

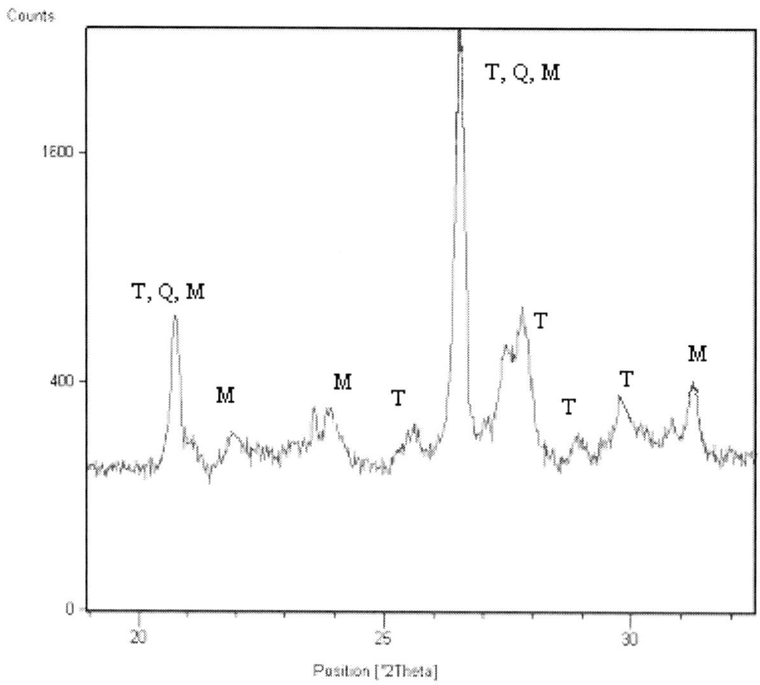

Figure 7. XRD spectrum of a fired ceramic brick of A composition with addition of 2.5% wt of galvanic sludge, after calcination for 24h at 950°C (Tarapacaite (T): K_2CrO_4, Microcline (M): $KAlSi_3O_8$, Quarz (Q): SiO_2).

Table 10 also shows that the addition of chromium sludges into the bricks obtained by fast thermal treatment (B base composition) does not modify the technological properties to a great extent. Looking at the main technological properties of the bricks, it is evident that the changes are within the industrial tolerances. Comparing with the slow thermal treatment, the fast one does not produce efflorescences which could be considered an important technological improvement. The experimental data indicates a substantial decrease in the mobility of chromium ions introduced into the mixture, compared with the mobility of these ions in slime.

Furthermore, Figure 8 shows the lack of Cr(VI) containing phases. This fact has been confirmed from the low ppm values of Cr(VI) obtained from the leaching test (<0.02 mg/Kg) and these values are in agreement with the legal limits.

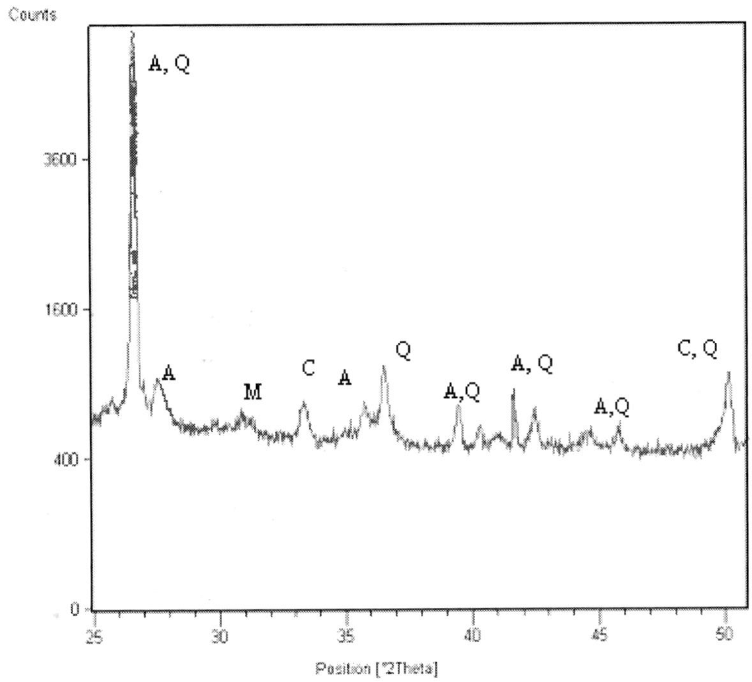

Figure 8. XRD spectrum of a fired ceramic brick manufactured from batch B with 2.5%wt of galvanic sludge addition after calcinations for 2h at 1150°C (Eskolaite (C): Cr_2O_3, Anorthite (A): $CaAl_2Si_2O_8$, Microcline (M): $KAlSi_3O_8$, Quarz (Q): SiO_2).

The chemical composition of the ceramic matrices also plays an important role in the inertization of galvanic sludge. The higher is the CaO content in the ceramic matrix, lower is the possibility of obtaining a successful immobilization. High levels of CaO or $CaCO_3$ could promote chromium oxidation and the formation of $CaO·CrO_3$ increases the hazardous character of the final products [22]. The fact that the bricks obtained from the batch based on the formulation B are free of Cr(VI) containing phases could be due to the above mentioned reason (low CaO content and to the use of a fast firing cycle at higher temperature). Furthermore, the firing temperature and the extension of the thermal treatment seem to have a determinant role in the immobilization degree of chromium in bricks. From the two thermal treatment experimented it appears evident that in the slow treatment addition of chromium mixture up to 2.5 wt% does not vary significantly the technological properties of the bricks. However, Cr leaching exists even at low percentage of $Cr(OH)_3$, in particular Cr(III) is partially oxidized into hazardous Cr(VI). Hence the stabilization of $Cr(OH)_3$ in building ceramic bricks following the slow process is not safe for the environment. On the other hand, the faster thermal treatment at higher temperature seems to assure metal immobilization, since it induces reactions between sludge and clay particles leading to formation of more insoluble trivalent chromium phases. This result could be considered a promising way to immobilize this galvanic waste. Hence, this fast firing technique presents interesting potential for wastes encapsulation purposes, since an effective barrier is created that minimizes the contact between hazardous substances and the outside. Since the encapsulation in the ceramic matrix identified with B and the fast firing technique seem to give encouraging results both considering technological and environmental aspects, optimization of each processing step should be considered.

8. Valorisation of Galvanic Sludge: Synthesis of Chromium Containing Pigments

The screening results of the scientific activity conducted on the possibility to inertise and valorise chromium (III) containing galvanic sludge as a chromium precursor for ceramic pigments such as $CaCr_{0.04}Sn_{0.97}SiO_5$ and $Ca_3Cr_2(SiO_4)_3$ are reported in this section. The sludge valorisation can be obtained by achievement of thermal and chemical stable crystal structures, which are able to colour ceramic material. Ceramic pigments were synthesized by solid state reactions and the colour developed in a suitable ceramic glaze was investigated in comparison with the colour developed by the commercial pigments prepared from pure Cr_2O_3. The synthesis of pink $CaCr_{0.04}Sn_{0.97}SiO_5$ structure was performed using the three different galvanic sludges with chemical composition reported in Table 3, while the green $Ca_3Cr_2(SiO_4)_3$ pigment was synthesized using only the industrial chromium sludge. Both of the pigments were also synthesized starting from an industrial Cr_2O_3 (Cr oxide content ≥98 wt%) as chromium precursors. The samples obtained with pure chromium oxide were used as a reference in order to establish a comparison of the color strength with a traditional pigment composition. A ceramic glaze, having a typical composition for high temperature (Table 11), was used for pigment application. The pigments, added to the composition at 4 wt%, were wet mixed with the ceramic frit and the obtained slips were deposited on ceramic tiles obtaining a uniform glaze layer that was fired at 1140°C for 35 min following an industrial cycle.

Table 11. Chemical composition of industrial frit used

Oxide	Frit (wt%)	Oxide	Frit (wt%)
SiO_2	61.95	TiO_2	0.46
Al_2O_3	11.80	NiO	/
CaO	12.40	CuO	/
MgO	10.40	MnO	/
Na_2O	0.93	CdO	/
K_2O	1.03	CoO	/
Cr_2O_3	/	SnO_2	0.7
Fe_2O_3	0.30	I.L	/

Figure 9 reported the XRD patterns of the obtained pink powders. The mineralogical comparison indicates no variation of the main crystalline phases when galvanic powders (for simplicity only the XRD pattern of industrial galvanic sludge is reported) or pure Cr_2O_3 are employed. In fact, Figure 10 shows the presence of malayaite $CaSnSiO_5$ (ICDD 25-0176), wollastonite $CaSiO_3$ (ICDD 27-0088) and non reacted cassiterite SnO_2 (ICDD 27-0088) as main crystalline phases in both pigments.

From the comparative analysis of the XRD patterns, it can be concluded that from the mineralogical point of view the sludge used in the place of the oxide of chrome does not involve any meaningful changes in the reactivity of the system. After calcination the formed phases in the pigments from the three different sludges are entirely similar to that formed using the pure Cr_2O_3. Only in the pigment obtained from sludge by NaOH the intensity of the peaks suggests that the coloring phase is not the principal phase: the not reacted cassiterite, in fact, has the more intense peaks. Moreover, in all the XRD patterns there is no presence of chromium containing phases and hence it can be hypothesized that chromium forms a solid solution in $CaSnSiO_5$ structure.

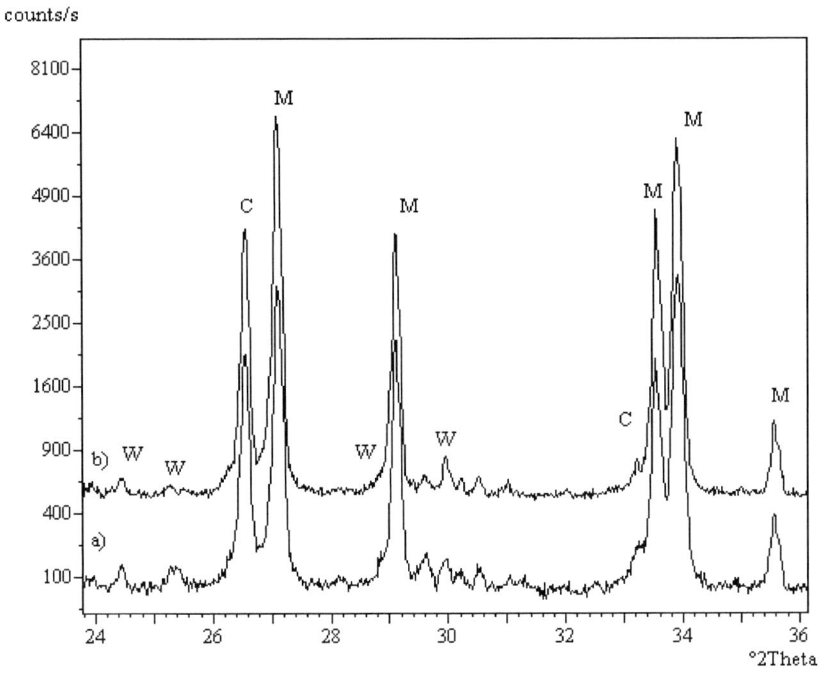

Figure 9. XRD patterns of the pink pigments obtained starting from both industrial a) galvanic waste and b) pure Cr_2O_3. (Cassiterite (C): SnO_2, Malayaite (M): $CaSnSiO_5$, Wollastonite (W): $CaSiO_3$).

The XRD patterns of the green pigments put in evidence slight differences between the powder obtained starting from the chromium sludge (Figure 10a) and that obtained starting from pure chromium oxide (Figure 10b). In fact, beside uvarovite ($Ca_3Cr_2(SiO_4)_3$, ICDD 01-087-1007), which is the main colored structure in pattern a), a spinel phase $(Mg,Fe)Cr_2O_4$ (ICDD 01-071-1256) not present in the pattern of powder obtained starting from pure chromium oxide, was also identified.

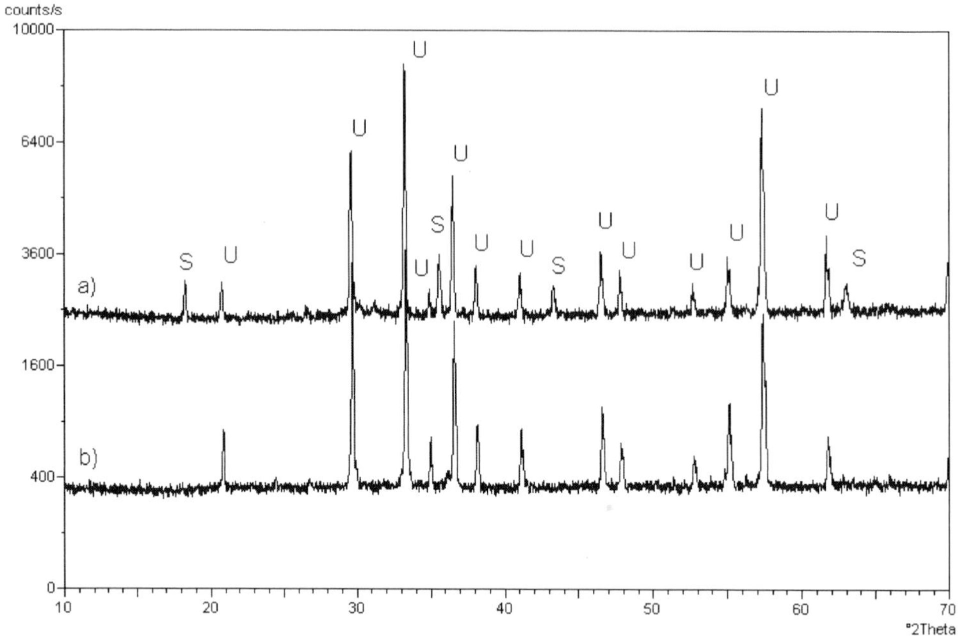

Figure 10. XRD patterns of the green pigments obtained starting from both industrial a) galvanic waste and b) pure Cr_2O_3. (Uvarovite (U): $Ca_3Cr_2(SiO_4)_3$, Spinel phase (S): $(Mg,Fe)Cr_2O_4$).

5 μm.

(A) **(B)**

Figure 11 (Continued)

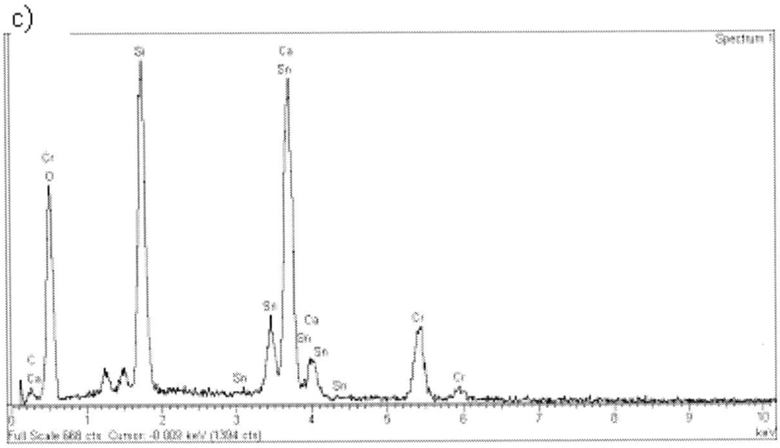

(C)

Figure 11. SEM images of the pink pigments obtained starting from both a) pure Cr_2O_3 and b) galvanic waste and c) EDS spectra of the area labelled A.

SEM observations, reported in Figure 11 a and b, indicate that the microstructure of the pink pigment powders does not vary using either Cr_2O_3 or galvanic waste as chromium precursors. In fact, independently of the chromium precursor used, the images show the formation of Ca-Sn-Si rich crystals, as evidenced by EDS analysis (Figure 11c) with an homogenous and narrow grain size distribution centred at around 10μm. In particular, Figure 11 c also shows the presence of needle-shape crystals of wollastonite, which is in accordance with XRD patterns.

(A) **(B)**

Figure 12. SEM images of the green pigments obtained starting from both a) pure Cr_2O_3 and b) galvanic waste.

Figure 12 shows the microstructures of the green pigments. The pigment obtained starting from pure chromium oxide shows a homogeneous microstructure constituted by particles with a narrow grain size distribution centred at around 10μm. The pigment obtained by sludge, instead, presents a wide grain size distribution (Figure 12b) with hard agglomerates.

The successful formation of the coloring phases was also confirmed by colorimetric analysis. Color measurements were performed on both pigments and glazes by UV–Vis spectroscopy (Perkin Elmer®, model Lambda 19™) using the CIElab method in order to obtain L^*, a^* and b^* values [23]. The method defines a color through three parameters, L^*, a^*, and b^*, measuring brightness, red/green and yellow/blue color intensities, respectively [24]. The method allows, moreover, to define a color difference as ΔE*, based on the relationship:

$$\Delta E^* = \left[(\Delta L^*)^2 + (\Delta a^*)^2 + (\Delta b^*)^2\right]^{\frac{1}{2}}$$

where ΔL*, Δa* and Δb* measure the differences in luminosity and in chromaticity between two colors. For the synthesized pigment, the a^* parameter, in relation to the scale extending from green ($-a^*$) to red ($+a^*$), is undoubtedly the most significant parameter for defining the powders color. Table 12 showed that the pink powders have a high $+a^*$ value independently to the chromium oxide precursor used. The other color parameters, L^* and b^*, are also quite similar. The pigment obtained with NaOH sludge, even if shows the higher a* value, develops instead a very clear (high value of L*) red / pink coloration because of the high amount of non reacted cassiterite in the pigment. The green powders have $-a^*$ values very different with the pigments obtained starting from pure eskolaite that developed a more intense hue, as expected by the XRD results.

Table 12. CIELab values of the obtained pigments

PIGMENT	L^*	a^*	b^*
Pink from industrial sludge	49.67	11.69	2.52
Pink from NaOH sludge	56.04	16.94	8.03
Pink from NH_4OH sludge	51,36	11,42	3,19
Pink from Cr_2O_3	50.07	11.85	1.84
Green from sludge	46.64	-1.40	3.36
Green from Cr_2O_3	43.67	-3.92	5.00

The color developed by these pigments added at 4 wt% in an industrial glaze for high temperature (Figure 13 and 14) is reported in Table 13. Regarding the pink pigments, the table shows that the obtained pigments develop a good red shade in the glaze, irrespective of the raw materials used, indicating the chemical and thermal stability of the synthesized powders. Instead, the powder obtained using NaOH as precipitating agent presents the lower a* values as expected considering the different mineralogical composition with a high amount of non reacted cassiterite.

Concerning the L^*, a^*, b^* parameters of the green glazes demonstrate a significant difference in color. In this case, the glaze hue variation is very high (ΔE* = 7.75) and it can be attributed to the crystalline phases developed as evidenced in the XRD patterns.

Table 13. CIELab values of the obtained glazes

GLAZE	L*	a*	b*
Pink from industrial sludge	48.06	24.34	8.52
Pink from Cr_2O_3	47.73	23.52	8.20
Pink from sludge (NaOH)	64,34	19,59	9,43
Pink from sludge (NH_4OH)	56,00	21,46	9,30
Green from sludge	40.24	-2.85	8.34
Green from Cr_2O_3	35.04	-7.40	11.87

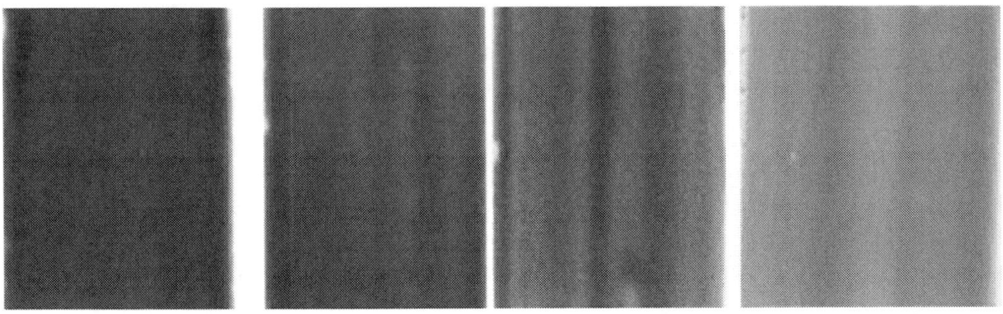

(A) (B) (C) (D)

Figure 13. Pink colored glazed tiles obtained with synthetized pingments: a) industrial sludge, b) pure Cr_2O_3, c) sludge from NH_4OH, d) sludge from NaOH.

(A) (B)

Figure 14. Colored glazed tiles obtained with the synthesized pigments. a) sludge; b) pure Cr_2O_3.

Stabilization efficiency was assessed by water-leaching test (DIN 38414, liquid/solid ratio 10:1, leaching time 24 hrs), TCLP test (EPA method n° 1311 aqueous solution of acetic acid, pH 2.88 ± 0.05, liquid/solid ratio 20:1, leaching time 18 hrs). Moreover leaching experiments with sulphuric acid 1N (liquid/solid ratio 10:1) were carried out at room temperature for 24 hrs in order to evaluate a very strong acid condition. A magnetic stirrer was used to ensure the full suspension of particles and improve liquid-to-solid phase contact. The heavy metals content of the solutions was determined by ICP spectroscopy. All leaching tests performed on powdered glazes indicate levels of toxic elements clearly under the legal safety limits (amounts lower than 0.05 ppm). In summary, the results confirm the possibility to use galvanic sludge as chromium oxide precursor. The characterization carried out corroborates the thermal and chemical stability of the synthesized powders and, especially for the Cr-Sn pink pigment, the powders obtained from industrial galvanic sludge developed an intense color that is very similar to the color developed by the pigments obtained starting from pure Cr_2O_3. The pigment obtained from mud using NaOH to pH 10, instead, developed a color which is not comparable with that industrial. A future experimental work could be directed to optimize, also for this sludge, the conditions of optimal (time, temperature, mineraliser) synthesis of this pigment.

Concerning the green pigments, the L*, a*, b* parameters of the glazes are significantly dissimilar due to the differences emerged by XRD. For this system, thus, further work should be done to optimize the synthesis parameter such as calcination temperature and time. Nevertheless, in the synthesis of these pigments, chromium (III) was stabilized within an inorganic crystalline matrix at high temperatures, avoiding oxidation to toxic chromium (VI).

CONCLUSION

In recent years, because of the global environmental problem, the realization of products containing sludge from industrial production processes can represent an alternative method for waste management with respect to disposal in dump. In particular in this feasibility study, the incorporation of galvanic waste in 3 different matrices was attempted as an effective low expense technique for the fixation of chromium in usable products (glasses, bricks and pigments) and for decreasing the residue volume for further disposal.

Regarding the use of different amount of galvanic sludge containing Cr_2O_3 added to a glass cullet, the results indicate that all the prepared glasses were inert and the chromium was immobilized. Compared to the raw sludge, the sludge-vitrified glasses displayed lower leaching concentrations (Cr amount lower than 0.05 ppm) fulfilling the legal limit.

Furthermore, the incorporation of galvanic waste in clay based fired matrices was attempted as an effective low expense technique for the fixation of chromium in usable products and for decreasing the residue volume for further disposal. The results suggest that the effectiveness of this inertization technique depends on several experimental parameters, such as inertization matrix composition, the physical aspect of the sample (powder and/or pressed pellets), amount of sludge, mixing time, etc. It was found that both the firing temperature and time have a determinant role in the immobilization degree of chromium in bricks. From the two thermal treatments experimented on it appears evident that in the slow treatment Cr^{6+} leaching occurs even at low percentage of $Cr(OH)_3$. Hence the stabilization of $Cr(OH)_3$ in building ceramic bricks following the slow process is not safe for the environment. On the other hand, the faster thermal treatment at higher temperature seems to assure metal immobilization, since it induces reactions between sludge and clay particles leading to the formation of more insoluble trivalent chromium phases. This latter result could be considered a promising way to immobilize this galvanic waste.

Finally, the use galvanic sludge as chromium oxide precursor for the development of Cr-based ceramic pigments was also investigated. The characterization carried out corroborates the thermal and chemical stability of the synthesized structures. In particular, an intense pink color very similar to the color developed by the pigments obtained starting from pure Cr_2O_3 was obtained in the case of Cr-Sn pink pigment.

The formations of Cr- based glasses, bricks and pigments deeply described in this work appear as a promising solution for the valorization and recycling of these residues. This represents an important technological result both from an economical and environmental point of view considering the strict limitations and the expense of the disposal of chromium (III) .

REFERENCES

[1] Mandich, N. V. *Plat Surf Finish* 1998, 85(12), 91-95.

[2] Surviliene, S.; Jasulaitiene, V.; Nivinskiene, O.; Cesuniene, A. *Appl Surf Sci* 2007, 253, 6738-6743.

[3] Surviliene, S.; Bellozor, S.; Kurtinaitiene, M.; Safonov, V. A. *Surf Coat Tech* 2004, 176, 193-201.

[4] Kok, Y. N.; Hovsepian, P. Eh. *Surf Coat Tech* 2006, 201, 3596-3605.

[5] Petukhov, I. V.; Shcherban', M. G.; Kichigin, V. I.; Zaitseva, A. V.; Mikhailov, E. V.; Zavodchikov, S. V. *Prot Metals* 2006, 42(4), 378-388.

[6] Fink, C.G. 1926, U.S. Pat. 1581188.

[7] U.S. Environmental Protection Agency. Toxicological Review of Hexavalent Chromium. National Center for Environmental Assessment, Office of Research and Development, Washington, DC. 1998.

[8] Fontanesi, C.; Giovanardi, R.; Cannio, M.; Soragni, E. *J Appl Electrochem* 2008, 38, 425–436.

[9] Odle, R. R.; Martinez, I.; Deets, L. A. *JOM* 1991, June, 28–31.

[10] Andreola, F.; Barbieri, L.; Bondioli, F.; Cannio, M.; Ferrari, A. M.; Lancellotti, I. *J Haz Mat* 2008, 156(1-3), 466-471.

[11] Andreola, F.; Barbieri, L.; Cannio, M.; Lancellotti, I.; Siligardi, C.; Soragni, E. *WIT Transact Ecology Environment* 2006, 92, 23-30.

[12] Silva, A. C.; Mello-Castanho, S.; Guitian, F.; Montero, I.; Esteban-Cubillo, A.; Sobrados, I.; Sanz, J.; Moya, J. S. *J Am Ceram Soc* 2008, 91(4), 1300-1305.

[13] Andreola, F.; Barbieri, L.; Hreglich, S.; Lancellotti, I.; Morselli, L.; Passarini, F.; Vassura, I. *J Haz Mat* 2008, 153(3), 1270-1274.

[14] Davim, E.; Fernandes, M. H. V.; Monteiro, R. C. C. *Glass Tech-Euro J Glass Sci Tech, Part A* 2007, 48(4), 164-167.

[15] Wang, G.; Wang, S.; Wang, Q. *World J Eng* 2008, 5(1), 71-83.

[16] Leonelli, C.; Veronesi, P.; Boccaccini, D. N.; Rivasi, M. R.; Barbieri, L.; Andreola, F.; Lancellotti, I.; Rabitti, D.; Pellacani, G. C. *J Haz Mat* 2006, 135(1-3), 149-155.

[17] Fernandes, H. R.; Ferreira, J. M. F. *J Euro Ceram Soc* 2007, 27(16), 4657-4663.

[18] Mello-Castanho, S. R. H.; Da Silva, A. C.; Esteban-Cubillo, A.; Pecharroman, C.; Moya, J. S. *Bol Soc Esp Ceram V* 2006, 45(1), 52-57.

[19] Karlovic, E. S.; Dalmacija, B. D.; Tamas, Z. S.; Prica, M. D. J.; Ranogajec, J. G. *J Environ Sci Health A Tox Hazard Subst Environ Eng* 2008, 43(5), 528-537.

[20] Lazau, R. I.; Pacurariu, C.; Becherescu, D.; Ianos, R. *J Euro Ceram Soc* 2006, 27(2-3), 1899-1903.

[21] Boccaccini, D. N.; Cannio, M.; Romagnoli, M.; Veronesi, P.; Leonelli, C.; Boccaccini, A. R. *J Am Ceram Soc*2010, 33(5), 1355-1363.
[22] Lee, Y.; Nassarala, C. L. *Metall Mater Trans* 1998, B29, 405-410.
[23] CIE, Recommendations of Uniform Color Spaces, Color Difference Equations, Phychometrics Color Terms. Supplement no. 2 of CIE Publ. No. 15 (E1-1.31) 1971, Bureau Central de la CIE, Paris 1978.
[24] Hunter, R. S. *J Opt Soc Am* 1958, 48, 985-995.

In: Management of Hazardous Residues Containing Cr(VI) ISBN: 978-1-61470-637-3
Ed: Maria José Balart Murria

Chapter 10

INCORPORATION OF CHROMIUM-TANNED LEATHER RESIDUE IN CIVIL ENGINEERING: CHALLENGES AND APPLICATIONS

Viviane Mattje, M.Sc[1], Karine Krummenauer, M.Sc[2] and Jairo José de Oliveira Andrade[3*]

[1] Civil Engineer

[2] Architect

[3] Associate Professor of Engineering School of Catholic University of Rio Grande do Sul (PUCRS).

Av. Ipiranga, 6681, Prédio 30, Bloco F, Sala 212. Porto Alegre – RS – Brazil. 90619-900

ABSTRACT

It is widely known that nowadays there's an urge to establish sustainable conditions for all kinds of industrial processes. In the leather industry, tanning is the main process used to protect the material from environmental effects such as heat and microbial degradation. In Brazil, 60% of the leather processed comes from an area in the state of Rio Grande do Sul called Vale dos Sinos, that has approximately 90 tanneries and accounts for 30% of the Brazilian leather exportation. Studies show that 30% of the leather processed in tanneries is rejected, especially after the shaving process, in which many chemical products are used. The chromium-tanned leather has been used for many decades in the manufacturing of a large variety of personal use leather products. This process generates liquid and solid waste containing chromium salts that are discharged as effluents into the environment, mainly in water and ground. Some solutions can be used to minimize the amount of chromium residue, as the increase of material fixation due the control of pH, recycling of chrome tanning liquors to reduce the discharge into waste water, pyrolysis of material for the production of activated carbon and use in Portland cement clinker production. In order to contribute with an alternative for this residue disposal, two MSc dissertations were developed under my supervision about the use of

* E-mail: jairo.andrade@pucrs.br

chromium-tanned leather residue in Civil Engineering. The first MSc dissertation evaluated the residue incorporation in mortars. Three mortar mixture proportions were studied, each one with 5 levels of sand replacement by residue. Tests were conducted in order to verify the mortars properties, in fresh and hardened states, and the evaluation of total and chrome VI. The second MSc dissertation was about the use of chromium-tanned leather in asphalt micro-surface layer. A laboratory design for asphalt micro-surface was performed and the best mix proportion was experimentally applied to a 320-meter stretch of Hélio M. Salles Highway in the town of São Sebastião da Grama - in the state of São Paulo, Brazil. In both works the results showed the potentiality of use of chromium-tanned leather residue in Civil Engineering, in which technical, economical and environmental benefits can be obtained.

1. Introduction

The tanning process used by about 90% of the leather processing industries is mineral tanning using chromium salts, generating residues with the presence of the chromium metal which, according to National Brazilian Standard (NBR) 10004 (1987) are classified as Class I Residues - Dangerous, requiring special treatment and disposal (Teixeira et al., 1999).

Among these residues is the chromium-tanned leather residue, a bulky material, impregnated with toxic salts, generated in the shaving operation, which standardizes the leather thickness.

Each tanned leather generates an average of 4 kg of residue, which represented in 2002 about 86.64 thousand tons of chromium-tanned leather residue in Brazil (AICSUL, 2004). The main issues about this situation are the volume occupied by the residue in landfills and its inadequate disposal, contaminating the environment by the slow leaching of trivalent chrome. Being a slowly biodegradable product, it stays for a long time in the environment, which allows its oxidation into hexavalent chromium, a hundred to a thousand times more toxic than its reduced form (Oliveira, 1999).

According to Basegio and Bergmann (1999), there are several studies intended to attempt to substitute the chromium used in tanning, i.e., vegetal tanning. However, until recently there had been no tanning element that was able to rival the physicochemical characteristics conferred to leather by the chromium salts.

Recycling through reuse or recovery of residues or its constituents is one of the most attractive ways to solve the issues of final destination of these materials, both from the business and the environmental protection agencies perspective (ROCCA et al., apud GARLET and GREVEN, 1996).

One of the sectors with greater potential to absorb industrial solid waste is the construction industry. This is basically due to the need of reducing construction costs, the large housing deficit and the low income levels. For these reasons, several materials have been tested in order to optimize the properties and reduce the cost of construction materials. These tests allow the development of composites, generated by the incorporation of residues originated from local industrial processes, which constitute abundant and low cost raw material, that can lead to the possession of new cheap and durable construction resources.

2. Use of Tanning Industry Solid Residues in the Construction Field

Teixeira et al. (1999) and Oliveira (1999) reports that chromium-tanned leather residue was industrially used for the first time in Germany, during World War II. Although processing leather using lactic acid was expensive, the drying process was an important source of food, being cooked in order to produce gelatin.

In Rio Grande do Sul, most of this residue has been accumulating in landfills, called ARIP (acronym for Aterro de Resíduos Industriais Perigosos, Dangerous Industrial Residues Landfill). As for the ARIPs, there are many opinions against their use, due to the space occupied by them and the need for constant monitoring of groundwater levels, besides the impossibility of future use of their physical sites.

Due to its low density, chromium-tanned leather residue (CTL) occupies a large volume, making it very difficult to transport and final disposal, besides damaging the environment when improperly discarded.

Given this reality, in the last few years a growing number of researches about the reuse, processing and disposal of this residue have taken place. The main disposal and/or use alternatives for CTL that have been studied are presented below.

2.1. Aggregated Leather

According to Teixeira et al. (1999), the use of CTL for the manufacturing of aggregated leather does not require previous untanning of the residue and, basically, has a simple obtention process, consisting fundamentally in mixing the previously ground tanned leather residues, scraps and pieces obtained from the footwear and leather goods industries in general with binders such as natural latex or synthetic resins. The product obtained can, in some cases, be used as an insulating material, but more commonly it is destined for the footwear industry and its components, which uses it in the manufacture of inner soles.

2.2. Regenerated Leather

Over twenty years ago there had been researches aiming at the obtention of collagen-artificial leather types that would resemble as closely as possible real leather, reproducing in an artificial way a fibrous texture similar to leather's (Teixeira et al., 1999). Comte (1982) described uses for chromium-tanned leather residues such as synderm regenerated leather, designed in order to obtain a new material, with good physical performance, through the mix of moist leather fibers and its thermal fusion. The material obtained can be used as inner soles, and as leather goods, in which it can be used as reinforcement, lining or separation material.

2.3. Composition For Soles

Serrano et al. (1993) conducted a study of the influence of adding CTL residue as filler in compositions. A traditional polymer of the sole industry was used, the styrene-butadiene rubber (SBR) copolymer. The based composition suffered the addition of chromium-tanned leather residue as filler. The conclusion obtained was that CTL use as filler was possible up until a 25% concentration, without modifying the basic characteristics of the elastomeric composition.

2.4. Manufacture of Ceramic Materials

Basegio and Bergmann (1999) conducted a study on the use of CTL ash to obtain refractory materials. The results showed that it was possible to obtain refractory materials employing CTL ashes up to 30% of the total weight in formulations with magnesium oxide and alumina with qualities compatible with existing products on the market. Chemical analysis of the leachate did not detect any chromium.

Keller et al. (1997) and Oliveira (1999) conducted studies aimed at the immobilization of the chromium contained in CTL in ceramic materials. The organic component is burnt and chromium becomes unavailable within the solid. The mechanical resistance of the materials resulting from the addition of up to 6% of solid waste met the requirements of technical standards for ceramic products. There was no significant change in color and general appearance of the material.

Keller et al. (1997) and Oliveira (1999) cited the following advantages for the incorporation of CTL residue: the fact that the material is burned inside the ceramic, avoiding gas emissions; within the ceramic body is formed a reducing environment that can prevent the oxidation of trivalent chromium into hexavalent; the chromium gets stabilized within the ceramic material; and the residue - chromium-tanned leather residue - is transformed into a product.

Fujikawa (2002) conducted a CTL incorporation study in oven burnt clay bricks. The mix with a 5% CTL addition on the clay mass, which corresponded to the highest amount of CTL added to the brick, showed compressive strength of 3.5 MPa and water absorption of 21.60%. The bricks proved to be viable regarding resistance to compression and water absorption, however, the solubilization and leaching tests proved to be Class II - Non-inert, if compared with the residues.

Similar investigation was conducted by Aguiar et al. (2002), where the authors produced prismatic bricks incorporating 1, 3 and 5 mass-% of leather residues in clay. The bricks were tested regarding their flexure, compression and leaching. The results showed that chromium toxicity disappeared in the bricks, but the mechanical tests showed a decrease in strength in specimens with leather residue.

2.5. Use in Concrete

Fujikawa (2002) also conducted an experiment in which the incorporation of CTL residue in concrete was proved feasible for non-structural concrete, obtaining compression resistance of 5.40 MPa at an age of 28 days, and being totally inert - Class III according to the solubilization and leaching tests.

Oliveira (1999) examined the feasibility of adding CTL to concrete. CTL's characterization revealed a material capable of absorbing large quantities of water. There was a small increase in density of the hardened concrete, as a result of CTL's property of filling the voids. CTL also caused a small increase in the concrete resistance to both compression and surface impact, which was deduced to have occurred because of the decrease of free water. This work did not include residues leaching tests.

Thus, in order to contribute to an appropriate destination for the residue, two alternative employments in Civil Engineering were evaluated: the incorporation in mortar and the use in a microsurface pavement layer.

3. Use of Chromium-Tanned Leather Residue in Mortar

This study, which sought to evaluate the potential of using chromium-tanned leather residue in mortar, was the same as developed by Mattje (2005). In order to achieve this goal, a set of characterization tests of the material was performed.

3.1. Materials

3.1.1. Cement

The cement variety used was pozzolanic Portland, similar to Type IP ASTM C 595 cement. Table 1 shows the physicochemical characteristics of the cement used.

Table 1. Properties of cement used

Parameter	Brazilian Standard	Value		NBR 5736 Specification (1991)
Density	NBR NM 23 (2001)	2.82 g/cm^3		-
Retained # 200	NBR 11579 (1991)	0.15%		$\leq$ 8.0 %
Retained # 325	NBR 12826 (1993)	0.86%		-
Specific Surface	NBR NM 76 (1998)	527 m^2/kg		-
Hot Expandability	NBR 11582 (1991)	0.08 mm		< 5 mm
Initial Setting time	NBR 11581 (1991)	403 min		> 60 min
Loss on ignition	NBR 5743 (1989)	2.27%		< 4.5%
Insoluble residue	NBR 8347 (1991)	39.60%		-
SO_3 content	NBR 5745 (1989)	2.31%		< 4.0%
Compressive strength	NBR 5745(1989)	3 days	19 MPa	> 10 MPa
		7 days	23 MPa	> 20 MPa
		28 days	34 MPa	> 32 MPa

3.1.2. Sand

Natural sand was used. Table 2 presents the results of the material's characteristics.

Table 2. Characterization of sand

	Sieve (mm)	Retained weight (g)			Percentage	
		1st	2nd	Average	Retained	Accumulated
Grain size distribution – NBR NM 248 (2003)	2.4	0	0	0	0	0
	1.2	9	8	8.5	3	3
	0.6	26	25	25.5	8	11
	0.3	138	139	138.5	46	58
	0.15	118	120	119	40	97
	< 0.15	9	8	8.5	3	100
	Fineness modulus = 1.69			Classification – Fine sand		
	Maximum size = 1.2 mm					
Bulk density - NBR 7251 (1982)				$1.48\ g/cm^3$		
Apparent specific gravity - NBR 9776 (1987)				$2.62\ g/cm^3$		

3.1.3. Chromium-Tanned Leather Residue (CTL)

The chromium-tanned leather residue (CTL) employed in this research was collected from a tannery facility and, after gathered, was left exposed to the sun until it became free of any trace of moisture. In order to obtain a homogenous grain size distribution, the CTL used in all the tests was previously screened.

The material used was the one that passed the 2.4 mm sieve but was retained on the 0.075 mm sieve. After the screening, the CTL was stored in an appropriate place.

3.1.3.1. CTL Characterization

To perform the grain size distribution test, two samples of CTL with 50 grams each were collected. Due to the large volume occupied by material, it was not possible to use the minimum amount required by the NBR NM 248 (2003), which corresponds to samples of 500 grams each. The results of grain size distribution of CTL are shown in Table 3.

Table 3. Physical characterization of residue

	Sieve (mm)	Retained weight (g)			Percentage	
		1st	2nd	Average	Retained	Accumulated
Grain size distribution – NBR NM 248 (2003)	2.4	0	0	0	0	0
	1.2	7	8	7.5	15	15
	0.6	16	18	17	34	49
	0.3	16	16	16	32	81
	0.15	9	7	8	16	97
	0.075	2	1	1.5	3	100
	< 0.075	0	0	0	0	100
	Fineness modulus = 1.69			Classification – Average grain size distribution		
	Maximum size = 2.4 mm					

The bulk density of the CTL was equal to 0.13 g/cm^3. This value does not differ much from those mentioned in researches from Krummenauer (2005) and Oliveira (1999), who obtained values between 0.064 and 0.156 g/cm^3 for such material. The apparent specific density was of 1.71 g/cm^3.

3.1.3.2. Chemical Analysis

The values of CTL chemical analysis are presented at Table 4.

Table 4. Chemical properties of CTL used

Parameter	Result
Volatile matter content	62.5%
Sulphated total ash	11%
pH value	3.6
Difference figure	0.8
Chromium oxide content	3.62%

3.1.3.3. Water Absorption

Due to CTL's structure, the material is highly water-absorbing. In order to avoid problems with the mortar workability, the amount of water absorbed by the CTL was determined. Through tests using a hydrostatic scale, it was found that the material showed total water absorption after one minute of immersion, not presenting significant increases in weight after this period. In this study the maximum absorption values reached 550%, mainly due to the presence of protein fibers that have, among other properties, large water absorption coefficients after tanning.

3.2. Method

First of all, the definition of the mix proportions for the study of chromium-tanned leather residue incorporation in mortar took place. The mix proportions, in volume (cement: CTL + sand) adopted in this study were of 1:3, 1:4 and 1:5, with 0%, 3%, 5%, 7%, 15% and 25% of CTL in replacement of sand in volume, as shown in Table 5.

Table 5. Mortar mix proportions adopted

Mix proportion (in volume)	% in volume of CTL use	Mix proportion unit in volume		
		Cement	CTL	Sand
1:3	0	1.00	0.00	3.00
	3	1.00	0.09	2.91

	5	1.00	0.15	2.85
	7	1.00	0.21	2.79
	15	1.00	0.45	2.55
	25	1.00	0.75	2.25
1:4	0	1.00	0.00	4.00
	3	1.00	0.12	3.88
	5	1.00	0.20	3.80
	7	1.00	0.28	3.72
	15	1.00	0.60	3.40
	25	1.00	1.00	3.00
1:5	0	1.00	0.00	5.00
	3	1.00	0.15	4.85
	5	1.00	0.25	4.75
	7	1.00	0.35	4.65
	15	1.00	0.75	4.25
	25	1.00	1.25	3.75

Table 6 presents the tests adopted for the characterization of the fresh mortar, according to the Brazilian standards.

Table 6. Tests for the characterization of fresh mortar

Test	**Test method**
Mortar consistency index	NBR 13276 (1995)
Apparent mass density and air void formation	NBR 13278 (1995)
Water Retention	NBR 13277 (1995)

Table 7. Tests for the characterization of mortar in the hardened state

Test	Age	Test method
Water absorption by immersion, void rate and specific mass	28 days	NBR 9778 (1987)
Capillarity water absorption	7 and 28 days	NBR 9779 (1995)
Apparent mass density	28 days	NBR 13280 (1995)
Compressive strength	3, 7 and 28 days	NBR 13279 (1995)
Diametral compression test	7 and 28 days	NBR 7222 (1994)
Total and hexavalent chromium determination	after 120 days	Specific for each test

The tests conducted to characterize the mortar in the hardened state are presented in Table 7.

3.3. Results and Discussions

3.3.1. Mortar Consistency Index

The standard consistency index determined by NBR 13276 (1995) is 255 ± 10 mm. Figure 1 and Figure 2 show, respectively, the values for consistency index and water percentage added to the mix in relation to the mortar's anhydrous components' (dry material) total mass.

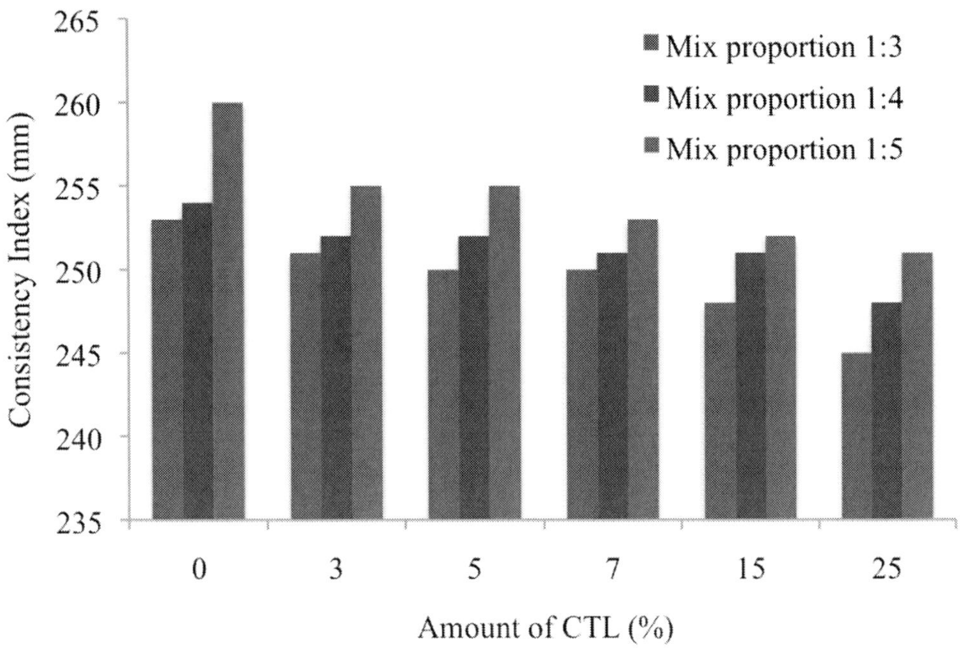

Figure 1. Mortar consistency index.

In Figure 1, the scale corresponding to the index values started at 235 mm, in order to favor the results visualization. It is observed that, as previously established, all mortar consistency index values were within the range allowed by NBR 13276 (1995).

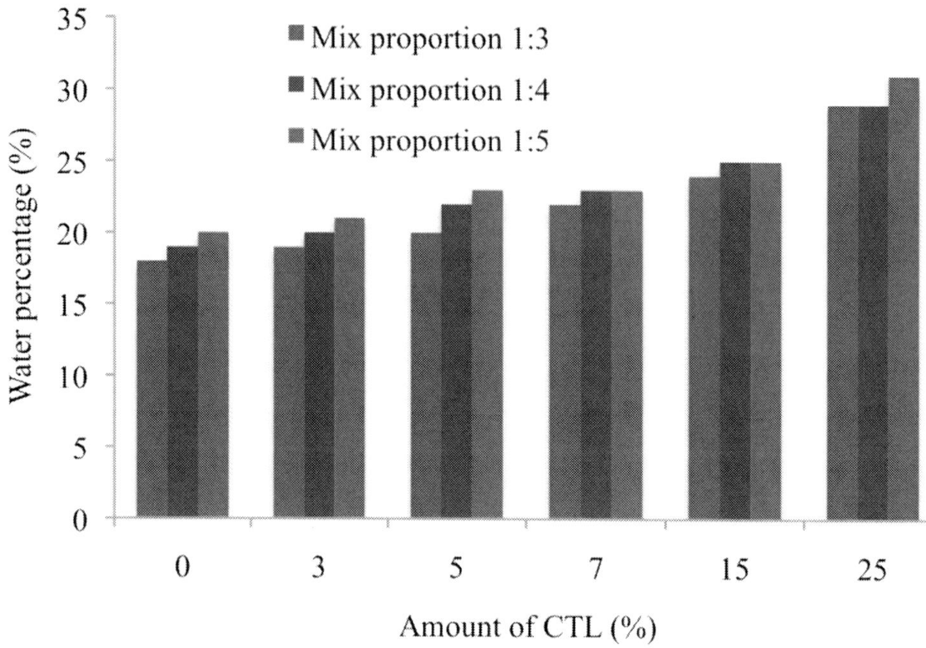

Figure 2. Percentage of water added to the mortar.

It can be observed that as the percentage of chromium-tanned leather residue increases, the amount of water added to the mortar increases as well, as opposed to the consistency index, which decreases. In regular mortars, the increment of water content is expected to lead to an increase of the consistency index, according to NBR 13276 (1995). This behavior was not observed in the mortars studies in this work, due to CTL's water absorption capacity caused by the presence of collagen fibers in its composition. CTL water absorption decreases the amount of free water, reducing the workability.

Figure 3 illustrates the consistency test for the 1:3 mortar trace, by volume, with the use of 25% CTL in place of sand (mix 1:0,75:2,25), which obtained the consistency index of 245 mm (minimum allowed by NBR 13276 (1995)).

Figure 3. Consistency index of the 1:3 mortar mix with 25% of CTL.

Although all mixes with 25% of CTL reached a standard level of consistency, a lot of exudation occurred during the mortars preparing, as shown in Figure 4.

Figure 4. Exudation of the 1:3 mortar mix with 25% of CTL.

Mortar exudation occurs because the excess water cannot be retained within the mix by the other constituent materials (Mattos, 2001). It was expected that with a high CTL level exudation would not occur due to CTL's capacity to absorb water. However, some factors should be taken into consideration:

The water percentage for the 1:3 mortar with 25% of CTL was 29%, with the standard consistency index (SCI) of 245 mm. An amount of water inferior than 29% was not tested, which might obtain the same consistency index value;

Regarding the 1:4 and 1:5 mix proportions with 25% of CTL, 29% and 31% water were used respectively, resulting in consistency rates of 248 mm and 251 mm. In this case, inferior percentages of water that might result in lower consistency indexes (but within SCI range) were not tested as well.

3.3.2. Apparent Mass Density and Void Rate

The results for the apparent mass density and void rate are presented, respectively, in Figure 5 and in Figure 6.

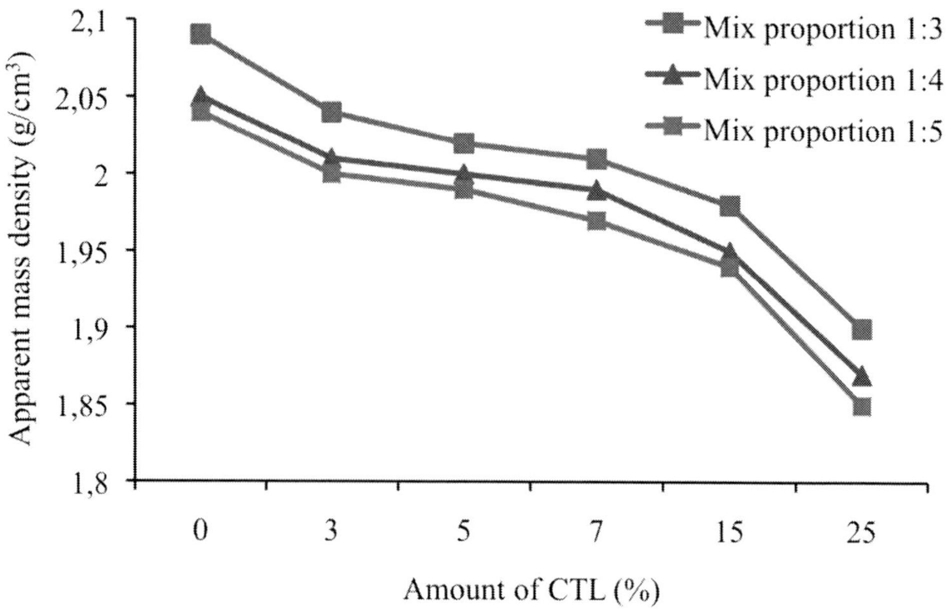

Figure 5. Apparent mass density.

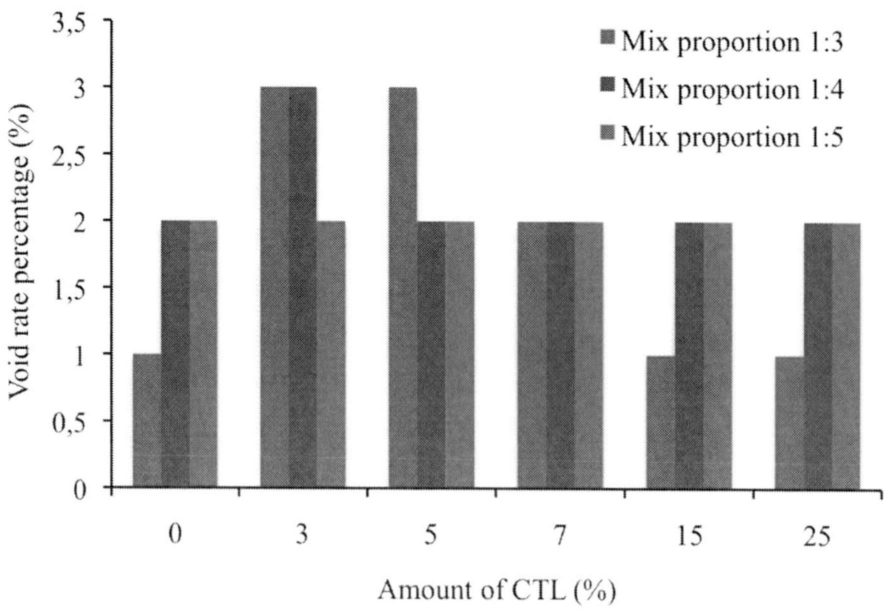

Figure 6. Void rate percentage.

It shows that the higher the percentage of CTL in mortar is, the lower the apparent mass density values. The void rate percentage in all of the mortars did not exceed 3%. According Nakakura (2003), during mixing incorporated air reaches a maximum of 3% to 4% of the total volume, a number that can be increased with the use of air-entraining additives. Thus, it is observed that the apparent density reduction did not occur due to the increase in air content, but due to the low-density characteristic of the CTL.

3.3.3. Water Retention

The results of the water retention determination test of the mortar are presented in Table 8.

Table 8. Water retention

CTL (%)	Water retention (%) NBR 13277 (1995)		
	1:3	1:4	1:5
0	84	84	85
3	83	83	83
5	85	84	83
7	83	83	83
15	83	83	81
25	83	83	83

The water retention in all mix proportions was between 80% and 90%, values considered by NBR 13281 (2001) as normal. As CTL absorbs a great amount of water, it could be expected that with the CTL content increase, water retention would increase as well, which did not occur. On the other hand, according Cincotto et al. (1995), water retention can be augmented because of the increase of the constituent materials specific surface, a characteristic that was not determined for CTL in this work. It is known that CTL presented a greater fineness modulus than sand, which theoretically indicates a lower specific area, which would reduce water retention.

3.3.4. Water Absorption by Immersion, Void Index and Specific Mass

Figure 7 and Figure 8 show respectively the results for water absorption by immersion and void index.

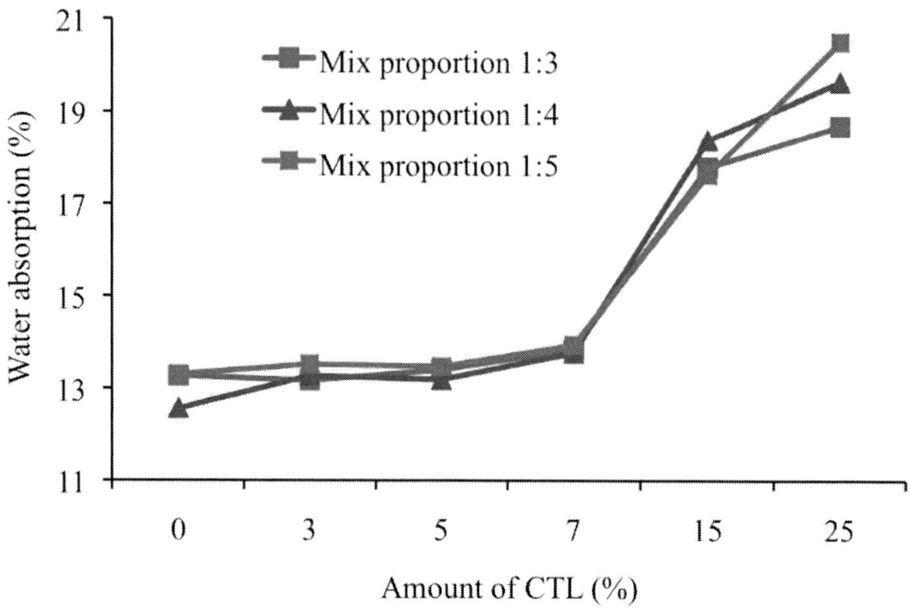

Figure 7. Water absorption by immersion at age 28 days.

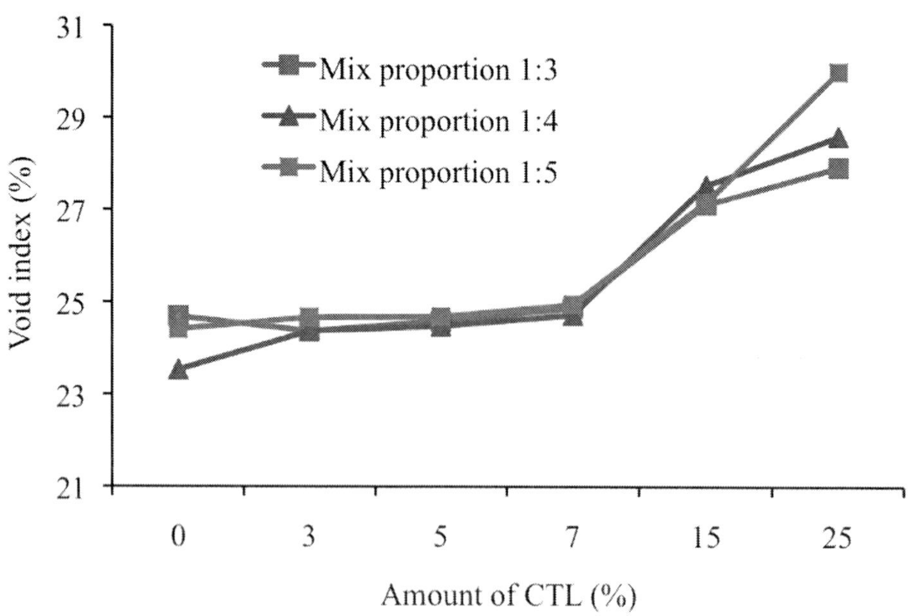

Figure 8. Mortar void index at age 28 days.

It is possible to observe that as far as a 7% CTL use, water absorption by immersion kept virtually constant for all mix proportions, around 13.5%. The same behavior can be seen in the void index results, in which the values up to a 7% percentage remained practically the same, around 24.5%.

It can be observed that at the 15% and 25% percentages, water absorption by immersion increased due to greater presence of CTL in the mortar and also due to a greater number of voids in these mix proportions. In the 15% and 25% percentages the void index also showed an increase, which may be due to the spaces left after the evaporation of excess water in the mortar. In relation to the values found for the 25% percentage in all mixes, it should be noted that mortar exudated during the molding process, which may have contributed to the increase of voids in the hardened mortar.

Figure 9 presents the results for the determination of apparent mass density at 28 days.

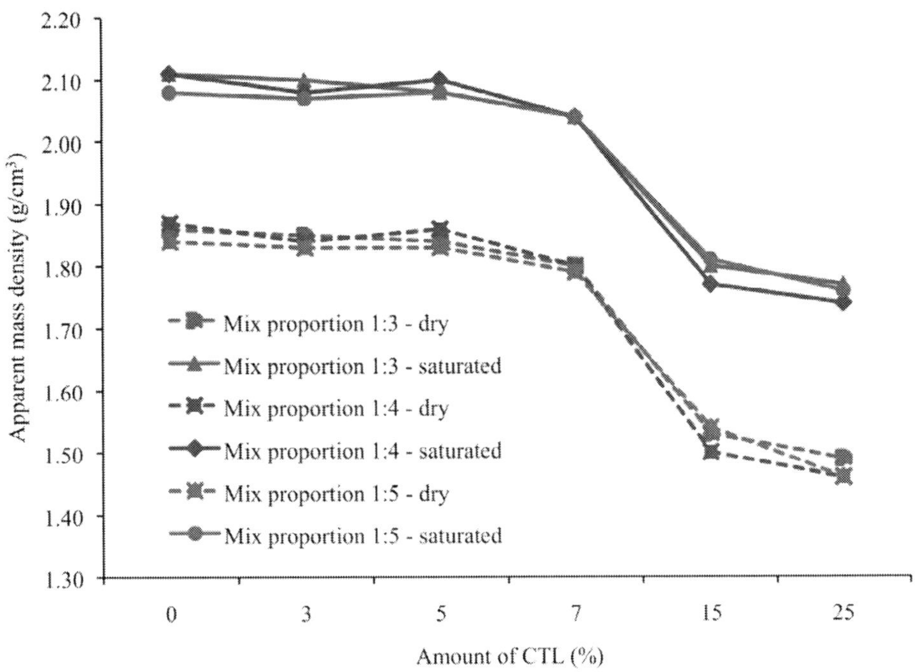

Figure 9. Apparent mass density of mortar at 28 days.

It is evident that the apparent mass density of the samples is lower in the dry condition, increasing when saturated. It can be observed that up to a 7% percentage, the results remained virtually constant. In the 15% and 25% percentages, the apparent mass density values decreased due to the greater CTL presence in the mortar as well as to the greater number of voids.

Comparing the 1:3, 1:4 and 1:5 mix proportions, it is possible to verify that there is virtually no variation in the values found.

3.3.5. Capillarity Water Absorption

It was noticed through visual observation that the maximum capillary rise was always inferior to the test specimen height, being flagrant that the water absorption by capillary action continues to increase during the determinations for at least the 48 hours of testing. It could be expected that with the increase of the CTL amount, the absorption by capillarity would increase as soon as the first few hours, due to CTL's characteristic of rapidly absorbing water. However, capillarity water absorption also depends on other factors, such as the presence and distribution of voids in the mortar, the grain size distribution of all constituent materials and the CTL distribution within the test specimen.

Figure 10 shows the results obtained at the end of the test, after 72 hours, of the mix proportions studied at the age of 28 days.

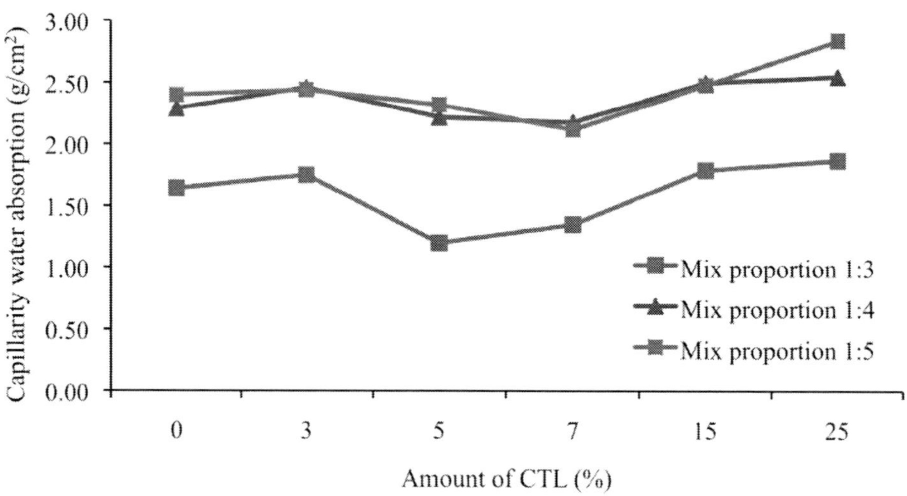

Figure 10. Capillarity water absorption at age 28 days after 72 hours of testing.

It can be observed that capillarity water absorption aged 28 days showed the lowest results in all 1:3 mix proportions compared to the 1:4 and 1:5 mixes. This fact was also reported by Vargas and Comba (1984) and Mattos (2001) who found that the increase in the cement/aggregate ratio reduced the mortar's absorption by capillarity. In addition, there is an increased absorption of water in relation to the reference at age 28 days, in the 3% CTL mixes and a reduction in water absorption, at mortars with 5% and 7% CTL incorporation. In the 15% and 25% percentages, water absorption by capillary action at age 28 days was higher than the absorption in the reference mortar (without CTL).

Another fact observed is that the absorption of water by capillary action of the mix proportions at age 28 days is generally lower than that of mixes at 7 days of age, as the absorption of water by capillary decreases as the hardening of the mortar (Cincotto et al., 1995).

3.3.6. Apparent Mass Density

The results obtained at the apparent mass density determination test in the mortars at age 28 days are presented at Figure 11.

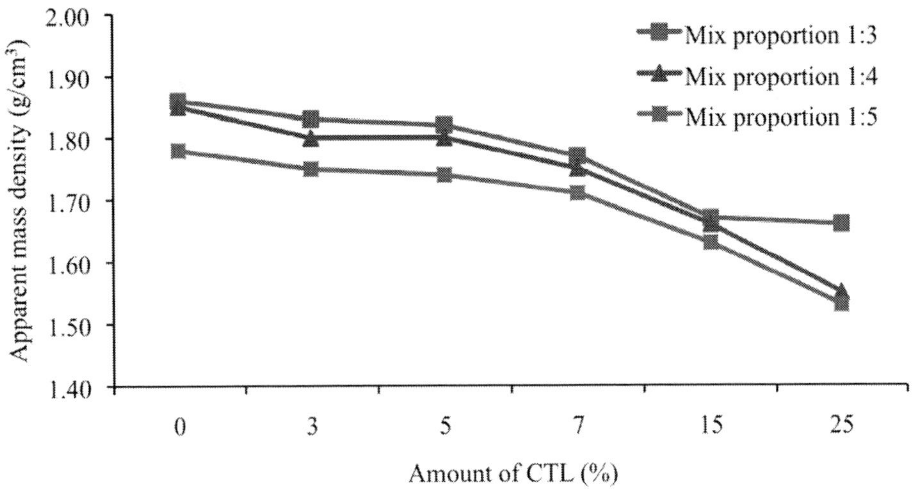

Figure 11. Apparent mass density of mortars at age 28 days.

As previously observed, the higher the use of chromium-tanned leather residue in mortar, the lower are the density values obtained, mainly due to the CTL low specific mass and the increase on the amount of voids.

3.3.7. Compressive Strength

The resistance values obtained for the 1:3, 1:4 and 1:5 mix proportions at ages 3, 7 and 28 days are shown, respectively, at Figure 12, Figure 13 and Figure 14.

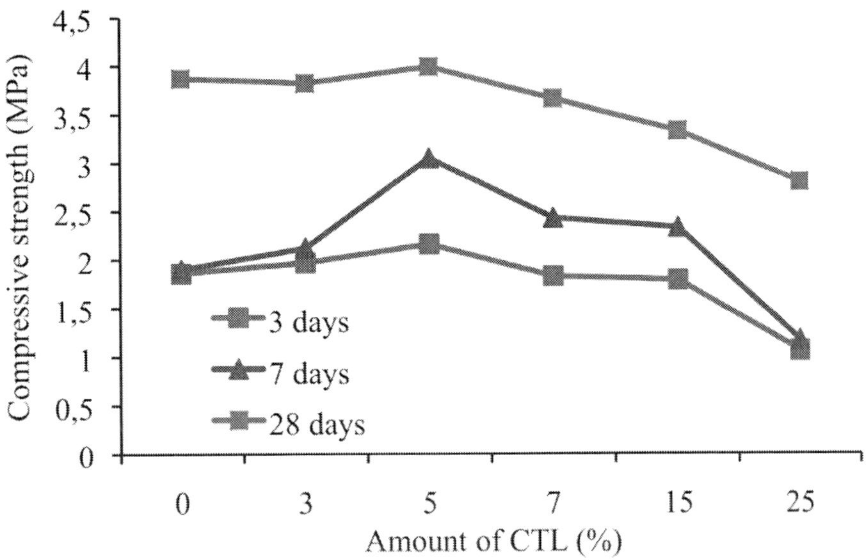

Figure 12. Compressive strength of the 1:3 mix proportion.

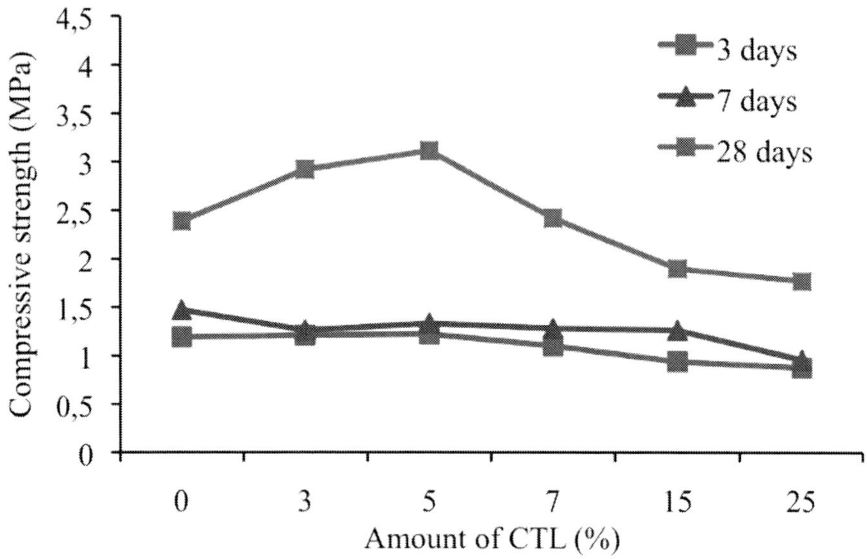

Figure 13. Compressive strength of the 1:4 mix proportion.

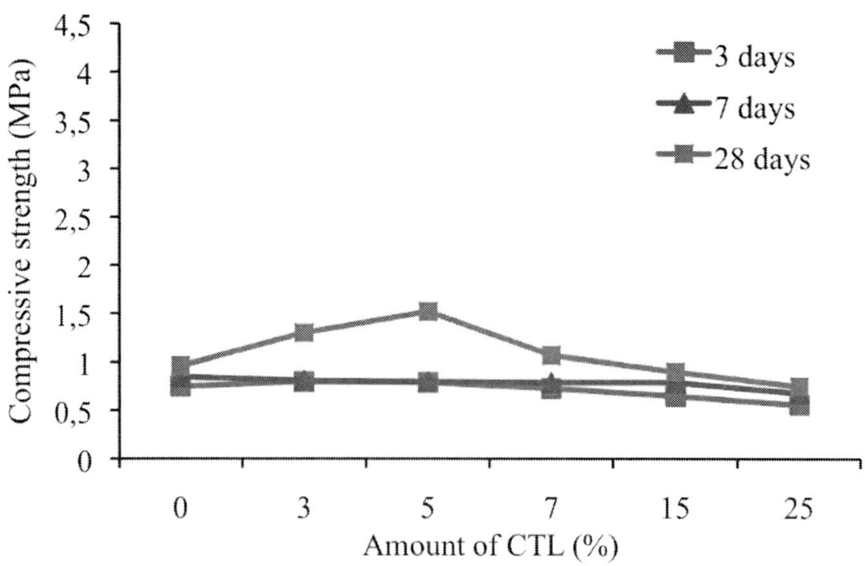

Figure 14. Compressive strength of the 1:5 mix proportion.

It can be observed that the mix proportions with the higher cement/aggregates ratios obtained higher resistance values. As with the 1:5 mix with 25% of CTL, comparing the results obtained at ages 3, 7 and 28 days, it is found that the axial compression resistance did not vary much with increase of the age of the test, which may indicate a delay in hydration, as a result of the mix having a lower cement/aggregate ratio (weak mix proportion) and a high CTL level.

Analyzing the data obtained for the axial compression resistance of mortars at the age of 28 days, it can be verified that CTL levels of 3% and 5% showed an increased resistance in comparison with the reference level (0%), with the exception of the 1:3 mortar mix with 3% of CTL, which showed a small decrease in value compared to the reference. It can also be observed that of all 28 days of age mix proportions, the 5% amount of CTL was characterized as the best performance percentage, resulting in higher resistance values than the reference amount (0%). This behavior may have occurred because of the fiber effect that CTL provided the mortar with, through the intertwining of the constituent material's grains.

The use of 7% of CTL obtained values similar to the ones of the reference mortar, and the 15% and 25% percentages showed an axial compressive resistance decrease when compared to CTL-free mortars. Oliveira (1999) found an axial compression resistance increase in concretes of 28 days of age with the addition of high amounts of CTL, and obtained the value of 20.53 MPa for the reference test specimens and 24.26 MPa for the concretes with 5% CTL. Fujikawa (2002) found a decrease in resistance in concretes at 28 days of age, with the addition of high levels of CTL (10% and 20%), obtaining the values of 20.52 MPa (reference), 5.39 MPa (10% of CTL) and 1.68 MPa (20% of CTL).

3.3.8. Diametral Compression Strength

Figures 15, 16 and 17 show the results of diametral compression for the 1:3, 1:4 and 1:5 mix proportions, respectively.

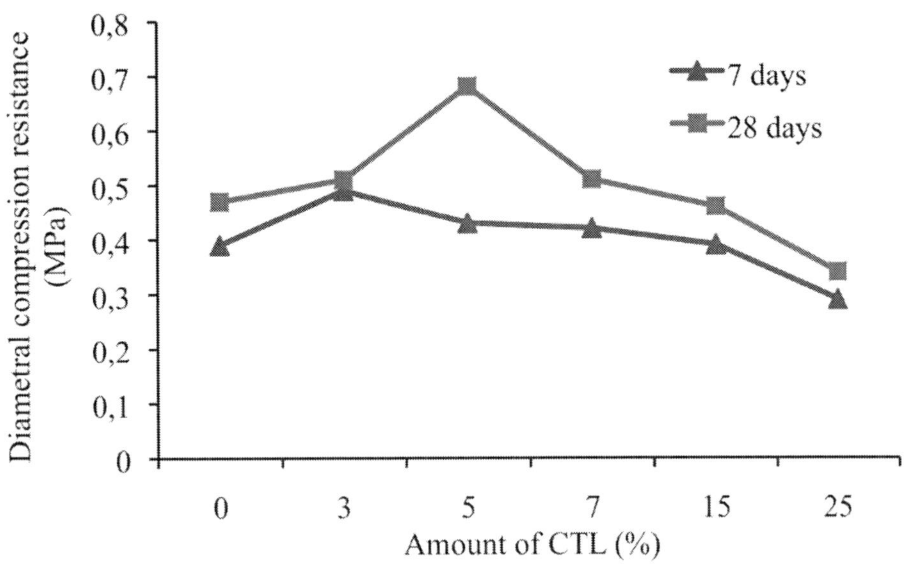

Figure 15. Diametral strength for the 1:3 mix proportion.

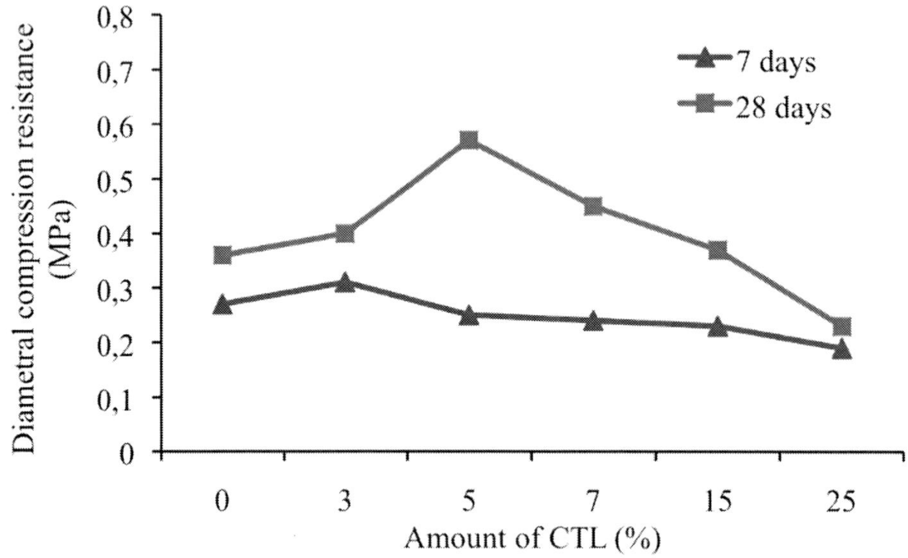

Figure 16. Diametral strength for the 1:4 mix proportion.

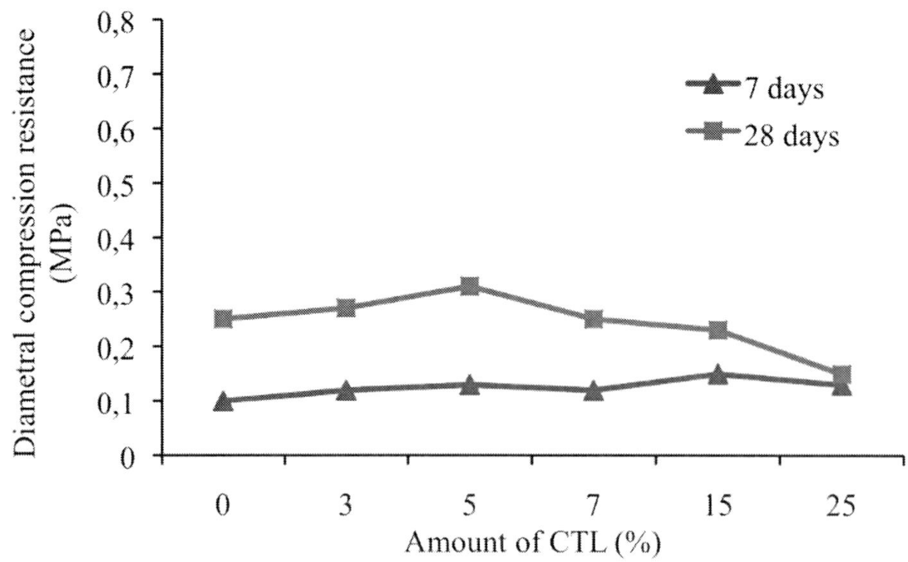

Figure 17. Diametral strength for the 1:5 mix proportion.

It can be observed that the mix proportions with the higher cement/aggregates ratio showed higher results, as happened to the axial compression. Analyzing the data obtained for the diametral strength compression of mortars at 28 days of age, it was shown that the use of 3%, 5% and 7% of CTL evidenced an increase in resistance compared to the reference (0%).

In all mix proportions, aged 28 days, the 5% level of CTL was characterized as the percentage of best performance, with resistance values superior to the mortars with no CTL. The 15% and 25% percentages showed a decrease in resistance when compared to the reference mortar. Such behavior could also be observed in the resistance to axial compression test.

3.3.9. Total and Hexavalent Chromium

These tests were performed on mortars with 5% CTL of each mix proportion, as a result of it having been characterized as the best performance percentage. An analysis was made in order to determine the total and hexavalent chromium in the samples and the leaching. No hexavalent chromium was detected in any sample. As of the total chromium, the highest value found was 56.7 mg/kg, for the 1:3 mix mortar with 5% of CTL. EN 71/3 (1994) specifies the maximum amount allowed for migration of certain elements/heavy metals originated by the migration from toys, being adapted to tinted materials and leather, with representative results. According to EN 71/3 (1994), the maximum chromium migration limit is 60 mg/kg. Thus, it is acknowledged that the results of total chromium determination were satisfactory, with values below 60 mg/kg.

As it was not detected hexavalent chromium in the samples, only the total chromium determination test was performed on the leaching. The higher results obtained for the determination of total chromium in the leachate test were of 0,81 mg/L, for the 1:3 mix mortar with 5% CTL. The maximum concentration of total chromium in leachate according to NBR 10004 (1987) is 5 mg/L. Thus, it is acknowledged that the results obtained at the leachate analysis were also acceptable, being below the maximum amount allowed by the standard.

Based on the results, it appears that the tests performed were satisfactory, since the chromium present in the CTL did not confer toxicity to the samples and leaching of mortars with 5% of CTL. Thus, promising data was obtained from this study. It should be noted, however, the importance of further research addressing the use of these mortars, and evaluating the behavior of chromium in relation to its toxicity and possible oxidation of trivalent chromium into hexavalent chromium.

Based on this study, it was found that the 5% level of CTL was characterized as the percentage of best performance in this work, revealing itself as of promising use in mortars. However, it is necessary to point out the large water absorption capacity of the CTL and the presence of chromium, factors that should be carefully evaluated in mortars in service conditions.

4. Use of Chromium-Tanned Leather Residue in Asphalt Micro-Surface Layer

Another alternative of reusing CTL residue was studied by Krummenauer (2005), who incorporated the material to cold asphalt micro-surface. Due to CTL's characteristics, a significant improvement of the asphalt mechanical properties was observed, as well as a decrease in the cracking of the pavement's superficial layers. A more thorough analysis of the subject is presented in Krummenauer (2005) and in Krummenauer and Andrade (2009).

4.1. Materials

4.1.1. Identification of the Materials Used

Table 9 shows data identifying the type and nature of the constituent materials of asphalt micro-surface layer.

Table 9. Identification of the constituent materials of asphalt micro-surface layer

Material	Type	Nature
Aggregates	Stone dust	Basalt mineral
	Fine aggregate	Basalt mineral
	Coarse aggregate	Basalt mineral
Filler	Calcium hydroxide	Inorganic with a 5% max content of CO_2
Fiber	Residue – Chromium tanned leather sawdust.	Organic material with mineral tanning, with a cationic load
Asphalt emulsion	Asphalt emulsion RL – 1C	Organic, cationic with slow setting, 3% SBS polymer (3% SBS. Polymer)
Additive	Break retardant	Organic material

4.1.2. Aggregates

Table 10 shows the values of aggregate tests. Specifications of aggregates for asphalt micro-surface layer were normalized according to the National Brazilian Standard (NBR) 14948 (ABNT, 2003).

Table 10. Values of the tests performed on the aggregates

Tests	**Stone dust**	**Fine aggregate**	**Coarse aggregate**	**Specifications** NBR 14948 (2003)
Sand equivalent (%)	70	-	-	min. 60%
Los Angeles abrasion test (%)	-	-	17	max.40%
Evaluation of durability using sodium sulfate or magnesium sulfate (%) - Soundness	-	6	8	max. 15%
Characterization of the fine fraction by methylene-blue absorption (mg/g)	4	-	-	max. 10
Bulk density (g/cm³)	1.585	1.410	1.358	-

4.1.3. Asphalt Emulsion

Table 11 shows the values of tests performed on asphalt emulsion and its residue, supplied by a report from the product manufacturer.

Table 11. Values of the asphalt emulsion tests

Tests	Results	Units	Limits NBR 14948 (2003)
Viscosity Saybolt Furol, 50°C	25	Seconds	max. 70
Sieve Test	0.02	%	0.15
Settlement	1.5	%	5
Particle charge	Positive	-	Positive
Residue by distillation	62.3	%	62.0
Elastic recovery in the ductilometer	77.5	%	60
Penetration, 25°C, 100g, 5s	66	mm	min. 45
Ring and ball softening point	56	°C	min. 55

4.2. Method

4.2.1. Determination of the Kind of Residue Incorporated to the Asphalt Micro-Surface Layer

The residue was used in the asphalt micro-surface layer mix composition, through the dry mix process, i.e., as part of the aggregate composition, referring to the process using fibers to increase asphalt micro-surface layer flexibility and resistance.

The job mix of the designs presented in Table 12 is performed based on the grain size test of the mix constituent materials. After this the appropriate percentages of these materials incorporated in the mix are adjusted until they fit the appropriate gradation required by the designer. The International Slurry Seal Association (ISSA) recommends gradation III to restore the paving surface layer.

Table 12. Composition of the asphalt micro-surface layer design job mix

MATERIALS	**DESIGNS**			
	1	**2**	**3**	**4**
	Reference	**0.3% residue**	**0.5% residue**	**0.7% residue**
Stone dust (%)	49.0	48.7	48.5	48.3
Fine aggregate (%)	25.0	25.0	25.0	25.0
Coarse aggregate (%)	25.0	25.0	25.0	25.0
Residue (%)	0.0	0.3	0.5	0.7
Lime (%)	1.0	1.0	1.0	1.0

4.2.2. Experimental Performance Test

4.2.2.1. Determination of Loss by Wet Abrasion – WTAT (NBR 14746, ABNT, 2001)

This is a performance test that determines the asphalt micro-surface layer wear under wet abrasion conditions in which, due to the difference in mass, loss of stone material from the test specimen occurs, where the optimum emulsion content of a given mixture is established, which makes up the asphalt micro-surface layer.

4.2.2.2. Determination of Excess Asphalt and Sand Adhesion(NBR 14841, ABNT 2002)

This test allows evaluating the compaction, the layers deformation of the micro-surface mix and the maximum limit of asphalt mix content, in order to avoid exudations under traffic, simulating the action of traffic in a Loaded Wheel Tester (LWT) equipment.

4.2.2.3. Determination of the Optimum Asphalt Emulsion Content

The optimum asphalt emulsion content to be applied in the asphalt micro-surface layer mix design is the point in which the curves intersect, from the values obtained in the wet abrasion loss tests and excess asphalt and sand adhesion compared with the percentage of emulsion incorporated, drawn on a graph.

4.2.2.4. Determination of the Cohesion and Characteristics of Cure Using the Cohesimeter (NBR 14798, ABNT 2002)

This test determines the resistance of an asphalt micro-surface layer mix to surface torque (cohesion) during the cure process, submitted to the cohesimeter.

4.3. Results and Discussions

4.3.1. Wet Abrasion Test Loss (WTAT)

Table 13 shows the results obtained in the test to determine wet abrasion loss for the four mix designs with four different percentages of asphalt emulsion addition.

Table 13. Values obtained in the wet abrasion loss test in the asphalt micro-surface layer designs

		Emulsion contents			
Design	**Type**	8 %	10%	12%	14%
Design – 1	Reference	1237 g/m^2	539 g/m^2	297 g/m^2	197 g/m^2
Design – 2	0.3% residue	658 g/m^2	384 g/m^2	268 g/m^2	150 g/m^2
Design – 3	0.5% residue	903 g/m^2	663 g/m^2	479 g/m^2	274 g/m^2
Design – 4	0.7% residue	2487 g/m^2	1347 g/m^2	921 g/m^2	561 g/m^2
Max specification g/m^2		500	500	500	500

The results of the wet abrasion test show that in the test variables, the higher the emulsion content added to the mix, the better the resistance of the material to wear. In this case, the mixes with 14% asphalt emulsion presented a better performance. Besides, as the residue addition increased, there was excessive absorption of the emulsion by the residue, leading to less aggregation of the constituent materials, and consequently less resistance to wear.

The addition of residue with a 0.3% content had better results compared to the mix without residue, and there was less loss of stone material in all emulsion contents tested. The wear values in the different mixtures present a similar behavior to the asphalt emulsion contents tested.

4.3.2. Excess Asphalt and Sand Adhesion

Table 14 shows the results of the test to determine excess asphalt and sand adhesion using the LWT machine, with the variation of the asphalt emulsion content incorporated to the 4 mix designs.

Table 14. Values obtained in determining excess asphalt and sand adhesion

Design	**Types**	**Emulsion Contents**			
		10%	12%	14%	16%
Design – 1	Reference	343 g/m^2	448 g/m^2	486 g/m^2	571 g/m^2
Design – 2	0.3% residue	314 g/m^2	448 g/m^2	505 g/m^2	571 g/m^2
Design – 3	0.5% residue	324 g/m^2	495 g/m^2	543 g/m^2	590 g/m^2
Design – 4	0.7% residue	429 g/m^2	467 g/m^2	543 g/m^2	600 g/m^2
Max specification g/m^2		Heavy traffic – 538 g/m^2			
		Light traffic – 646 g/m^2			

With the increased emulsion content, there is an increased mix plasticity, leading to greater deformity and with the different additions of residues the values obtained in the test were satisfactory for the light traffic specification, with values below 646 g/m^2 (limit value), prescribed by the Brazilian standard NBR 14798 (ABNT, 2002). Only mix designs with the addition of 0.3% of residue and no addition of residue fitted the specification for heavy traffic up to a percentage of 14% of asphalt emulsion.

4.3.3. Optimum Asphalt Emulsion Content

Table 15 shows the optimum recommended asphalt emulsion contents for each mix design, based on the results of wet abrasion loss and the determination of excess asphalt and sand adhesion tests by the LWT machine.

Table 15. Asphalt emulsion contents recommended for the corresponding designs

Design	**Type**	**Asphalt emulsion**	
		Optimum content %	Acceptable variation (±0,5%)
Design – 1	Reference	11.20	10.70 to 11.70
Design – 2	0.3% residue	10.40	9.90 to 10.90
Design – 3	0.5% residue	11.90	11.40 to 12.40
Design – 4	0.7% residue	14.30	13.80 to 14.80

Asphalt emulsion is the product that adds greater value to the cost of the asphalt micro-surface layer design, and Table 15 shows that the mix design made by adding 0.3% residue had the lowest cost because of the lower asphalt emulsion content needed in the mix.

4.3.4. Cure Test Using the Cohesimeter

Table 16 shows the results of the test to determine cure using the cohesimeter, for the different mixture designs with the incorporation of asphalt emulsion to its optimum content, with values presented in Table 15.

Table 16. Values obtained in the cure test using the cohesimeter

Design	**Type**	**Optimum emulsion content %**	**Torque (kg.cm)**			
			30 min	60 min	120 min	180 min
Design – **1**	Reference	11.20	10	14	16	21
Design – **2**	0.3% residue	10.40	10	13	16	21
Design – **3**	0.5% residue	11.90	11	12	16	25
Design – **4**	0.7% residue	14.30	10	12	15	16
Specification of NBR 14748 (2002): 30 minutes (for break) minimum 12 kg.cm 60 minutes (for traffic) minimum 20 kg.cm						

Based on the results shown in Table 16, it is possible to say that the increased percentage of residue added to the mix up to 0.5% improves material resistance to breaking, where the mix design with 0.5% of residue presented the best performance in the test. It is noted that none of the mix designs fits the Brazilian specifications concerning the break values in 30 minutes and opening for traffic within 1 hour.

4.3.5. Leaching Test - NBR 10005 (ABNT, 2004)

Total chromium and hexavalent chromium were analyzed in the bath of the wet abrasion loss test corresponding to the test specimen with addition of 0.7% residue and in the leachate of the test specimen with the addition of 0.3% of residue (Table 17).

Table 17. Results of total and hexavalent chromium

Specimen	**Analysis**	
	Cr Total	Cr VI
Wet abrasion loss test bath for the test specimen adding 0.7% residue	0.34 mg l^{-1}	Nd*
Leaching in the test specimen adding 0.3% residue.	0.22 mg l^{-1}	Nd*
Methodology	Atomic absorption	Visible UV (sensitivity 0.05 mg/l)

*Nd – not detected

The results show that in the bath test, it is necessary to study the normal conditions to which the paving is submitted, since the test simulated an asphalt layer submerged in water and undergoing constant abrasive wear. The total chromium values detected in the WTAT test bath were higher than those detected in the leachate of the specimen with an addition of 0.3% residue; even with higher total chromium content, no hexavalent chromium was detected in the specimen. Besides, no hexavalent chromium was detected in the test specimens with additions of 0.3% and 0.7% residue submitted to the test.

4.3.6. Implementation of the Road Stretch

The experimental road stretch with asphalt micro-surface layer was implemented on Hélio M. Salles highway, at km 215, in São Sebastião da Grama – SP, is located on the limits with the county of Divinolândia – MG. The object of this application was to restore the existing pavement by resurfacing it with asphalt micro-surface layer, because there were cracks in the lane (Figure 8). The asphalt micro-surface layer was applied using a truck with a mobile plant (Figure 9). This was done to a total thickness of 13 mm.

Figure 18. Cracking problems on the pavement.

Figure 19. Applying asphalt micro-surface layer on the road stretch.

In order to evaluate a similar condition, half the road width was demarcated (3.5 meters) to apply an asphalt micro-surface layer without residue addition, considering this as the test blank. At the same time, in parallel to the design lane without adding residue, the pavement was resurfaced with the asphalt micro-surface layer mix design to which 0.3% of residue had been added (Figure 10).

Figure 20. Lane on the right resurfaced without residue, and on the left with 0.3% residue.

The experimental stretch was built on January 13, 2004, and the photography and visual evaluation were performed on March 31, 2004, 77 days after implementation. Figure 11 shows how the paving was worn down by the traffic on the stretch done with asphalt micro-surface layer without residue addition. Recently, another evaluation was performed in the stretch, and the same present adequate conditions for traffic in area with residue incorporation.

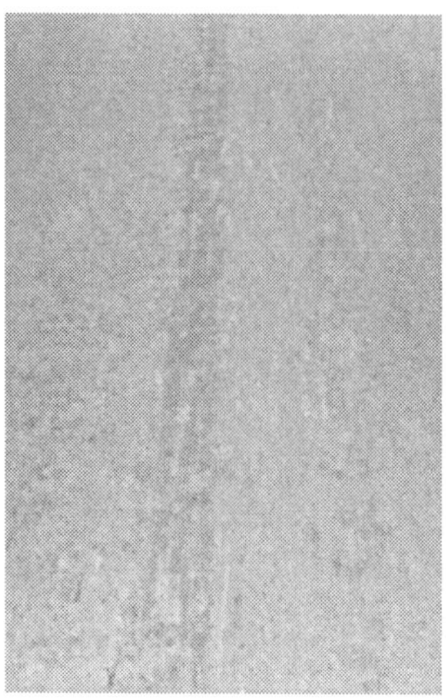

Figure 21. Detail of wear on the coating layer without residue.

According to the calculations based on percentages and respective densities of the constituent asphalt micro-surface layer materials, an application rate of 14.40 kg/m^2 was obtained, and 0.045 kg/m^2, of residue was incorporated to an amount of 162 kg on the experimental stretch (single lane).

According to the results presented up to the present time, it can be concluded that (Krummenauer and Andrade, 2009):

It was technically and economically feasible to incorporate residue to the asphalt micro-surface layer;

Around 0.3% of residue can be appropriately incorporated to the asphalt micro-surface layer design;

No hexavalent chromium was detected in the test specimens used with 0.3% and 0.7% additions of residue;

The asphalt micro-surface layer design with the results in performance tests was design 2, with the addition of 0.3% of residue;

The visual evaluation of the lanes after implementation showed that lanes to which fibers were added had a better visual aspect, since the lane without added fibers showed wear due to traffic.

Final Considerations

This chapter had the intent of showing alternatives for the use of chromium-tanned leather residue in Civil Engineering. The results showed were promising, but further investigation must take place, mainly due to the possibility of environmental issues provoked by chromium leaching. Looking at the existing waste management options for chromium containing wastes, it is interesting that so far no practical solution has been found which is wholly satisfactory. Prevention is possible only to a certain extent, because chrome-free tanning is not feasible for all types of leather (e.g. shoe leather production). Recycling of chromium by Cr-extraction from shavings and trimmings is not done for economical reasons and depends on the market situation. (Lorber 2003).

References

AGUIAR, J. B.; PIRES, M. J.; TAVARES, T. Reutilization of leather residue by incorporation in bricks. *Tile & Brick International*, v. 18, n. 1, 2002. p. 30-32.

ASSOCIAÇÃO BRASILEIRA DE NORMAS TÉCNICAS. NBR 10004: Resíduos sólidos. Rio de Janeiro: 1987. 63p.

ASSOCIAÇÃO BRASILEIRA DE NORMAS TÉCNICAS. NBR 10004: Resíduos sólidos. Rio de Janeiro: 1987. 63p.

ASSOCIAÇÃO BRASILEIRA DE NORMAS TÉCNICAS. NBR 10004: Resíduos sólidos. Rio de Janeiro: 1987. 63p.

ASSOCIAÇÃO BRASILEIRA DE NORMAS TÉCNICAS. NBR 10005: lixiviação de resíduos. Rio de Janeiro, 2004.

ASSOCIAÇÃO BRASILEIRA DE NORMAS TÉCNICAS. NBR 11579: Cimento Portland – Determinação da finura por meio da peneira 75 micrômetros (número 200). Rio de Janeiro: 1991.

ASSOCIAÇÃO BRASILEIRA DE NORMAS TÉCNICAS. NBR 11581: Cimento Portland – Determinação dos tempos de início e fim de pega. Rio de Janeiro: 1991.

ASSOCIAÇÃO BRASILEIRA DE NORMAS TÉCNICAS. NBR 11582: Cimento Portland – Determinação da expansibilidade de Le Chatelier. Rio de Janeiro: 1991.

ASSOCIAÇÃO BRASILEIRA DE NORMAS TÉCNICAS. NBR 12826: Cimento Portland e outros materiais em pó – Determinação do índice de finura por meio do peneirador aerodinâmico. Rio de Janeiro: 1993.

ASSOCIAÇÃO BRASILEIRA DE NORMAS TÉCNICAS. NBR 13276: Argamassa para assentamento de paredes e revestimento de paredes e tetos – Determinação do teor de água para obtenção do índice de consistência padrão. Rio de Janeiro: 1995. 2p.

ASSOCIAÇÃO BRASILEIRA DE NORMAS TÉCNICAS. NBR 13277: Argamassa para assentamento de paredes e revestimento de paredes e tetos – Determinação da retenção de água. Rio de Janeiro: 1995. 2p.

ASSOCIAÇÃO BRASILEIRA DE NORMAS TÉCNICAS. NBR 13278: Argamassa para assentamento de paredes e revestimento de paredes e tetos – Determinação da densidade de massa e do teor de ar incorporado. Rio de Janeiro: 1995. 2p.

ASSOCIAÇÃO BRASILEIRA DE NORMAS TÉCNICAS. NBR 13279: Argamassa para assentamento de paredes e revestimento de paredes e tetos – Determinação da resistência à compressão. Rio de Janeiro: 1995. 2p.

ASSOCIAÇÃO BRASILEIRA DE NORMAS TÉCNICAS. NBR 13280: Argamassa para assentamento de paredes e revestimento de paredes e tetos – Determinação da densidade de massa aparente no estado endurecido. Rio de Janeiro: 1995. 2p.

ASSOCIAÇÃO BRASILEIRA DE NORMAS TÉCNICAS. NBR 13281: Argamassa para assentamento de paredes e revestimento de paredes e tetos – Requisitos. Rio de Janeiro: 2001.

ASSOCIAÇÃO BRASILEIRA DE NORMAS TÉCNICAS. NBR 13525: Requisitos para aceitação de couro quanto à analise química. Rio de Janeiro: 1995. 63 p.

ASSOCIAÇÃO BRASILEIRA DE NORMAS TÉCNICAS. NBR 13528: Revestimento de paredes e tetos de argamassas inorgânicas - Determinação da resistência de aderência à tração. Rio de Janeiro: 1995.

ASSOCIAÇÃO BRASILEIRA DE NORMAS TÉCNICAS. NBR 13529: Revestimento de paredes e tetos de argamassas inorgânicas. Rio de Janeiro: 1995.

ASSOCIAÇÃO BRASILEIRA DE NORMAS TÉCNICAS. NBR 14746: microrrevestimento a frio e lama asfáltica – determinação de perda por abrasão úmida (WTAT). Rio de Janeiro, 2001. 4 p.

ASSOCIAÇÃO BRASILEIRA DE NORMAS TÉCNICAS. NBR 14798: microrrevestimento asfálticos – determinação da coesão e características da cura pelo coesímetro. Rio de Janeiro, 2002. 3 p

ASSOCIAÇÃO BRASILEIRA DE NORMAS TÉCNICAS. NBR 14841: microrrevestimento a frio – determinação de excesso de asfalto e adesão de areia pela máquina LWT. Rio de Janeiro, 2002. 4 p.

ASSOCIAÇÃO BRASILEIRA DE NORMAS TÉCNICAS. NBR 14948: microrrevestimento asfálticos a frio modificados por polímero – materiais, execução e desempenho. Rio de Janeiro, 2003. 8 p.

ASSOCIAÇÃO BRASILEIRA DE NORMAS TÉCNICAS. NBR 5743: Cimento Portland - Determinação de perda ao fogo. Rio de Janeiro: 1989.

ASSOCIAÇÃO BRASILEIRA DE NORMAS TÉCNICAS. NBR 5745: Cimento Portland - Determinação de anidrido sulfúrico. Rio de Janeiro: 1989.

ASSOCIAÇÃO BRASILEIRA DE NORMAS TÉCNICAS. NBR 6508: Grãos de solos que passam na peneira de 4,8 mm – Determinação da massa específica. Rio de Janeiro: 1982. 8p.

ASSOCIAÇÃO BRASILEIRA DE NORMAS TÉCNICAS. NBR 7200: Execução de revestimento de paredes e tetos de argamassas inorgânicas – Procedimento. Rio de Janeiro: 1998.

ASSOCIAÇÃO BRASILEIRA DE NORMAS TÉCNICAS. NBR 7215: Cimento Portland – Determinação da resistência à compressão. Rio de Janeiro: 1996. 7p.

ASSOCIAÇÃO BRASILEIRA DE NORMAS TÉCNICAS. NBR 7222: Argamassa e concreto – Determinação da resistência à tração por compressão diametral de corpos-de-prova cilíndricos. Rio de Janeiro: 1994. 3p.

ASSOCIAÇÃO BRASILEIRA DE NORMAS TÉCNICAS. NBR 7251: Agregado em estado solto – Determinação da massa unitária. Rio de Janeiro: 1982. 3p.

ASSOCIAÇÃO BRASILEIRA DE NORMAS TÉCNICAS. NBR 8347: Cimento Portland - Determinação do dióxido de silício. Rio de Janeiro: 1991.

ASSOCIAÇÃO BRASILEIRA DE NORMAS TÉCNICAS. NBR 9287: Argamassa de assentamento para alvenaria de bloco de concreto - Determinação da retenção de água. Rio de Janeiro: 1986.

ASSOCIAÇÃO BRASILEIRA DE NORMAS TÉCNICAS. NBR 9776: Agregados – Determinação da massa específica de agregados miúdos por meio do frasco Chapman. Rio de Janeiro: 1987. 3p.

ASSOCIAÇÃO BRASILEIRA DE NORMAS TÉCNICAS. NBR 9777: Agregados – Determinação da absorção de água em agregados miúdos. Rio de Janeiro: 1987. 3p.

ASSOCIAÇÃO BRASILEIRA DE NORMAS TÉCNICAS. NBR 9778: Argamassa e concreto endurecidos – Determinação da absorção de água por imersão – Índice de vazios e massa específica. Rio de Janeiro: 1987. 5p.

ASSOCIAÇÃO BRASILEIRA DE NORMAS TÉCNICAS. NBR 9779: Argamassa e concreto endurecidos – Determinação da absorção de água por capilaridade. Rio de Janeiro: 1995. 2p

ASSOCIAÇÃO BRASILEIRA DE NORMAS TÉCNICAS. NBR NM23: Cimento Portland e outros materiais em pó – Determinação de massa específica. Rio de Janeiro: 2001.

ASSOCIAÇÃO BRASILEIRA DE NORMAS TÉCNICAS. NBR NM248: Agregados - Determinação da composição granulométrica. Rio de Janeiro: 2003. 6p.

ASSOCIAÇÃO BRASILEIRA DE NORMAS TÉCNICAS. NBR NM76: Cimento Portland - Determinação da finura pelo método de permeabilidade ao ar (Método de Blaine). Rio de Janeiro: 1998.

ASSOCIAÇÃO DA INDÚSTRIA E CURTUMES DO RIO GRANDE DO SUL. Estimativa de mercado mundial e brasileiro de couros bovinos 1990/2002. Guia Brasileiro do Couro 2004, Estância Velha: ABQTIC, 2004. p. 10.

BASEGIO, T. M.; BERGMANN, C. P. Análise da influência da temperatura em materiais refratários produzidos a partir da cinza da serragem de couro curtido ao cromo. In: Encontro Nacional dos Químicos e Técnicos da Indústria do Couro, 14º, 1999, Florianópolis. Anais. Florianópolis: 1999. Não paginado.

CINCOTTO, M. A.; SILVA, M. A. C.; CASCUDO, H. C. Argamassas de revestimento: Características, propriedades e métodos de ensaio. São Paulo: IPT, 1995. 118p.

COMTE, P. Valorização de subprodutos de curtumes. *Tecnicouro*, Novo Hamburgo, v. 4, n. 6, p. 36-43, nov./dez. 1982.

EUROPEAN NORM. EN 71/3: Safety of toys – Part 3: Migration of certain elements. 1994.

FUJIKAWA, E. S. Incorporação do resíduo "serragem cromada" em materiais de construção civil. Bauru: UNESP, 2002. 80p. Dissertação (Mestrado em Engenharia Industrial), Faculdade de Engenharia, Universidade Estadual Paulista, 2002.

GARLET, G.; GREVEN, H. A. Aproveitamento de resíduos de E.V.A. da indústria calçadista na construção civil. In. Reciclagem e reutilização de resíduos como materiais de construção civil. São Paulo: ANTAC, 1996.

KELLER, W.; CHERNICHARO, C. A.; LAUFFER, J.; MOREIRA, R. V. O cromo no resíduo sólido de curtumes: imobilização em materiais cerâmicos. 19º Congresso Brasileiro de Engenharia Sanitária e Ambiental. *Anais*. Foz do Iguaçu, 1997. p. 1759-1772.

KRUMMENAUER, K. ANDRADE, J. J. O. Incorporation of chromium-tanned leather residue to asphalt micro-surface layer. *Construction and Building Materials*. 23, 2009. p. 574-581.

KRUMMENAUER, K. Estudo da incorporação da serragem de couro curtido ao cromo no microrrevestimento asfáltico. Dissertação (Mestrado em Engenharia), Universidade Luterana do Brasil. Canoas, 2005. 97p.

LORBER, K. E. Waste management and treatment of residues in the leather-tanning industry. In: (Ed.). Middle Pomeranian Scientific Society of the Environment Protection, 2003. Waste management and treatment of residues in the leather-tanning industry, p.7-28

MATTJE, V. Avaliação da potencialidade da utilização da serragem de couro curtido ao cromo em argamassas. Dissertação (Mestrado em Engenharia: Energia, Ambiente e Materiais) Universidade Luterana do Brasil. Canoas, 2005. 158p.

Mattos, L. R. S. Identificação e caracterização das argamassas para revestimento externo utilizadas na cidade de Belém-PA. Porto Alegre: UFRGS, 2001. 145p. Dissertação (Mestrado em Engenharia), Escola de Engenharia, Universidade Federal do Rio Grande do Sul, 2001.

Nakakura, E. H. Análise e classificação das argamassas industrializadas segundo a NBR 13281 e a MERUC. São Paulo: EPUSP, 2003. 198p. Dissertação (Mestrado em Engenharia), Departamento de Engenharia de Construção Civil, Escola Politécnica da Universidade de São Paulo, 2003.

Oliveira, J. A. C. Contribuição ao estudo do comportamento mecânico dos sistemas de revestimento à base de argamassa modificada com polímeros de base látex. Universidade de Brasília, 1999.

Serrano, C. L. R.; Rocha, E. C.; Mautone, A. Estudo da influência da adição de serragem de couro curtido ao cromo como carga de enchimento em composições elastoméricas. In: 2º Congresso Brasileiro de Polímeros. **Anais**. São Paulo, 1993. p. 25-29.

TEIXEIRA, R. C.; BASEGIO, T. M.; BERGMANN, C. P. Caracterização química de resíduos sólidos de curtume (serragem de couro ao cromo) e sua aplicação como carga em materiais cerâmicos. In: 14° Encontro Nacional dos Químicos e Técn icos da Indústria do Couro. *Anais*. Florianópolis, 1999. Não paginado.

VARGAS, J. L.; COMBA, G. J. M. Estúdio teórico y experimental sobre los morteros para muros resistentes de fábrica de ladrillo para la actualización de la Norma Básica MV 201-1972. *Materiales de Construcción*,. v. 34, n. 196, 1984. p. 3-41.

In: Management of Hazardous Residues Containing Cr(VI)   ISBN: 978-1-61470-637-3
Ed: Maria José Balart Murria

Chapter 11

USE OF INDUSTRIAL WASTE WITH CHROMIUM IN CONSTRUCTION

María Cecilia Panigatti[1*], Rosana Boglione[1], Carina Griffa[1], Hugo Begliardo[1], Jorge Sota[2] and David Cassina[1]
[1] Universidad Tecnológica Nacional. Facultad Regional Rafaela.
Bv Roca 989. Rafaela. Santa Fe. Argentina
TE: (054 3492) 432702. FAX: (054 3492) 432710
[2] Universidad Tecnológica Nacional. Facultad Regional La Plata

ABSTRACT

Waste disposal is a major industrial concern since it may cause environmental damage due to the mobility of toxic elements.

Chromium -a heavy metal used in metalworking companies- is found in two oxidation states, Cr (III) and Cr (VI) in aqueous solutions, the last one being the most toxic and dangerous.

After hexavalent chromium reduction and precipitation, a trivalent chromium hydroxide residue is obtained. The disposal of these dangerous wastes, thus, becomes a major problem due to the absence of adequate places to dispose them and to increasing environmental-standards demands.

The capacity of immobilization of Cr in Portland cement matrix depends on the state of oxidation of the element. The Cr (III) effectively incorporates in the solid phase, with low levels of Cr in the pore water. In contrast, Cr (VI) incorporates by phases, pore water thus containing high Cr levels.

Including Cr waste in construction works represents an important alternative both to ensure its use and to reduce its hazard. The solidification/stabilization (S/S) technology was developed to improve the handling of waste and reduce the mobility of its toxic components towards the surrounding environment and is used worldwide. It is an economic process for the final disposition of various types of waste, being applicable to treat the pollutant under study.

* Email: maria.panigatti@frra.utn.edu.ar

The identification of the residue using X-ray tube electrons was made. Physicochemical characterization and Cr (VI) and Cr (III) content were also determined once the leachate test was made. Based on preliminary tests, concrete dough with addition of different (3, 5, 7 and 10) percentages of waste were prepared. For each concentration, cylindrical concrete samples were made, being afterwards cured under water in different containers. During storage, pH, Cr (III) and Cr (VI) concentration in cured water were measured.

The proper dosage, adding as much waste as possible without affecting phyisico-mechanic concrete properties and its durability, while maintaining the values of leachate standards set by EPA, full-time test of the concrete, was studied.

The results allow analyzing not only advantages but disadvantages of setting the waste in a cement matrix.

1. Introduction

1.1. Chromium

Heavy metals represent an important ecological and health hazard due to their toxic effects and their accumulation throughout the food chain. Chromium is widespread in the environment and is derived from both natural and anthropogenic sources. With its great economic importance in industrial use, it is a major metal pollutant of the environment. Chromium is released into the environment by a large number of industrial operations such as electroplating, petroleum refining, pigments and allied products, chromate manufacturing, leather tanning and wood preservation.

Chromium can be found in three valence states: +2, +3 and +6. At low levels, chromium is an essential element for animal metabolism, but at levels close to 0.1 mg/g of the body weight, it can be lethal (Palomo et al., 2003). Chromium (III) and (VI) are the more frequent states and of major environmental significance due to their stability in the natural environment.

Chromium (VI) is studied as one of the most commonly occurring toxic pollutants; it is highly toxic, mobile, and has a long residence time in surface and groundwater. Hence, it is a severe toxic contaminant of natural waters, sediments and soils. Besides, Cr(VI) is carcinogenic, and mutagenic to animals as well as humans and it is associated with decreased plant growth and changes in plant morphology. Chromium (VI) is acidic, forming chromates $(CrO_4)^{2-}$ and the dichromate $(Cr_2O_7)^{2-}$, the other valences being basic. The effects of chromates are mainly on the skin and mucous membrane (Conner, 1990). In contrast, trivalent chromium, Cr(III), is relatively less toxic and less mobile.

Thus, metal-contaminated industrial effluents are treated, as a rule, by use of chemical and physicochemical methods, which usually involve pH adjustment and chemical reduction followed by flocculation or coagulation. The traditional technique of Cr (VI) treatment is initially to reduce it to Cr(III) in an acid medium using a reducing agent ($FeSO_4$. $7H_2O$, $Na_2S_2O_5$ or $Na_2S_2O_4$). Chromium (III) is then precipitated. In a basic medium, the hydroxide $Cr(OH)_3$, is almost completely insoluble. Then it should be dried and a Cr_2O_3 residue is obtained (Figure 1). In general, these methods are relatively chemical and energy intensive and the resultant metal-containing chemical sludge may be a potential source of metal pollution. So an additional treatment for final disposal must be sought.

In Argentina, this type of waste is not received in a landfill but in a safety one and there are very few of these.

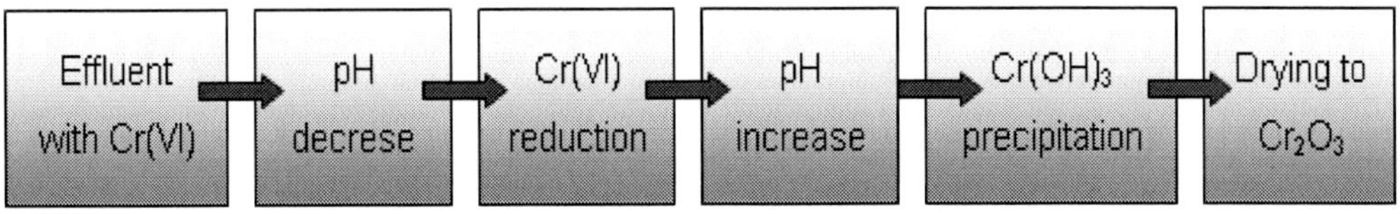

Figure 1. Physico-chemical treatment scheme of an effluent containing Cr (VI).

Due to their toxic effects on living systems, worldwide strict limits have been set for chromium discharge into the environment. The permissible industrial effluent discharge level of chromium into inland water for Santa Fe (Argentina), depending on dilution in the receiving body, can be seen in Table 1. The national law limit of hazardous residues is also shown.

Table 1. Regulations to control the discharge of liquid course of surface water

Legislation	Metal concentration	Dilution		
Santa Fe, Argentina Legislation 1089/82		100-360	361-1300	> 1300
	Cr (VI) ($mg.L^{-1}$)	0.10	0.15	0.20
	Cr (III) ($mg.L^{-1}$)	1.00	1.50	2.00
Hazardous Residues National Legislation 24051. Decree 831	Total Cr ($mg.L^{-1}$)	0.02		

In turn, there are limits to regulate the content of metals in the leaching of solid residue (Table 2).

Table 2. Regulation of chromium leachated from solid waste

Legislation	Metal	Concentration
Hazardous waste. National Legislation 24051. Decree 831	Total Cr ($mg.L^{-1}$)	5

Hazardous Waste Legislation N° 24051.

1.2. Solidification/ Stabilization Processes

Environmental pollution by heavy metals is mainly caused by industrial and agricultural processes. Removal of heavy metals from wastewater is usually achieved by physical and chemical processes which include reduction, precipitation, coagulation, membrane processes, ion exchange and adsorption. These processes generate waste that must be efficiently disposed.

S/S is accepted as a well established disposal technique for hazardous waste. As a result, many different hazardous wastes are treated with different binders. The S/S processes aim to carry out a mineralogical control of the metals to be stabilized. The processes of solidification/stabilization are classified in terms of the matrix used to encapsulate the waste, and can be divided as follows (Conner, 1990):

Inorganic-matrix processes (also known as cementation processes), where the matrix used is cement, lime, pozzolan.

Organic-matrix processes, where the matrix is composed of thermoplastic substances (bitumen paraffin, polyethylene) or organic polymers.

There is increasing public interest in finding ecological solutions for safe disposal of by-products that might be suitable for incorporation into cementitious materials and concrete (Mehta, 1998). The solidification/stabilization process utilizes chemically reactive formulations that, together with the water, components in sludges and other aqueous hazardous wastes, form stable solids. The material used for this process not only solidifies the hazardous waste by chemical means but also insolublizes, immobilizes, encapsulates, destroys, sorbs, or otherwise interacts with selected waste components. The results of these interactions are solids that are non-hazardous or less hazardous than the original waste.

S/S technology does not remove heavy metals from the polluted waste but has the purpose to physically as well as chemically fix them in the solid matrix in order to reduce their mobility, to minimize the threat to the environment, and to ensure compliance with existing regulatory standards (Malviya & Chaudhary, 2006; Parsa et al., 1996). The S/S products have different properties from waste and binders individually.

The effectiveness of S/S process is studied by physical, chemical and micro structural methods. The degree of effectiveness is defined basically by strength and the leach resistance. The continuing need to develop economical and improved waste management techniques has increased the potential importance of solidification technology through the world, in a process defined as "Best demonstrated available technology" (BDAT) (Conner, 1990; Barth, et al., 1989).

Cement is the most frequently used and usually studied binder due to its low cost and the resultant pH, rendering metal contaminants insoluble. Solidification with cement is widely understood and simple, is easily available and results in a stable product (Roy, et al., 1991). Important parameters to assess solidification are strength, setting time and extent of hydration. Mixing waste with cement can have either a positive and negative effects on the cement matrix properties. Binders other than cement are generally used as partial replacement of cement (Zhang, et al. 2008). In general other binders like sulphate resistant Portland cement (SRPC), white cement, kiln dust, lime, fly ash, pulverized fuel ash (PFA), silicates, etc., partially replace OPC, can have a positive or negative effect.

Pore solution is the solution that is contained within solid cement but is not combined into solid hydration product. Pore water is extracted using high compression devices. Pore solution of cement has a high concentration of alkali metal cations, which are balanced by OH ions. The OH-anions are responsible for high alkalinity. Other anions like aluminate and silicates are insoluble (Malviya & Chaudhary, 2006).

Compressive strength is dependent on the quality of the pore structure of the cementitious matrix. In turn, this quality depends on the kind and quantity of the constituents forming the pore structure, which are cement hydration products and their reaction products with admixtures. The pH of pore solution depends on the presence of waste species (Zhang, et al. 2008).

A significant role in the immobilization process is attributed to the so-called C–S–H phase (calcium silicate hydrates) the main component of hardened paste (Giergicznya & Król, 2008). Absorbability, low permeability preventing the migration of liquids and the presence of small gel pores (micropores) are among the decisive factors affecting high immobilization potential of C–S–H (Bhatty, 1987). The properties and amount of C–S–H are highly bound with the content of mineral additions in cementitious material.

The role of calcium aluminate and sulfoaluminate hydrates in the immobilization process should not be ignored. In the structure of calcium sulfoaluminates, the incorporation of mono- and bi-valent ions is possible (Glasser, 1992). The precipitated, insoluble hydroxides of heavy metals calcium sulfoaluminates play an exceptional role as Cr and Mo immobilizers, forming solid solutions with these elements (Glasser, 1994).

The solidification/stabilization (S/S) of hazardous wastes by pozzolan-based binders is also a most familiar technology that has been applied to many types of industrial wastes, mainly those containing heavy metals (Conner, 1990; Means et al., 1995).

In some cases, stabilization is a pre-landfill waste treatment process, which has been used for different types of industrial wastes, but is particularly suited to those containing heavy metals (Conner, 1990; Barth, et al., 1989).

1.3. Solidification/Stabilization of Wastes with Chromium

There are studies of S/S using both Cr (III) (insoluble hydroxides) and Cr (VI) (in solution or salts of dichromate or chromate anion). It has been stated, that the level of heavy metals immobilization is combined with composites composition.

Cr(III) is both less toxic and less mobile, being relatively insoluble as $Cr(OH)_3$ and also as mixed hydroxide phases containing some calcium (Omotoso et al., 1996).

The precipitated, insoluble hydroxides of heavy metals, such as Cr(III), Pb and Zn modify the porosity of the cement matrix, the total porosity is reduced and the gel pores ratio increases (Stephan, et al., 2003; Tama`s, et al., 1992; Thevenin & Pera, 1999). Some substitution into calcium (alumino) silicate hydrate (C–A– S–H) phases is also believed to be possible (Glasser, 1997; Kindness, et al., 1994).

Chromium (III) has an amphoteric behaviour like aluminium, zinc and lead, i.e. these elements are soluble in both strongly acidic and basic solution. The minimum solubility of chromium is at a pH value near 9. The increase in solubility at higher pH is generally due to the formation of an oxyanious complex such as chromite (Klemm, 1998).

On the other hand, several studies have shown that there is a lack of chemical fixation of the concentrations of Cr(VI) in cement based S/S technology (Barth, et al., 1989; Conner, 1997; Kindness, et al.,1994; Duchesne, 1999; Zamorani, et al.,1998) Thus Cr(VI), remain free in the contaminated solution when ordinary Portland cement (OPC) is used without a reducing agent.

Giergicznya & Król, (2008) concluded that heavy metals (Pb, Cu, Zn, Cd, Mn) added as salts are very effectively immobilized in hydrated matrices (from 99.82% to 99.99%); Cr is an exception immobilized with lower degree (from 85.97% to 93.33%). Immobilization is effected by the composition of solidifying binder and time of maturing.

Hexavalente chromium has limited solubility in pore fluid. The chromium concentration increases in the pore fluid with time. On spiking cement with 89,000 ppm Cr (III) the Cr content of the resulting pore solution increased from 0 to 0.24 to 2.2 ppm, at 0, 34 and 250 days, respectively. Increasing the Cr(III) content of the system does not increase its pore fluid concentration. On spiking with Cr(VI), the dissolved Cr remains high, up to 200 ppm even after 250 days and the dissolved Cr valence state remains Cr(VI) (Kindness, et al., 1994). pH of pore solution depend both on the presence of fillers and heavy metals.

Wang & Vipulanandan (2000) found that Cr(VI) addition tends to have deleterious effects on the strength development of Portland cement-based (OPC) systems. They studied Cr(VI) immobilization in OPC, and found that a maximum loading of 0.2% Cr as K_2CrO_4 was possible while complying with the requirements of the TCLP leach test. In OPC blends with higher Al content, CrO_4^{2-} substitution into the SO_4^{2-} sites of ettringite and hydrocalumite is also possible (Palmer, 2000), although the solubilities of these phases are unsatisfactorily high in terms of their use in immobilization (Perkins & Palmer, 2001).

It has been noted that reduction of Cr(VI) to Cr(III), either by addition of $FeSO_4$ or by the sulfide present in blast furnace slags, is beneficial for immobilization in Portland cement-based matrices (Allan & Kukacka, 1995; Glasser, 1997). Deja (2002) studied the immobilization of Cr(VI) in alkali-activated slag binders, and showed that not only Cr(VI) is quite effectively immobilized in these systems, but that it also leads to an improvement in mechanical strength development in both carbonate and silicate-activated systems. This work also indicated that reducing conditions in the hydrated slag pastes (i.e. available S^{2-} ions) play a positive role in the immobilization process by mechanisms such as partial reduction of Cr (VI) to Cr (III).

Chromium addition causes a wide variety of pore diameters to appear with the small 0.01 μm diameter pores dominating indicating involvement of this metal in silicate formation. Lead addition changes the relative amount of pores in the structure, while the Cr addition affects the pore formation mechanism (Chatterji, 1978)

Moreover, GGBFS (Granulated blast furnace slag) has a reducing potential that creates an environment where the Eh values (- 200 to - 400 mV) will be weaker than in a system composed mainly of Portland cement (100 to 200 mV) (Kindness, et al., 1994; Glasser, 1997). Indeed, the activity field (conditions of particular Eh and pH) of the GGBFS is found in the zone of stability of chromium, i.e. when chromium is found in the $Cr(OH)_3$ form. With regard to Portland cement, its field of activity is located in the chromates $(CrO_4)^{2-}$ zone which are a very soluble form of Cr(VI).

GGBFS seems to have the capacity to reduce Cr(VI) to Cr(III) and simultaneously solidify the residue. Then the Cr(III) will precipitate as the stable and insoluble form $Cr(OH)_3$. The majority of analyzed heavy metals, added to binders' composition in the form of heavy metal salts achieves high level of immobilization, in mortar based on binder with 85% GGBFS and 15% OPC. The lowest immobilization level was reached for hexavalent chromium added to hardening mortars as $Na_2Cr_2O_7 \cdot 2\ H_2O$.

Glasser (1994) points out that the precipitation of heavy metals under S/S conditions is often a two-step process. First, the metal constituent precipitates as hydrous oxide or hydroxide under the high pH conditions, but subsequently this metal constituent may react chemically with other cement components such as calcium, alumina, silica and OH^- ion to obtain a more complete precipitation (Klemm, 1998). Chromium (III) precipitation occurs immediately after mixing, as $Cr(OH)_3$, but this precipitation is not entirely complete due to the amphoteric nature of chromium. This behaviour is also conditioned by the oxidation state. In fact, the Cr(III), a slightly amphoteric ion, is readily precipitated as $Cr(OH)_3$ in an alkaline environment contrary to Cr(VI). The chromates $(CrO_4)^{2-}$, the more frequent form of Cr(VI) in solution, are relatively soluble at all values of pH (Macpheem & Glasser, 1993). However, with increasing time the $Cr(OH)_3$ is resorbed and substitutes for octahedrally coordinated Al in aluminate phases (Klemm, 1998). This explains the presence of chromium in calcium aluminate phases observed by SEM for samples of OPC at initial low concentrations of Cr(VI).

2. A Case Study

The S/S technique in a Portland cement matrix was applied to residues with Cr (III), obtained from reduction, precipitation and drying of Cr (VI) waste, as generated in two metalworking manufacturing processes in Rafaela, Argentina.

Figure 2. Residue 1 used in the study.

Figure 3. Residue 2 used in the study.

Since these industries employ similar treatment procedures for rinse water with Cr (VI) until its reduction to chromium oxide, the first step was to perform waste physical chemical characterization (called Residue 1 and 2, Figure 2 and 3 respectively). The results allowed selecting the residue to make the fixation experiences (Panigatti, et al. 2008).

The task involved the joint and coordinated action of both Chemistry and Civil Engineering Labs, Universidad Tecnológica Nacional (Rafaela).

2.1. Waste's Physicochemical Characterization

The waste elements identification was made through X-ray analysis with tube electrons. The results are shown in Table 3.

Table 3. Chemical elements (%) of the residue obtained by X-ray analysis

Sample	Relative Relations (% g/100 g)									
	Na	Al	Si	S	Cl	Ca	Cr	Fe	Ni	Cu
Residue 1	22	7	1	25	2	3	36	2	--	2
Residue 2	3	--	2	33	--	54	5	2	1	--

The results show that Residue 1 has high chromium, sulfur and sodium content. The company makes Cr(VI) reduction treatment with sodium sulfite and neutralization with sodium hydroxide. In contrast, in Residue 2 most important elements are calcium and sulfur, while chromium was found in low percentages. In this case, hexavalent chromium is reduced with sodium hydrogen sulphite and neutralized with sodium hydroxide and lime.

Also, physicochemical characterization was performed (Table 4) using the following techniques: pH (electrometric method), humidity (evaporation and drying in an heater at 105 °C), volatile solids (drying at 550 °C in muffle), soluble substances in ethyl ether (extraction with ethyl ether and evaporation), content of Cr, Pb, Cu, Fe and Ni (Atomic Absorption Spectrometry Flame).

The samples differ significantly mainly in pH, humidity and chromium and copper concentration, due to the different processes of wastewater treatment used in each company. Residue 1 was used on experiences as it has higher content of chromium.

Table 4. Residues physicochemical analysis

Parameter	Residue 1	Residue 2
pH	7.4	12.5
Soluble solids in ethyl ether (g/100 g wet sample)	3.35	--
Humidity (g/100 g)	28.5	2.1
Volatile solids (g/100 g wet sample)	14.4	--
Chromium (g Cr / 100 g sample)	18.6	1.3
Lead (g Pb / 100 g sample)	0.03	ND (0.006)
Copper (g Cu / 100 g sample)	1.3	0.009
Iron (g Fe / 100 g sample)	0.8	0.7
Nickel (g Ni / 100 g sample)	0.02	0.14

2.2. Design and Development of Concrete

Within the premises of the experience, the preparation of concrete containing 0%, 3%, 5%, 7% and 10% of chromium residue (referred to the weight of Portland cement), was included. The procedure considered the residue fraction as a part of fine aggregates.

For each type of concrete, three equal doses paste were prepared. Five cylindrical samples (15 x 30 cm) for each one were produced, in order to determine the standard deviation introduced in the repetitiveness of the production process. Also, prismatic samples (7 x 7 x 15 cm) for further cut and internal observation with stereoscopic magnifier were made.

The cylindrical concrete samples were placed, according to the waste`s concentration percentage, in different curing containers in water without the addition of lime to avoid changes in pH. A simple compression at 7, 14, 28, 60 and 90 days was tested. During storage, weekly measurements of pH, Cr (III) and Cr (VI) concentration in water curing of each container were made, extracting samples of 100 ml each time without adding additional water. Curing of concrete samples for testing was performed at 23 ± 2 °C and the measurements at 20 ± 3 °C.

Additionally, cylindrical samples were triturated and milled to simulate extreme conditions (Figure 4). Physicochemical characterization and Cr (VI) and Cr (III) content were also determined after the leachate test was made (Standard EPA 1310A).

Figure 4. Triturated and milled cylindrical samples.

2.3. Concrete Dosage

2.3.1. Preliminary Design of Standard Concrete (Concrete Comparison)

Concrete pattern was designed under the following assumptions:

Characteristic resistance of 30 MPa
Standard deviation: ± 4MPa
Expected Half Resistance 28 days: 37 MPa
Minimum cement content: 350 kg
Water-cement ratio of 0.35 (w/c), using super plasticizer additive to achieve a descent of approx. 6 cm in the Abrams cone.

2.3.2. Used Materials

Two granulometric sizes of granitic origin stone were used as coarse aggregate. Two sizes of sand, commercially classified as "medium" and "gross," were used as fine aggregate.

Since the manufacturing standards ensure homogeneous properties, Portland cement with limestone filler (CPF 40) was employed,

To achieve the expected slump, with the low water-cement ratio (w/c), a super plasticizer additive was used. The chrome residue granulometry was determined by an internal procedure of the Civil Engineering Laboratory, due to its powdery nature. The method consisted in washing the sample 5 (five) times through the sieve ASTM N° 200 (0.074 mm).

Since there is no standard procedure for determining residue density, Standard IRAM 1624 (cement density determination) was applied.

2.3.3. Used Dosage

The relationship fine/total aggregate fitting the ideal Füller curve, was 35% (Figure 5), obtained from a proper combination of aggregates available.

The appearance and performance settings applied to the paste test (keeping cement content the same) led to work with a 0.40 w/c value, this value remaining below the maximum set by regulations′ conditions for durability and strength.

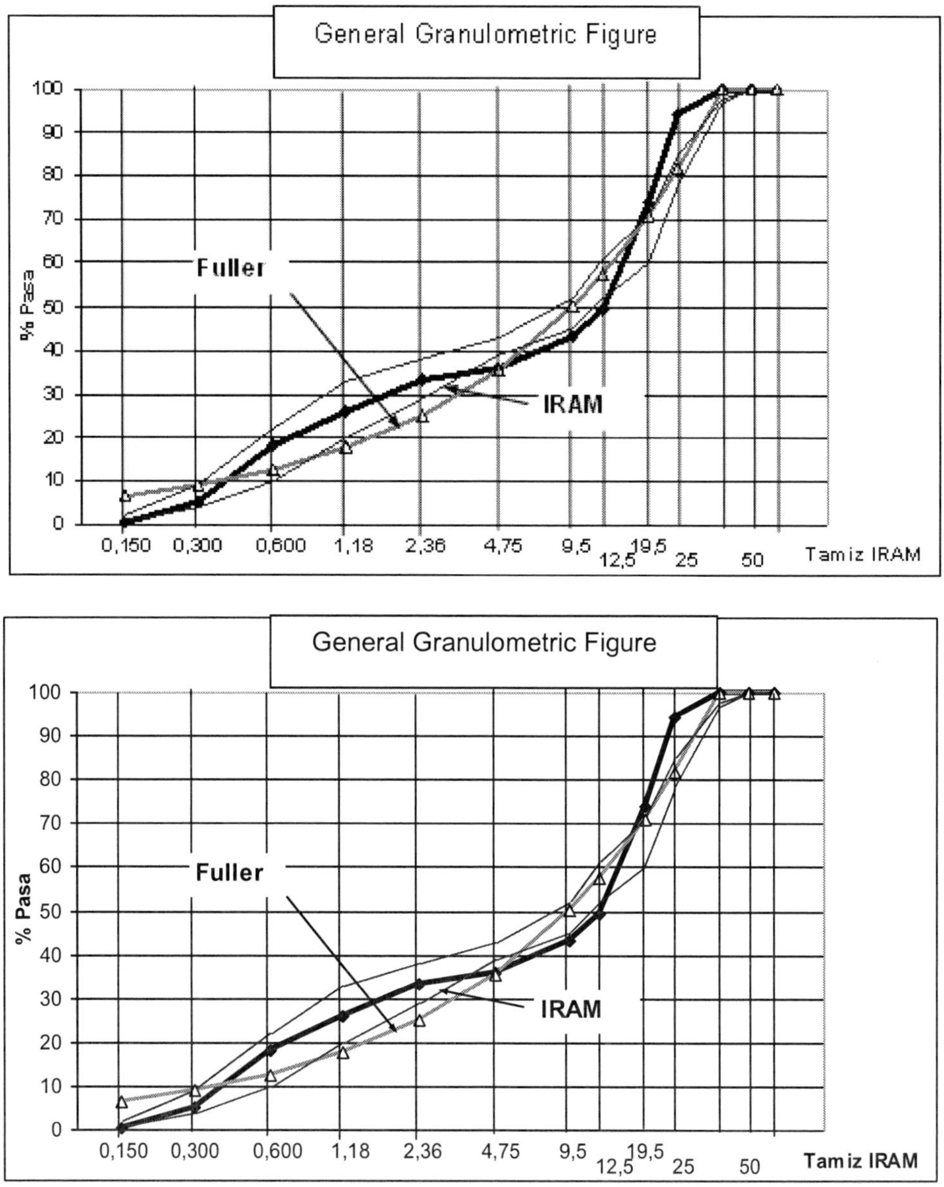

Figure 5. Granulometric Curves.

Based on the assumption that aggregates should be at saturated and surface dry (ssd) condition prior to the establishment of paste, moisture content of each fraction was measured, the amount of water incorporated into the concrete being then regulated up to the condition noted.

Besides the concrete pattern, four types of concretes were elaborated on by incorporating the residue, the four types being designated as follows:

Concrete 3 %: Base concrete pattern + 3 % of residue with chrome
Concrete 5 %: Base concrete pattern + 5 % of residue with chrome
Concrete 7 %: Base concrete pattern + 7 % of residue with chrome
Concrete 10 %: Base concrete pattern + 10 % of residue with chrome
Percentages refer to cement weight.

Table 5. Dosage concrete pattern

Concrete pattern		1	2	3	4	5
Material component		Absorption	Dry weight	Density	Solid volume	Weight s.s.d.
		w %	1 m^3 (kg)	(kg/dm^3)	(dm^3)	1 m^3 (kg)
1	Water		125.50	1.00	125.5	125.50
2	Cement Portland with filler CPF 40		350.00	3.10	112.9	350.00
3	Stone 10-30	0.71	578.50	2.65	218.3	582.61
4	Stone 6-10	1.04	578.50	2.71	213.5	584.52
5	Thick sand Cordoba	0.32	385.50	2.60	148.3	386.73
6	Medium sand Santa Fe	0.18	385.50	2.63	146.6	386.19
7	Additive super plasticizer		14.50	1.151	12.6	14.50
8	Residue with chrome		0.00	1.826	0.0	0.00
	Σ		2418.00			2430.05

Table 6. Dosage concrete 7 %

Concrete with 7 % of residue with chrome		1	2	3	4	5
Material component		Absorption	Dry weight	Density	Solid volume	Weight s.s.d.
		W %	1 m^3 (kg)	(kg/dm^3)	(dm^3)	1 m^3 (kg)
1	Water		125.50	1.00	125.5	125.50
2	Cement Portland with filler CPF 40		350.00	3.10	112.9	350.00
3	Stone 10-30	0.71	578.50	2.65	218.3	582.61
4	Stone 6-10	1.04	578.50	2.71	213.5	584.52
5	Thick sand Cordoba	0.32	385.50	2.60	148.3	386.73
6	Medium sand Santa Fe	0.18	361.00	2.63	137.3	361.65
7	Additive super plasticizer		14.50	1.151	12.6	14.50
8	Residue with chrome (7 %)		24.50	1.826	13.4	24.50
	Σ		2418.00			2430.01

Note: in order to maintain the w/c, additive volume was considered as water volume.

Dosing adopted for the concrete pattern was recorded (Table 5), the concrete with 7 % of residue being incorporated as example (Table 6).

2.4. Results

pH measurements were made during the treatment of cylindrical samples, since at pH values higher than 11, the residue trivalent chrome could dissolve as chromate into solution. The results obtained, corresponding to the cylindrical samples elaborated with 3, 5, 7 and 10 % of the residue, can be observed in Figure 6. In all cases, the values ranged between 9.8 and 11.8 and a slight decrease was observed throughout the time, due the effect of the sodium hydroxide carbonation.

When compared to the initial used water (pH = 7.50), a major pH increase was observed mainly due to matter exchange between the water and the cylindrical samples′ concrete.

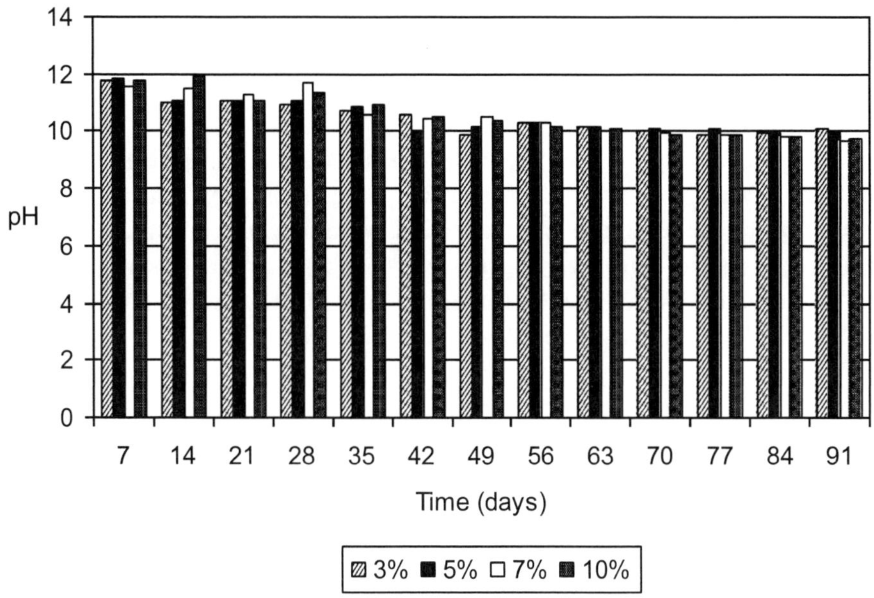

Figure 6. pH variation depending on time for cylindrical samples concrete treated with 3, 5, 7 and 10 % of residue.

During three months of control, tri- and hexavalente chrome were not detected in the treated (3, 5, 7 and 10 %) solutions analyzed.

After different treatment periods, the leached test on triturated and milled cylindrical samples (simulating extreme conditions) was performed. Figure 7 shows total chromium concentrations average, as a function of time, for different residue concentrations. Chromium (VI) was not detected in leached of analyzed conditions; hence total chromium concentration corresponds to Cr (III). Variations in chromium concentrations were observed. These could be due to work conditions (temperature, humidity, human factor, etc.) as well as to the inhomogeneity in the triturated samples.

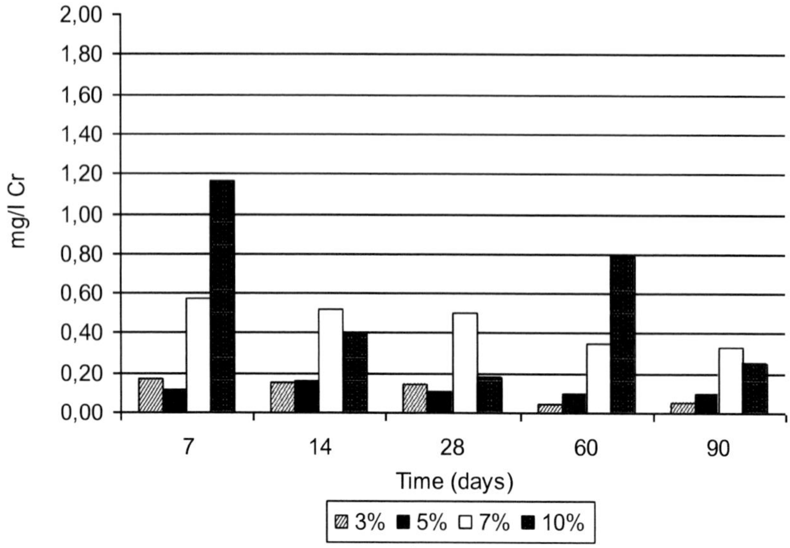

Figure 7. Concentrations of Total Chromium leached depending on the matrix age.

The highest concentrations of chromium leached were obtained using 7 and 10 % of residue. Working with 3, 5 and 7 % of residue, Cr (III) concentration in leached diminished in time, which may be due to a higher stabilization of the residue in the matrix.

In the 10 % residue cylindrical samples, major variability was observed in the results, since at the mentioned concentration there is a high probability of finding residues not retained by the matrix, the sample being more heterogeneous.

According to results, it is important to emphasize the Cr (VI) absence in leached, which is the most toxic and harmful oxidation state for the environment. It was observed that there was no oxidation of trivalent chromium during preparation and curing of cylindrical samples. Chromium (III) concentrations obtained are low and meet Argentine Dangerous Residues Law 24051 (less than 5 mg/l).

Stereoscopic magnifier observation on polished cuts of prismatic samples (7 x 7 x 15 cm) in different concretes was practiced. In most of the samples, very good residue dispersion in the matrix was verified, which would indicate high residue solubility in the paste (Figure 8 to 11).

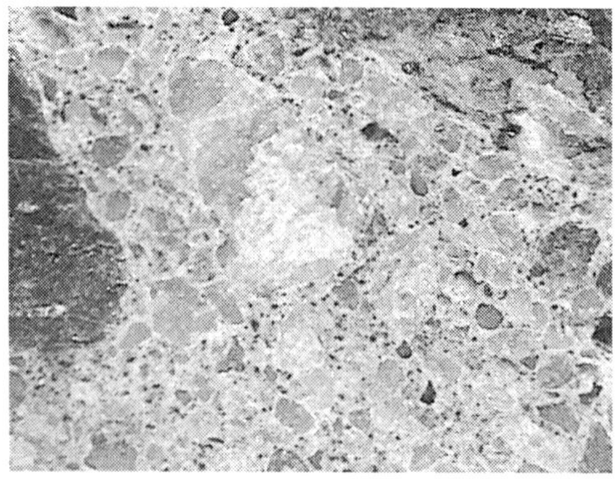

Figure 8. Concrete with 3 % addition of chromium waste.

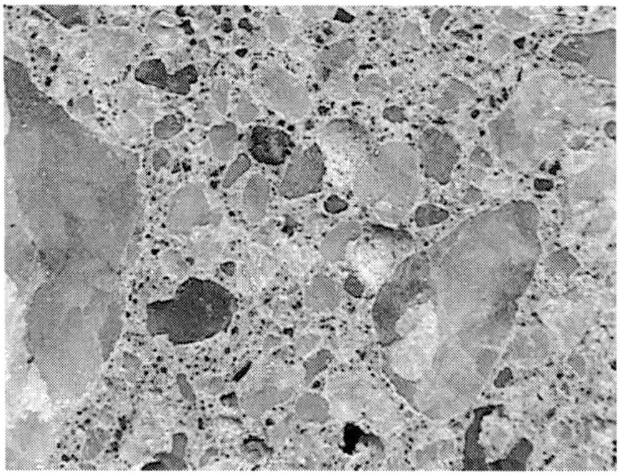

Figure 9. Concrete with 5 % addition of chromium waste.

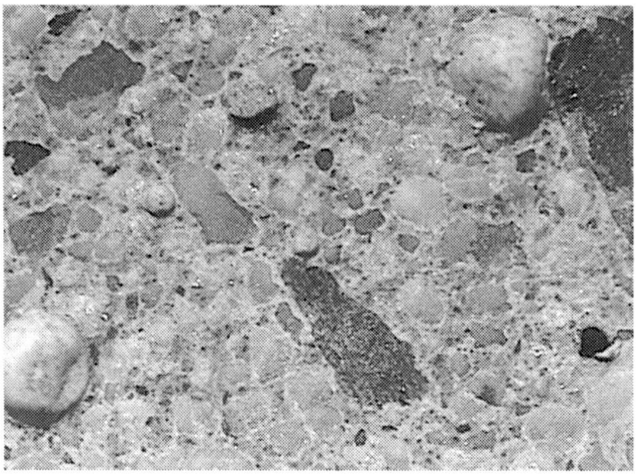

Figure 10. Concrete with 7% addition of chromium waste.

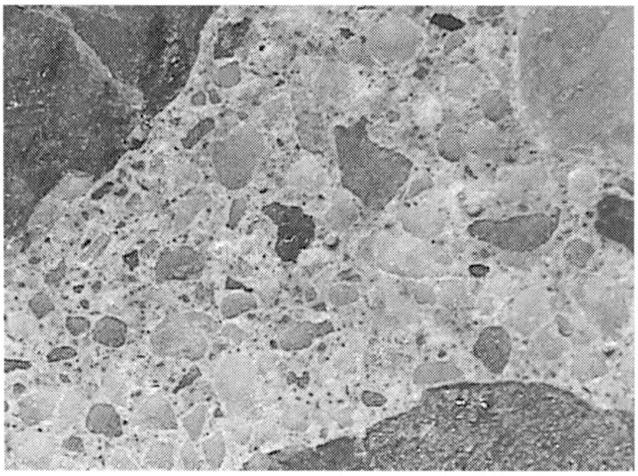

Figure 11: Concrete with 10% addition of chromium waste.

Figure 12 shows simple compression tests of cylindrical samples, at different ages and concretes. The average resistances obtained at 28 days, reached the values expected according to the premises of reference concrete design; the same occurred for the 3% waste sample.

Slightly low values were registered for the remaining concentrations, which was predictable due to the residue substitution on part of the thin fraction of the aggregates, in the composition of the respective concretes.

The obtained standard deviation, including the process repetitiveness, were below 4 MPa applied for the design of the concrete pattern, this resulting a typical value in some regulations for precise elaboration conditions with controls.

In all cases, fast resistances developed at an early 7 day period, reaching 80 - 87 % of the resistance at 28 days. This is due to used cement and the properties incorporated by the additive. The trend was a resistance increase along test age, and a decrease with the increase in residue concentration, matching our expectations.

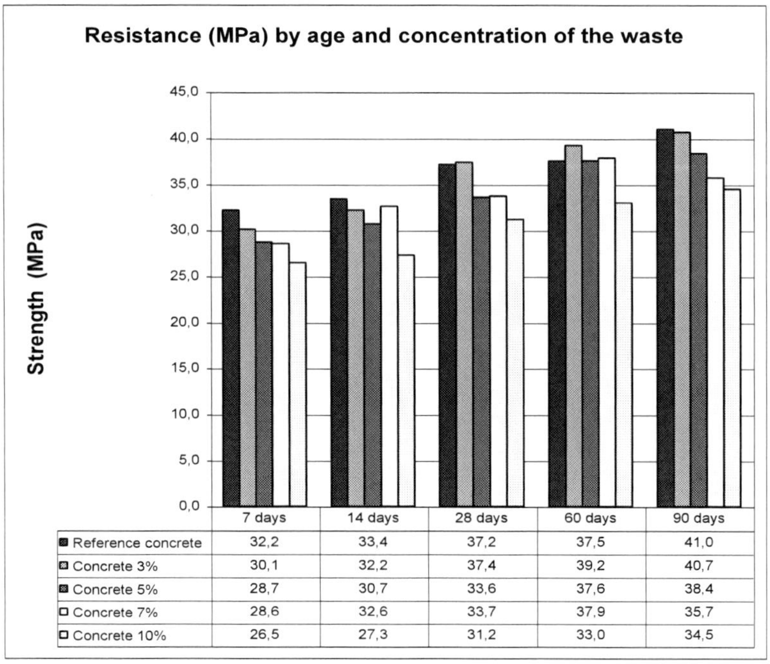

Figure 12. Resistance comparisons of simple compression, for age and residue concentration.

In Table 7, the residue and chrome masses contained in 1 m^3 of elaborated concrete for each of the residues percentages used in the present study can be observed. Although the added chromium masses are not high, it is a viable alternative to dispose this waste.

Table 7. Residue and chrome (III) content for m^3 of concrete in the different work conditions

	Percentages of residue			
	3 %	5 %	7 %	10 %
Residue (kg/m^3)	10,5	17,5	24,5	35,0
Cr(III) (kg/m^3)	2,0	3,3	4,6	6,6

CONCLUSION

Solidification/stabilization technology is a very effective tool and is becoming a standard process in hazardous waste treatment and disposal.

The solidification of heavy metal compounds in matrices is beneficial from the natural environment point of view in several aspects. The wastes from the industry, metallurgy particularly, can be very effectively adapted for this purpose.

Solidification of heavy metals containing inorganic materials in mortars and concretes produced using OPC (Ordinary Portland Cement) is an effective way of immobilization. The mechanism of the processes occurring on immobilization, as well as the assessment of the effectiveness of solidification, has been studied and discussed in many reports.

The capacity of immobilization of Cr in Portland cement matrix depends on the element oxidation state. Cr (III) is effectively incorporated in the solid phase, with low levels of Cr in the pore water. In contrast, Cr (VI) incorporates by phases, pore water thus containing high Cr levels.

From experiences carried out with waste containing Cr (III), there is the possibility of inclusion of such waste in cementitious matrix.

A very good dispersion of the residue was observed in the cement matrix. The leached tests, under extreme triturated and milled conditions of the concrete with residue, after 90 days, have demonstrated low chromium (III) concentration and chromium (VI) absence. It is concluded that a good fixation of the residue in the matrix occurs with 3, 5 and 7 % concentrations. Cr (III) in leached obeys with Argentinean Dangerous Residues Law 24051 (minor to 5 mg/l).

The compression resistances have been satisfactory. Concrete has achieved good end-resistance for its employment in pavements or vehicles maneuvers parking, among other uses of common order.

The results are short-term, so it should be evaluated what happens with the concrete with chromium residues, in medium and long term scales.

The possibility of including waste of processes in construction works represents an important alternative to ensure their use and to reduce their hazard. The materials thus produced are not detrimental for environment and can be stored or practically implemented, e.g. in sub base and in road making.

ACKNOWLEDGMENTS

The authors thank M. Belen Schierano and Ernesto Meier for the English correction.

REFERENCES

Allan, M.L. & Kukacka, L.E. (1995). Blast furnace slag-modified grouts for in situ stabilization of chromium-contaminated soil. *Waste Management, Vol. 15,* (3) 193–202.

Astrup, T., Mosbæk, H. W. & Christensen. T. H. (2006). Assessment of long-term leaching from waste incineration air-pollution-control residues. *Waste Management, Vol. 26*, 803 – 814.

Barth, E., Percin, P., Arozarena, M., Zieleinswski, J., Dosani, M., Maxey, H., Hokanson, S., Pryately, C., Whipple, T., Kravitz, R., Cullinane, M., Jones, L., & Malone, P. (1989). *Stabilization and Solidification of Hazardous Wastes*, Noyes Data Corporation, New Jersey.

Bhatty, M.S.Y. (1987). Fixation of metallic ions in Portland cement. *Proceedings of the Fourth National Conference Hazardous Wastes and Hazardous Materials*, Washington. 140 – 145.

Chatterji, S. (1978) Mechanism of the $CaCl_2$ attacks on Portland cement concrete cement, *Cement and Concrete Research, Vol. 8*, 461–467.

Conner, J.R. (1990). *Chemical Fixation and Solidification of Hazardous Wastes*. New York, Van Nostrand Reinhold.

Deja, J. (2002). Immobilization of Cr^{6+}, Cd^{2+}, Zn^{2+} and Pb^{2+} in alkali-activated slag binders, *Cement and Concrete Research, Vol 32,* (12), 1971–1979.

Duchesne, J. (1999) Réduction des concentrations en Cr et en Mo dans des lixiviats générées par des poussières de four de cimenterie. In: J. Méhu, G. Keck and A. Navarro, Editors, *Stabilisation des Déchets and Environnement*, Société alpine de publications, Grenoble, Vol. 99, 21–25.

Giergicznya, Z. & Król, A. (2008). Immobilization of heavy metals (Pb, Cu, Cr, Zn, Cd, Mn) in the mineral additions containing concrete composites. *Journal of Hazardous Materials,* Vol. 160, (2-3), 247-255.

Glasser, F.P. (1992). Application of cements to the treatment and conditioning of toxic wastes. *Ninth International Congress on the Chemistry of Cement New Delhi*, 114–118.

Glasser, F.P. (1994) Environmental uses of cement in waste conditioning. M.W. Grutzeck, S.L. Sarkar. *Engineering Foundation Conference, Proceedings, Advance and Cement and Concrete, American Society of Civil Engineers,* New York. 499–507.

Glasser, F.P. (1994) Immobilization potential of cementious materials. *Proceeding of the International Conference on Environmental Implication of Construction Materials and Technology Developments Maastricht*, 77–86.

Glasser, F. P. (1997). Fundamental aspects of cement solidification and stabilisation. *Journal of Hazardous Materials, Vol. 52,* 151 – 170.

Kindness, A., Macias, A. & Glasser, F.P. (1994). Immobilization of chromium in cement matrices. *Waste Management, Vol.1,* 3–11.

Klemm, W.A. (1998). Ettringite and oxyanion-substituted ettringites—their characterization and application in the fixation of heavy metals: a synthesis of the literature. *Portland Cement Association, Research and Development Bulletin RD116.*

Macphee, D.E. & Glasser, F.P. (1993) Immobilization science of cement systems, *MRS Bulletin*. 66 – 71.

Malviya, R. & Chaudhary, R. (2006). Evaluation of leaching characteristics and environmental compatibility of solidified/stabilized industrial waste, *Journal of Material Cycles and Waste Management. Vol. 8*, 78 – 87.

Malviya, R. & Chaudhary, R. (2006). Factors affecting hazardous waste solidification/stabilization: A review. *Journal of Hazardous Material.* B 137 267 – 276.

Means, J., Smith, L., Nehring, K., Brauning, S., Gavaskar, A., Sass, B., Wiles, C. & Mashni, C. (1995) *The Application of Solidification/Stabilization to Waste Materials*, Lewis Publishers, CRC Press, Inc., Boca Raton, FL.

Mehta, P.K. (1998). Role of pozzolanic and cementious material in sustainable development of the concrete industry, fly ash, silica fume, slag and natural pozzolans in concrete. *Sixth CANMET/ACI International Conference*, Bangkok, Thailand. 1– 20.

Omotoso, O.E., Ivey, D.G. & Mikula, R. (1996). Quantitative X-ray diffraction analysis of chromium(III) doped tricalcium silicate pastes. *Cement and Concrete Research, Vol. 26*, (9), 1369–1379.

Palmer, C.D. (2000). Precipitates in a Cr(VI)-contaminated concrete. *Environ. Sci. Technol., Vol. 34,* (19), 4185–4192.

Palomo, A. & Palacios, M. (2003). Alkali-activated cementitious materials: alternative matrices for immobilisation of hazardous wastes: Part II. Stabilization of chromium and lead. *Cement and Concrete Research, Vol. 33*, 289 – 295.

Panigatti, M. C.; Begliardo, H.; Sota, J.; Boglione, R.; Griffa, C. y Cassina, D. (2008). *"Incorporación de residuos industriales con cromo en matrices cementicias"*. Memorias III Congreso Internacional y 17° Reunión Técnica de la Asoc. Arg. de Tecnología del Hormigón, Vol. 1, 15-22.

Parsa, J., Munson-McGee, S.H. & Steiner, R. (1996). Stabilization/solidification of hazardous waste using fly ash. *Journal Environmental Engineering, Vol. 122*, 935 – 940.

Perkins, R.B. & Palmer, C.D. (2001) Solubility of chromate hydrocalumite ($3CaO{\cdot}Al_2O_3{\cdot}CaCrO_4{\cdot}nH_2O$) 5–75 °C, *Cement and Concrete Research, Vol. 31,* (7), 983–992.

Roy, A., Eaton, H.C., Cartledge, F.K. & Tittlebaum, M.E. (1991). Solidification/stabilization of heavy metal sludge by a Portland cement/fly ash binding mixture hazard. *Waste Hazardous Mater., Vol. 8*, 33–39.

Stephan, D., Knöfel, D. & Härdtl, R. (2003). Influence of heavy metals on the properties of cement and concrete-binding mechanisms and fixation. *Proceedings of the 11th International Congress on the Chemistry of Cement "Cement's Contri-bution on the Development in the 21st Century"*, Durban, South Africa, 2178-2186.

Tama´s, F.D., Csete`nyi, L. & Tritthart, J. (1992). Effect of adsorbents on the leachability of cement bonded electroplating wastes. *Cement and Concrete Research, Vol. 22,* 399 – 404.

Thevenin, G. & Pera, J. (1999). Interactions between lead and different binders. *Cement and Concrete Research, Vol. 29*, 1605 – 1610.

Wang, S. & Vipulanandan, C. (2000). Solidification/stabilization of Cr(VI) with cement. Leachability and XRD analyses, *Cement and Concrete Research, Vol.* 30, 385–389.

Zamorani, E., Brambilla, G., Serrini, G., Toussaint, N. & Cazzaniga, E. (1992). Particle fabrication by chemical processing for hazardous waste management. *Cement and Concrete Research, Vol. 22*, 331– 338.

Zhang, J., Provis, J., Feng, D. & van Deventer, J. (2008) .The role of sulfide in the immobilization of Cr(VI) in fly ash geopolymers. *Cement and Concrete Research, Vol.* 38, 681–688.

In: Management of Hazardous Residues Containing Cr(VI) ISBN: 978-1-61470-637-3
Ed: Maria José Balart Murria

Chapter 12

RECOVERY AND RE-USE OF CHROMIUM FROM THE WASTE GENERATED BY LEATHER INDUSTRY

K. Kolomaznik*

Tomas Bata University in Zlin, Faculty of Applied Informatics, Nad Stranemi 4511, 760 05, Zlín Czech Republic

ABSTRACT

Leather industry processes hides of slaughtered and game animals; meaning by-products of the meat industry. The main output is stabilized leather. The conversion of raw hide into leather comprises a complex series of chemical, physical-chemical and mechanical operations. Approximately 80% of leather is stabilized (tanned) by trivalent chromium (CrIII) salts. The processing of 1,000 kg of raw hide into leather results in the creation of 200 kg of leather with chromium content of 3 kg, 250 kg of non-tanned protein waste and 200 kg of tanned waste with chromium content of 3 kg. The tanning process generates up to 50,000 kg of waste water containing 5 kg of chromium. Consequently, only 20 % of collagen raw material is valorized by leather manufacturing processes and about 73 % of chromium ends as waste (primary leather waste). Another kind of leather waste is the so-called manipulation waste (secondary waste) generated during processing of the stabilized leather into final leather products, e.g. shoes and other leather goods. These products, when worn-out and disposed of, represent the final stage of leather waste (tertiary waste). Due to the content of chromium, all the above named kinds of waste are potentially hazardous. While chromium III is harmless, chromium VI is highly toxic and carcinogenic. It is considered a causative agent of cancer of urinary bladder, renal pelvis and kidney. The main problem of all kind of chrome-tanned leather waste lies in a possibility of spontaneous oxidation from CrIII to CrVI. The answer is the free enthalpy change of the oxidation reaction. In the case of the oxidation from CrIII to CrVI the value of free enthalpy is negative, which implies that the spontaneous oxidation can occur. While there are many ways of disposal of the primary waste and we can consider this problem solved, disposal or further processing of tertiary waste, especially old shoes, still represent a technological challenge. In co-operation with the USDA

* Emails: kolomaznik@fai.utb.cz, uhlirova@fai.utb.cz

Eastern Regional Research Center, Wyndmoor and the Texas Technical University in Lubbock we have developed chemical methods for separation of chromium from the collagen part in worn-out leather products. We have also found applications for both the protein part and separated chromium.

1. Introduction

Compounds of trivalent and hexavalent chromium are widely used in many areas of human activity. Directly and most frequently we meet chromium compounds when using products of shoe-manufacturing and textile industry.

People use products of the leather-processing industry on a daily basis. These include especially shoes, leather and textile goods; we normally encounter leather products even in both public and private transport. Although the majority of chromium present in leather products is in its trivalent form, numerous reports [1-8] point out the possibility of the presence of hexavalent chromium in leather goods. A considerable risk for human health lies especially in wearing shoes which are in direct contact with human skin. For this reason it is necessary to deal with the minimization and waste management of hazardous residues containing Cr (VI) in a complex way, it means to focus on the recycling of both trivalent and hexavalent chromium due to a considerable thermodynamic instability of trivalent chromium compounds (see further in the text).

The primary raw materials for final products is hide from animals from slaughter houses and hide from game – i.e. waste from the meat industry, which is processed in tanneries and turned into leather. Therefore, the tanning industry can be considered one of the first industries to use and recycle secondary raw materials.

Although the tanning industry is environmentally important as a principal user of meat industry waste, the industry is perceived as a consumer of resources and a producer of pollutants. Processing one metric ton of raw hide generates 200 kg of final leather product (containing 3 kg of chromium), 250 kg of non-tanned solid waste, 200 kg of tanned waste (containing 3 kg of chromium), and 50,000 kg of waste water (containing 5 kg of chromium) [1]. Thus, only 20% of the raw material is converted into leather, and more than 60% of the chromium ends up as solid or liquid waste.

During the production of leather goods, especially shoes, manipulation waste is produced, which makes about 15-20% of the entry material – leather. The last kind of waste are used leather products which have lost their utility value.

2. The Possibility of Oxidation of Cr (III) to Cr (VI)

The basic question is the possible oxidation reaction from chromium III to chromium VI. In basic solutions, the oxidation of CrIII to CrVI by oxidants such as peroxides and hypohalide occurs with ease [2]. Such strong oxidation conditions are realized in the process of the sterilization of drinking water. This is the first threat to human health and life. Rain (especially acid rain) can leach chromium III from waste dumps, and soluble salts can then reach sources of drinking water. During the sterilization process by ozone or hypochloride, chromium III is converted into chromium VI and reacts with magnesium and calcium ions occurring in drinking water to produce carcinogenic magnesium and calcium chromate or dichromate salts. Another problem concerns the possibility of oxidation of CrIII into CrVI in gentle conditions by air in the wide range of pH. Principally, oxidation can be realized after the following equations:

$$2Cr_2O_3 + 8OH^- + 3O_2 \rightarrow 4CrO_4^{2-} + 4H_2O$$

in alkali medium, and

$$2Cr_2O_3 + 2H_2O + 3O_2 \rightarrow 2Cr_2O_7^{2-} + 4H^+$$

for acid medium.

The probability of a spontaneous oxidizing process of CrIII into CrVI is real as the Gibbs Energy values for both reactions show. The published thermodynamics functions (Gibbs energies of formation) are shown in Table 1 [3]:

Table 1. Gibbs free energy of reaction components

Component	**Gibbs Energies of Formation** $\Delta Gf^{\circ}\,kJmol^{-1}$
Cr_2O_3	-1053
OH^-	-157.28
O_2	0
CrO_4^{2-}	-727.85
H_2O	-228.61
$Cr_2O_7^{2-}$	-1301.2
H^+	0

We calculated Gibbs Energy for both reactions:

$$\Delta Gf^0 = -4\times 727.85 - 4\times 228.61 - (-2\times 1053 - 8\times 157.28) = -461.6 \quad kJmol^{-1}$$

for alkali medium and

$$\Delta Gf^0 = -2\times 1301.2 - (-2\times 1053 - 2\times 228.61) = -39.18 \quad kJmol^{-1}$$

for acid medium.

The negative values of both thermodynamic functions prove the possibility of spontaneous oxidation within a wide range of pH. This fact is more dangerous than in the previous case because we do not know under which conditions CrIII may be converted into CrVI. It is a question of the kinetics of the above mentioned reactions.

Leather and leather products may sometimes contain traces of hazardous CrVI, although only CrIII compounds are used in the tanning process. There are some possible sources for the occurrence of CrVI in the leather-tanning process. CrVI may possibly be present as a contaminant in the tanning agent. Fat-liquoring with unsaturated fatty acids and fish oils may, after photo-ageing with UV light or thermal treatment (dry heating over 80°C), possibly lead to the oxidation of CrIII. Storage of fat-liquored leather at a relative humidity above 35% may result in CrVI formation. Use of alkaline glues in shoe production may also provoke the formation of CrVI.

3. Potential Dangers of Hexavalent Chromium in Connection with Leather Goods

The primary routes of entry for animal exposure to chromium compounds are inhalation, ingestion, and for hexavalent compounds absorption through the skin. This last route is the most significant. Most hexavalent chromium compounds are readily absorbed, are more easily soluble than trivalent chromium in the pH range 5-7, and react with cell membranes. The target organ for acute systematic toxicity is the kidney. Usually, poisoning by CrVI results in acute tubular necrosis of the kidney, a reported cause of death [2]. Prolonged contact with certain chromium components may produce allergic reactions and dermatitis in individuals [2]. There is evidence that hexavalent chromium may be carcinogenic. The National Institute for Occupational Safety and Health has classified the chromate and dichromate salts of calcium and magnesium (elements occurring in soil and drinking water) as carcinogenic compounds [2]. The key to the hexavalent chromium's mutagenicity and possible carcinogenicity is the ability of this oxidation state to penetrate the cell membrane.

There is another aspect of CrVI that may have both health and environmental impact. Although only minor attention has been paid to this problem so far, there is increasing evidence that in the near future this problem should be taken into account. As it has already been mentioned above, one of the potential entrances of CrVI into human body is absorption through the skin. This may possibly happen, for example, when wearing shoes on bare feet. Usually, shoe production is under strict control, but a considerable amount of shoes have been imported within the last two decades from countries in which the chromium content is not always carefully monitored, and this import is rapidly increasing. These shoes thus represent not only a potential threat to the users, but also become an environmental issue when worn out and dumped in the open air.

As a reported carcinogen affecting kidney and urinary tract [2], CrVI can be possibly connected with some cases of neoplasms. The fact that malignant tumours of these organs may be caused by CrVI compounds is also pointed out in literature [4,5]. This, together with a potential entrance of CrVI into human body through skin, makes a combination that should not be ignored.

The following figures (Figures 1-6) [5], presented with kind permission of the Institute of Health Information and Statistics of the Czech Republic, show time developments of the incidence of urinary bladder, renal pelvis and kidney neoplasms. The symbol ASR – *age standardized rate* - in the y-axis of the figures means a summary measure of a rate that a population would have if it had a standard age structure. Standardization is necessary when comparing several populations that differ with respect to age because age has such a powerful influence on the risk of cancer. The symbol (W) expresses that the numbers have been adjusted to the World Standardized incidence rate. It is expressed per 100,000 persons.

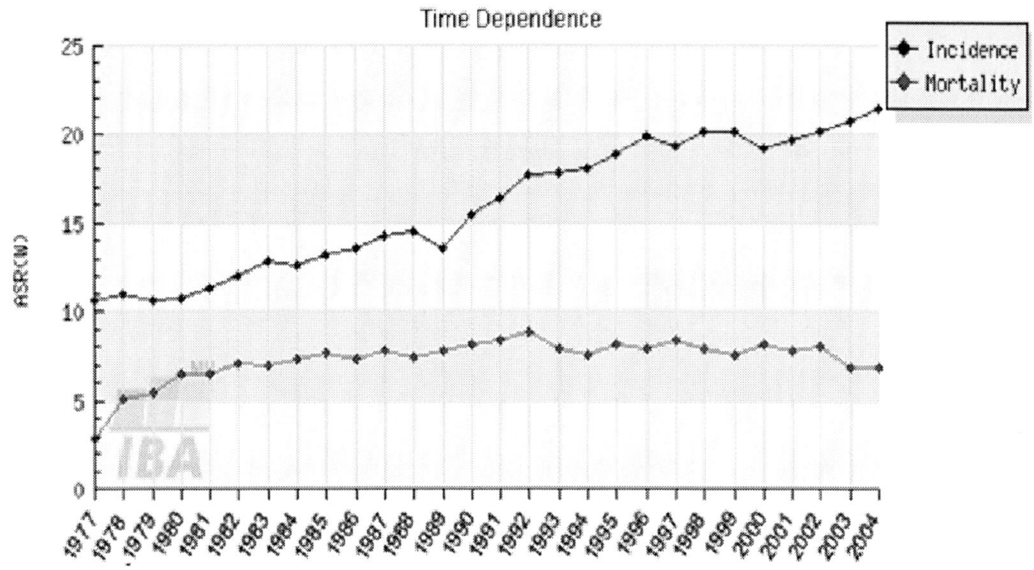

Figure 1. Time development of mortality rate and incidence in the urinary bladder (men).

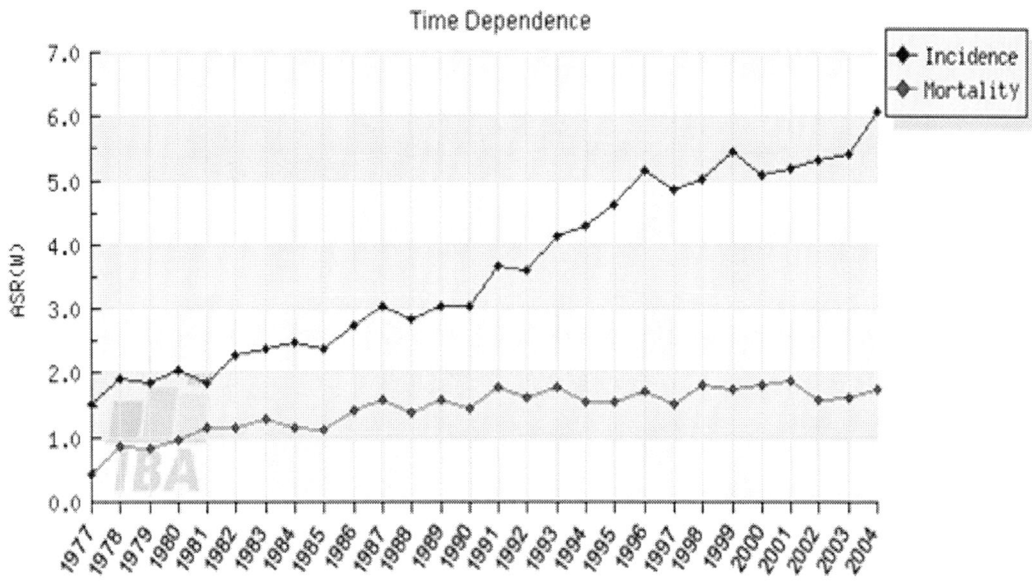

Figure 2. Time development of mortality rate and incidence in the urinary bladder (women).

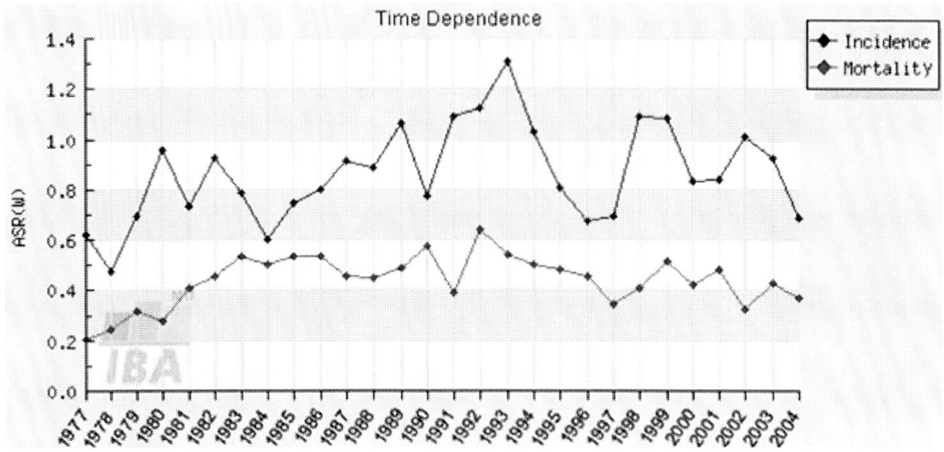

Figure 3. Time development of mortality rate and incidence in the renal pelvis (men).

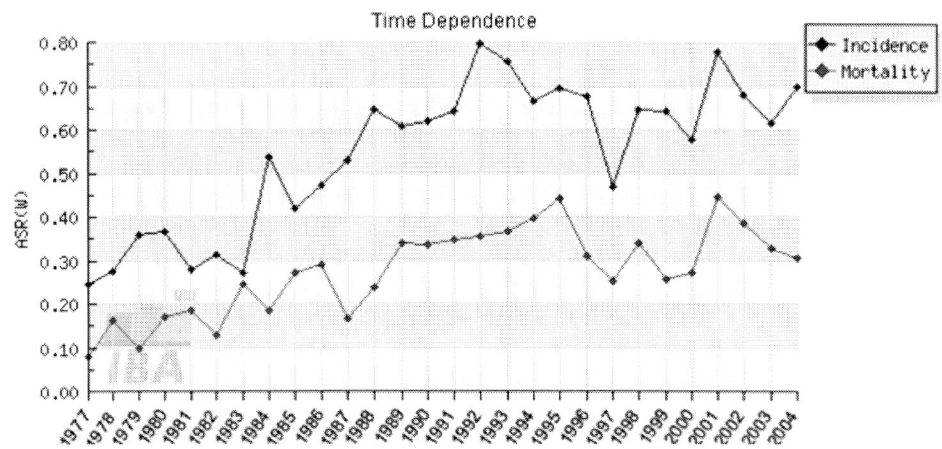

Figure 4. Time development of mortality rate and incidence in the renal pelvis (women).

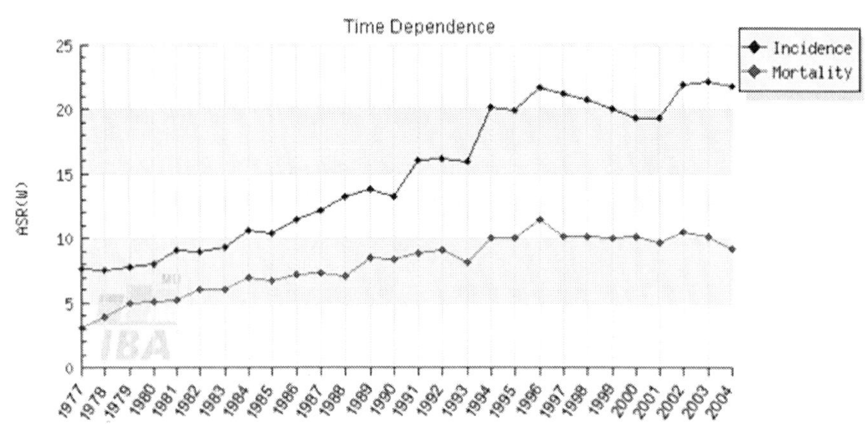

Figure 5. Time development of mortality and incidence in the kidney, excluding the renal pelvis (men).

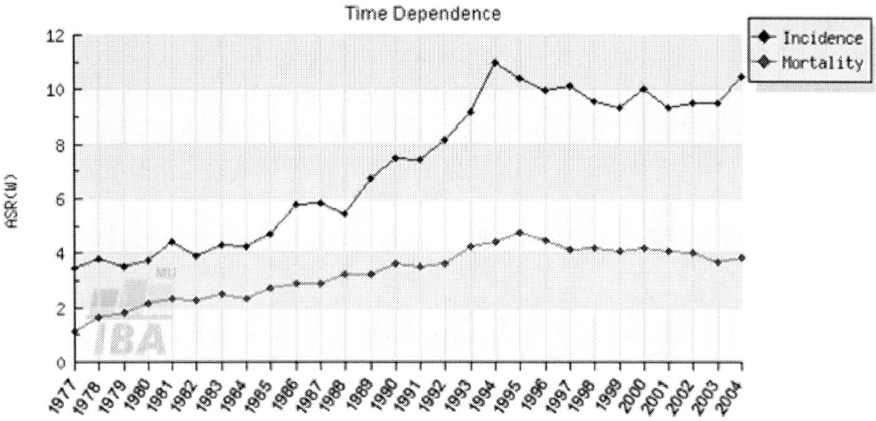

Figure 6. Time development of mortality and incidence in the kidney, excluding the renal pelvis (women).

The following Figures 7-8 show the overview of footwear production, import and consumption during the last two decades in the USA and the Czech Republic, respectively.

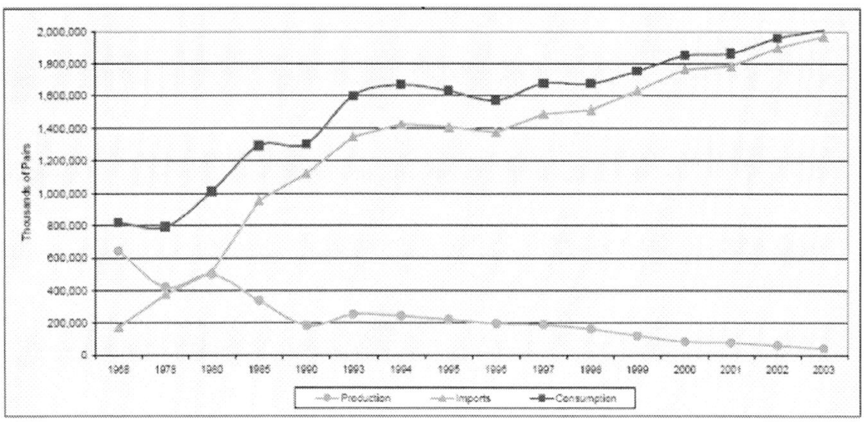

Figure 7. The overview of footwear import in the USA in the years 1990-2004 [6].

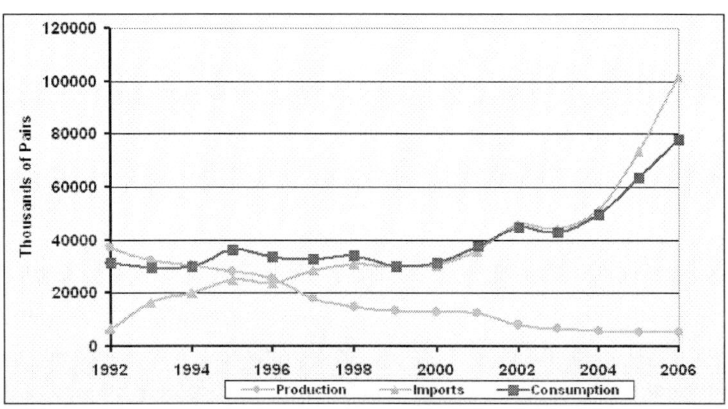

Figure 8. The overview of footwear import in the Czech Republic in the years 1990-2006 [7].

It is interesting to compare the percentage growth in the number of neoplasms in the urinary bladder, the renal pelvis and the kidney excluding the renal pelvis. As shown in Figures 1-6, the growth is 2-3 times higher in women than men. One of the possible explanations might be the fact that women use socks 10 times less often, while socks act as a barrier against the sorption of trace amount of hexavalent chromium from leather by the skin surface.

4. Existing Solutions for Chromium Tanned Waste

A) Waste Dumps

The simplest solution is to put the waste from the leather-processing industry on waste dumps. For example, in the Czech Republic, over 1.5 million tons of chrome-tanned waste is stored on various dumping sites. Considering the danger of the leaching of chromium salts by acid rain, with a potential danger of leakage into drinking water sources and subsequent oxidation of CrIII into CrVI during disinfection, this solution is the least suitable from the point of view of environmental protection.

B) Combustion

An advantage of combustion of chrome-tanned waste from the tanning industry is the relatively high content of organic base which produces heat in the process of combustion. A disadvantage is the total oxidation of trivalent chromium into hexavalent chromium, whose calcium chromate is a carcinogenic compound. The issues of combusting chrome-tanned wastes are also pointed out in Ref. [8]. A perfect separation of combustion gasses and ash is necessary as well as a safe disposal of the ashes as a highly hazardous waste. It is also necessary to pay attention to the increased content of nitrogen oxide which is created during the combustion of collagen proteins.

C) Pyrolysis or Carbonization

High-temperature decomposition of chrome-tanned wastes that produces chrome green of high quality is presently being intensely researched. However, practical industrial implementation has not been carried out yet.

D) Chemical Treatment

There are basically two possible ways of chemical treatment of chrome shavings – by oxidation or hydrolyses. In the oxidation process, in alkaline environment, chromium salt is liquidized by oxidation by hydrogen peroxide, and the chromate that is created is leached from the heterogenic reaction compound by water. The main disadvantage of this method, apart from the above mentioned creation of carcinogenic chromium salt, is also the production of a considerable amount of diluted solutions of chromium salt, which makes the process non-economical. However, an advantage is a high-quality gelling protein with high bloom value. Hydrolysis of chrome-tanned wastes in acid or alkaline conditions is more environmentally-friendly as no hexavalent chromium is created due to the reaction time.

One of the numerous possible solutions to the problem of chrome-tanned wastes is their enzymatic digestion which has been dealt with in detail in a number of publications [9-12]. When researching processing of chrome-tanned wastes in the Czech Republic, we decided to utilize the enzymatic dechromation technology worked out at the USDA in Philadelphia [13]. The technology was adapted to conditions in a factory that manufactures auxiliary tanning agents. Furthermore, we modified the American process by using volatile organic bases such as isopropylamine, diisopropylamine, cyclohexylamine, ammonia, and others. The use of volatile amines produces a higher quality of protein hydrolysates and chromium sludge.

E) Direct Application of Solid Tannery Waste

Due to a relatively large specific surface of solid tannery waste (approximately 10 $m^2.g^{-1}$ – [14]), it is possible to use this waste for decontamination of textile industry waste water, which contains organic dyes. This method has been proposed in [15] to remove methylene blue and reactive red textile dye from aqueous residues. High sorption capacity of solid tannery waste (80 $mg.g^{-1}$ and 163 $mg.g^{-1}$ for methylene blue and red textile dye, respectively) makes this method an interesting alternative to commercial adsorbent materials.

Another promising application of solid tannery waste has been published in [16] – the waste serves as an adsorbent for removal of toxic Cr(VI) and As(V) from aqueous media. The sorption capacity is described by Langmuir and Freundlich isotherms. The amount of adsorbed Cr(V) and As(V) (133 $mg.g^{-1}$ and 26 $mg.g^{-1}$, respectively) shows a great potential for future use. The sorption capacity of tannery waste (chrome shavings) has been also used in [17] for removal of motor oils, other oily waste and hydrocarbons from water. The shavings are capable of adsorbing approximately 7 times their weight in oils and hydrocarbons.

5. Processing Manipulation Waste and Used Leather Goods

However, the chemical procedures elaborated for processing primary wastes fail in cases dealing with secondary wastes from the production of leather products, especially shoes. The presence of CrVI in worn shoes was brought up in works [18]; more than 90% of leather goods are tanned with alkaline chromium salts [19]. We were looking for the solution to this problem in cooperation with USDA, ERRC – Wyndmoor U.S.A. and the Texas Technical University in Lubbock for manipulation wastes in the NIKE Company. We elaborated a hybrid technology in three steps. In the first step, synthetic resin (polyurethane, polyacrylate) is decomposed in a strongly alkaline environment of potassium hydroxide (pH~12) resulting in a strongly stabile emulsion, which cannot be broken up due to the stabilization by hydrolysing products. In the second step, after adjustment to pH ~ 9 by phosphoric acid, a proteolytic enzyme is dosed and enzymatic hydrolysis is carried out, decreasing the molecular weight from the original 100 kDa to 10 kDa. In the last step, pH is adjusted again to the neutral level 7 and the resulting suspension is relatively easy to filtrate. The filtrate is vacuum-evaporated to 20-40% solid matter content and used as NK (Nitrogen-Potassium) or NPK (Nitrogen-Phosphorus-Potassium) – organic fertilizer. The titanium-chromium sludge can be used as inorganic pigment in glass or ceramic industry.

Used leather products, especially old shoes, are at present most often taken to waste dumps together with communal waste or burnt, which is, as we have shown, not an ideal solution. Much more suitable is selective collection of used leather goods, just as is the case with used paper, plastics, glass and metal. This solution would enable separate processing especially of the chrome-tanned wastes.

In order to prevent health risks, several legislative regulations have been issued, limiting the CrVI content in leather products to the maximum of 2 to 10 $mg.kg^{-1}$, while for eco-labelled leather, the CrVI concentration should not exceed 0.5 $mg.kg^{-1}$. To comply with these legislative requirements, reliable measurement of CrVI content in leather is of extreme importance. However, with increasing shoe imports in the last few decades, it seems that CrVI health threats should be again taken into account. The imported shoes do not always comply with the required quality and chromium content is not always carefully monitored. In our study, we point out the possible connection between CrVI and some types of cancer. The goal of that part of our paper is to show that the role of CrVI as carcinogen cannot be neglected, and when there is an increasing incidence of some kinds of cancer, it is necessary to count all the possible causes.

The second part of our paper deals with the still topical need for economical and environmentally friendly processing of chrome tanned leather waste. We particularly focus on manipulation waste generated during leather processing in the final products of leather industry goods and used leather goods, such as old shoes. As there has not yet been a satisfactory technology developed for manipulation of wastes so far, this paper represents a pilot study in this field. We have proposed and industrially tested a hybrid three-step technology, the products of which can be applied to the agriculture and ceramic industry. The filtrate especially seems to be very economically interesting. Due to the addition of potassium hydroxide and phosphoric acid during the hybrid process, the concentrated and dried filtrate represents a good quality NPK fertilizer. In the semi-pilot testing, the dried filter cake containing chromium oxide, titanium oxide (Cr_2O_3, TiO_2) (in the case of the NIKE Company) and organic residue was tested for the dying of enamel in Vietnam with promising results.

In addition to the manipulation waste, the hybrid technology seems to be an effective solution also for used leather products, e.g. old shoes.

CONCLUSION

This study concerns the potential health risks of hexavalent chromium in the waste generated by leather industry and the possibilities of economical and environmentally friendly processing of this waste. Special attention is given to so-called secondary waste, i.e. manipulation waste generated during processing leather into the final products of leather industry and used leather goods. A new three-step hybrid technology of processing manipulation waste is presented and tested under laboratory, pilot-scale and industrial conditions. The filtrate obtained by this technology can be used in agriculture as a good quality NP or NPK fertilizer. The solid product can serve as an inorganic pigment in the glass and ceramic industry. The technology also seems to be an effective solution for hydrolysable parts of used leather products, e.g. old shoes. For this reason, the authors propose selective collection of this kind of waste.

ACKNOWLEDGMENT

The project was financed by the Ministry of Education of the Czech Republic, VZ MSM 7088352102.

REFERENCES

[1] Hüffer, S. & Taeger, T. (2004). Sustainable leather manufacturing - a topic with growing importance. *J Am Leather Chem Assoc,* 99, 423–428.

[2] Kirk, O. (1992). *Encyclopedia of Chemical Technology* (4th edition), John Wiley & Sons Inc., New York, 1992.

[3] Dean, A.J. (1992). *Lange's Handbook of Chemistry* (14th edition). Mc. Graw-Hill, Inc., New York.

[4] NCBI (National Center for Biotechnology Information), 2007. http://www.ncbi.nlm.ni h.gov/sites/entrez?db=pubmed.

[5] UZIS CR (Institute of Health Information and Statistics of the Czech Republic), 2007. Cancer Incidence 2004 in the Czech Republic. UZIS CR in cooperation with National Oncological Register of the Czech Republic (NOR CR).

[6] Mangione, P.T. Footwear Distributors and Retailers of America. *World Footwear Congress*, Logroño, Spain, 2007.

[7] Czech Statistical Office, The Overview of Production, Export, Import and Consumption of Footwear in the Czech Republic in the Years 1990-2006, Prague, 2006.

[8] Singh, J., Carlisle, D.L., Pritchard, D.E. & Patierno, S.R. (1998). Chromium-induced genotoxicity and apoptosis: relationship to chromium carcinogenesis (review). *Oncol Rep*, 5, 1307-1318.

[9] Kolomazník, K., Langmaier, F., Mládek, M. & Taylor, M.M. Industrial Treatment of Chrome Tanned Solid Waste. *IULTCS Centenary Congress*, London, 1997.

[10] Kolomazník, K., Mládek, M., Langmaier, F., Taylor, M.M., Diefendorf, E.J., Marmer, W.N., Brown, E.M. & Tribula, L. CR Patent 280 655, February 2nd, 1996.

[11] Taylor, M.M., Diefendorf, E.J., Marmer, W.N. & Brown, E.M. U.S. Patent 5, 271, 912. December 21st, 1993.

[12] Taylor, M.M., Diefendorf, E.J., Na, G.C. & Marmer, W.N. U.S. Patent 5, 094, 946, March 10th, 1992.

[13] Kolomazník, K., Taylor, M.M., Diefendorf, E.J. & Brown, E.M. Optimization of Time and Energy Consumption in the Enzymatic Treatment of Chrome Tanned Solid Waste. *XXIIIth IULTCS Congress*, Friedrichshafen, 1995.

[14] F. Langmaier, verbal communication. Tomas Bata University in Zlin, Faculty of Technology, Zlin, Czech Republic, E-mail: langmaier@ft.utb.cz.

[15] Oliveira, L.C.A., Gonçalves, M., Oliveira, D.Q.L., Guerreiro, M.C., Guilherme, L.R.G. & Dallago, R.M. (2007). Solid waste from leather industry as adsorbent of organic dyes in aqueous-medium, *J Haz Mat*, 141, 344-347.

[16] Oliveira, D.Q.L., Gonçalves, M., Oliveira, L.C.A. & Guilherme, L.R.G. (2008). Removal of As(V) and Cr(VI) from aqueous solutions using solid waste from leather industry. *J Haz Mat*, 151, 280-284.

[17] Gammoun, A., Tahiri, S., Albizane, Azzi, A.M., Moros, J., Garrigues, S. & de la Guardia, M. (2007). Separation of motor oils, oily wastes and hydrocarbons from contaminated water by sorption on chrome shavings. *J Haz Mat*, 145, 148-153.

[18] J. Font, J., Cuadros, R., Reyes, Mª R. & Marsal, A. (1999). Factors which influence the formation of Cr (VI) during ageing. *Leder Piel*, 17, 54-59.

[19] Soler, J., Morera, J.M., Almendros, J.F. & Martínez, M. Variaciones en la curtición al cromo [Variations in the Chrome Tanning]. *Proceedings from the Symposium Internacional de Curtición al Cromo*, p. 73, Igualada, Spain, 1992.

In: Management of Hazardous Residues Containing Cr(VI) ISBN: 978-1-61470-637-3
Ed: Maria José Balart Murria

Chapter 13

STUDY OF STABILIZATION/SOLIDIFICATION PROCESS OF TANNERY WASTE CONTAINING CHROMIUM

Carolina Afonso Pinto[1]; Pedro Mauricio Büchler[1] and Jo Dweck[2]
[1] Chemical Engineering Department, Polytechnic School, University of São Paulo, Brazil
[2] Chemistry School, Federal University of Rio de Janeiro, Brazil

ABSTRACT

The research on treatment of industrial wastes has been increased, as is also the case of the tannery industry, which generates considerable quantity of effluents with high concentrations of chromium, a hazardous waste that requires treatment. Nowadays, the mostwell-known treatments that have been used are either not efficient or have a high cost. The research developed by the authors proposes the treatment by stabilization solidification process in cement. For a more complete study, clays were also used with cement due to their property of adsorbing heavy metals containing products present in the waste. The process allows the junction of three actions: waste components are adsorbed by the clays decreasing their mobility; the heavy metals are transformed in respective hydroxides decreasing significantly their solubility in water and increasing their stability with respect to lixiviation by acidic rain and the whole system remains encapsulated in the solidified hydrated cement. The tannery waste solidification/stabilization study was performed with Portland cement original and modified Brazilian clay and commercial clays. In the experimental research waste characterization was done by solid-state nuclear magnetic resonance spectroscopy, and X ray diffraction. Analyses of the solidified samples and of the solidification process were done by thermogravimetry and by conventional and non-conventional differential thermal analysis, leaching and compressive strength tests. The behavior of the waste in the cement matrix was analyzed. The developed process shows a good efficiency, as shown by leaching tests. The chromium contents in lixiviates in different operating conditions are below the limit demanded by the Brazilian environmental regulation, which indicates that the stabilization/solidification process has potential to be applied to tannery waste containing chromium.

1. INTRODUCTION

A waste is defined as unwanted byproducts of a manufacturing process, most often with no value. The classification of a solid waste as either nonhazardous or hazardous as defined under the Resource Conservation and Recovery Act (RCRA) has a significant impact on the cost of disposal. Organic solid wastes when incinerated present a higher cost than stabilization + landfill for inorganic solid wastes [Mulholland and Dyer, 1999].

The tannery wastewater contains organic and inorganic compounds. Among the inorganic compounds present in high concentrations, there are ammonia, sulfides and chromium compounds that are inhibition factors for any kind of biological treatment [Vlyssides and Israilides, 1997] These come from the compounds used in the process to prevent decomposition of the leather and have been removed from the wastewater in decanters after precipitation

The trivalent chromium is less toxic than hexavalent chromium, classified as a human carcinogen if ingested. According to Avudainayagam et al. [2001], there is a little potential for hexavalent chromium to be present in ground waters from tannery waste disposal or sewage sludge to soils. Normally, it is expected to find trivalent chromium. Trivalent chromium may oxidize to hexavalent chromium only in the presence of Mn(IV) in soils [Bartlett and James apud Avudainayagam et al., 2001; Milacic and Stupar, 1995].

Dermal contact by chromium can cause primary irritation and ulceration as well as allergic eczema. Inhalation can cause nasal irritation, and septal perforation.Pulmonary irritation, and bronchogenic carcinoma, may result from breathing chromate dust. Ingestion causes violent gastro-intestinal irritation with vomiting, diarrhea [The Merck, 2001].

Ates et al. [1997] studied the characteristics of two major tannery complexes in different districts in Turkey monitored for more than one year. The first district, containing 107 tanneries processing cattle hide and sheepskin, presents a wastewater flow of 13000-14,000$m^3.d^-$. The second district contains 87 tanneries processing predominantly sheepskin and with an effluent with a similar magnitude as the first district. Table 2 shows the characterization results for some analyzed parameters.

Cr(III) is more stable and exhibits a tendency to form inert complexes and is an essential trace nutrient for humans involved in glucose metabolism [Balasubramanian and Pugalenthi, 1999].

There are a number of treatment technologies available for soils that have elevated levels of Cr other than conventional remediation by excavation and off-site disposal. The technologies applicable to a particular Cr site depend on the cleanup goals, the forms of Cr present in it, and the volume and physical/chemical properties of the Cr-containing soils [Pera et al., 1997].

Stepniewska and Bucior [2001] studied the tannery waste lagoon in Lubartów, Poland a source of chromium contamination. The tannery waste lagoon was not covered from the top and its bottom was not sealed. The annual precipitation in this region is about 600mm. Soil and plant samples were analyzed and water from a farm situated about 1-2km from the lagoon as well. The results showed high concentrations of chromium in drinking water from the farm. The chromium concentration in plants was not uniform. The concentration in stalks was 10 times higher than in the ears. The chromium absorption from the soil by plants depended of the plant species.

A study of animal's contamination by chromium was performed by Walsh and O'Halloran [1998]. They studied the behavior of depuration, digestion at a cellular level of a population of mussels transported to an estuary, which was receiving tannery effluents from river Colligan, Ireland. The chromium speciation in the effluent from the tannery consists entirely of Cr (III), with no evidence of oxidation in the estuary. The tannery industries that discharge their effluent in the studied site remove chromium from the primary precipitation treatment. Even with this removal, chromium is found in the effluent. The study showed high concentrations of chromium in the mussels transported to the estuary after 12 months. Depuration was slower but even with the contamination part of the chromium was digested and eliminated from the organism.

Lai et al. [2001] performed a study of soils, plants and human hair impacts of metals like cadmium, cobalt, chromium, copper, manganese, nickel, lead and zinc from anthropogenic activities in Nam Dinh Province, Vietnam. The results showed that high concentrations of theses metals were found in human hair and plants, indicating the toxic dimensions of pollution on people and vegetables at this area.

In leather tanning wastewater much of the chromium appears complexed with proteins and as inorganic compounds like oxide, sulfate or hydroxide [Avudainayagam et al., 2001 and Vieira et al., 2001]. In Brazil, this solid waste after removal from the wastewater consists of a hazardous material that needs treatment and appropriate disposal and has been stored waiting for a proper solution. It contains organic and inorganic compounds and several studies have been performed on characterization, treatments and contamination of tannery waste [Stepniewska and Bucior, 2001; Walsh and O'Halloran, 1998; Lai, et al., 2001; Dweck, 2000]. The application of stabilization/solidification process in cement intends to be an effective treatment for its final disposal and it is a potential method for preventing Cr leaching from contaminated soils, which aims to either bind or complex the components of hazardous wastes within the solid cement matrix.

Some treatments have been studied to application of tannery waste. Dweck et al. [2000] have studied, by differential thermal analysis, the behavior of sodium bentonite and the waste in cement. Pinto et al. [2005a, 2005b, 2004a] studied the mechanical behavior, the chromium leached from the system when 10% of tannery waste is applied to cement by differential thermal analysis.

Besides chromium influence in the cement hydration there are some studies referring to other compounds that may interfere in these reactions, like ammonium chloride, sulphate and carbonate that can react with the concrete [Jauberthie and Rendell, 2003; Irassar et al., 2003; Hills and Pollard, 1997].

Solidification/stabilization (S/S) is a potential method for preventing Cr leaching from contaminated soil.

Stabilization can be defined, according to Conner and Hoeffner [1998] as a technique that reduces the hazard potential of a waste by converting the contaminants into their least soluble, mobile or toxic forms. Solidification corresponds to technique that encapsulates the waste in a solid of high structural integrity.

Biczok apud Jauberthie and Rendell [2003] reported that ammonium chloride, sulphate and carbonate can react with the concrete. This influence of sulphate attack was also studied by Irassar et al. [2003]. The sulphate attack normally occurs when the system structures is exposed to environments like seawaters, rivers or groundwater [Tikalsky et al. 2002].

The carbonation in the cement system is capable of inducing setting and strength development in waste solidified bodies where hydration is significantly retarded. Slegers and Rouxhet apud Hills and Pollard [1997] showed that the process where carbonation proceeds depends on the retardation of hydration and the type of used waste. Carbonation reactions can occur with ettringite, calcium silicate hydrate (C-S-H gel), forming silica gel and calcium carbonate, and calcium hydroxide (CH).

According to Lea [1971], the products obtained from Portland cement blends are complex to be characterized due to significant changes in physical and mechanical properties that can be taking place over periods of many months or even years.

The chemistry related to the cement hydration in a S/S process with waste is complex as well. Some authors believe that when cement is mixed with water, the aluminate phases are capable of producing ettringite in less than 10 minutes after mixing cement with water. A gelatinous layer is formed around cement particles and at this step the waste modifies the system. These modifications can be by complexation or precipitation and a reduction of production of calcium silicate hydrate (C-S-H) and calcium hidroxyde (CH) may occur. The effects in the cement can be accelerating or retarding the hydration depending of the waste applied at the matrix [Hills and Pollard, 1997].

Weng et al. [1996] studied some chemical interactions between Cr(VI) and concrete particles in an aqueous system. They used ordinary Portland cement concrete blocks after being crushed and milled to a size fraction between 0,18-0,5mm. The waste was a chromite ore-processing residue (COPR), containing 2-5% of chromium. According to the authors, the concrete adsorbs Cr(VI). The pH is considered an important parameter affecting the chemical interactions. Below pH 3.5 the removal of Cr(VI) is controlled by a reduction reaction obtaining a Cr adsorption range of 0.1-0.8mg/g at pH 3.5. Between pH 4-6 there is the highest adsorption from 0.4 to 1.4mg/g. At alkaline pH the degree of Cr(VI) adsorption decreases due to weakening surface complex formation.

Research referring to the development of different types of cement has been performed continuously. Elkhadiri et al. [2002] showed the influence in the mechanical properties of cement when mixed with fly ash and limestone. According to the authors, the addition of fly ash extends the setting time of the cement while the limestone, when substitutes clinker, presents better behavior in the early days of hydration. Li and Wenquan Liang [2002], developed new cement materials with blast furnace slag, fly ash and limestone. Tsivilis et al. [2000 and 1999] studied the cement behavior when clinker, gypsum and limestone are mixed with different fineness and contents. The results showed that up to 10% of limestone addition the compressive strength is similar with samples of pure cement and one of the reasons is the particle size.

The use of clays in the sorption of hazardous waste has been studied. Some recent researches are reported in the work of Pinto et al. [2004a]; Vianna et al. [2004]; Souza Santos [1992]; Boyd et al. [2004]; Tillman et al. [2004]; Jose et al. [2003]; Chun et al. [2003], Celik et al. [2000]; Valenzuela-Diaz [2001]; Nennemann et al. [2001]; Nzengung et al. [1997, 1996] e Lo [1996].

Some studies performed by Volzone [1999, 1998, 1997] showed the efficiency of chromium adsorption by clays. In a work conducted in 1998 the retention capacity of chromium by kaolinite from Georgia was analyzed, also vermiculite from Arizona and sodium bentonite from Wyoming. The results showed that sodium bentonite presented more retention of chromium in its structure than kaolinite and vermiculite. In her previous work,

Volzone [1995] studied the chromium adsorption in smectites by specific surface area analysis, which showed the increase of SSA by the increase of chromium concentration in the mixture; by XRD analysis the d_{001} basal spacing was increased with the chromium presence showing that the intercalation occurred. It was also observed that in supernatants there were sodium, potassium and calcium evidencing the cation exchange of these metals by chromium in the clay.

Calvanese et al. [2002] studied the S/S process of tannery waste using organoclay as a pre-solidification adsorbent. The organoclay was prepared in laboratory, from an Italian commercial bentonite. The organic material was initially removed and the final sludge was kept in contact with the organoclay. The system was then solidified in two different materials: pozzolanic cement and a mixture of pozzolanic cement and blast furnace slag in a mass ratio of 3:7. According to the authors, the organoclay presence increased the efficiency of the S/S process and the system with pozzolanic cement had better results from environmental and technological points of view when compared to the mixture with blast furnace slag.

Some additives of organoclays and sodium bentonite were added to the cement matrix based on previous researches. [Pinto et al., 2005, 2004b; Boyd et al., 2004; Tillman et al., 2004; Valenzuela-Diaz, 2001; Nennemann et al., 2001; Dweck et al., 2001 and Volzone, 1999].

According to U.S. EPA [1992] the TCLP method consists of trying to determine the mobility of both organic and inorganic analytes present in liquid, solid, and multiphase wastes.

Metal ion leachability is controlled by solubility, diffusion and advection transport rates through connected porosity. Cement solidified wastes also produce a pH environment in which some metals have low solubilities [Conner apud Cheeseman et al. 1998].

Leaching is a phenomenon where hazardous or other undesirable soluble constituents are removed from the waste into the environment via the leachate. This phenomenon is quantified in terms of the final concentration of the constituent in the leachate. This is because the concentration determines the constituent's effects on living organisms, especially humans [Spence, 1992].

Leaching tests are invariably accelerated examinations of waste stability and are usually aggressive, providing conservative assessments of leaching performance to protect the environment [Hills and Pollard, 1997].

Solid-state ^{29}Si nuclear magnetic resonance has been shown to be an important mechanism in the chemistry study of solidified hazardous wastes [Ortego et al. 1991; Janusa et al. 1993].

According to Lambert and Mazzola [2004] and Mackenzie and Smith [2002] NMR analysis exploits the magnetic properties of nuclei to provide information on nuclear structure. An NMR experiment consists of three parts, the preparation of the nuclear spin system by placing it in an external magnetic field, its perturbation by applying a pulse of radiation, and the detection of phenomena accompanying its return to the initial state when the perturbation is removed. The resonance frequencies measured by NMR are normally reported as chemical shifts relative to an external standard compound. The Fourier transform is the mathematical process that converts the time domain function (FID) into a frequency domain function (the spectrum).

Silicon in inorganic materials is most commonly bonded to oxygen. The ^{29}Si chemical shifts of Si-O units are sensitive to the coordination number, those of Si(VI) units occurring in the range –60 to –120ppm, with respect to tetramethylsilane (TMS). The chemical shifts of ^{29}Si spectra are commonly quoted with respect to TMS. Silicate structures may be regarded as being built up to tetrahedral units with varying degrees of polymerization. In NMR it is described with a "Q" notation, which means silicon bonded to four oxygen atoms. The superscript number, from 0 to 4, corresponds to a specific degree that identifies the degree of polymerization. The Q^4 represents silicon bonded through oxygen to four other silicons [Mackenzie and Smith, 2002]. Table 3 summarizes the silicate structures in the ^{29}Si NMR spectra.

Portland cement setting occurs when the cement is hydrated and all the components show varying degrees of reactivity with water, but the most significant hydraulic activity is associated with the tricalcium silicate, which forms a cohesive mixture of calcium hydroxide and calcium silicate hydrate (C-S-H) with water. 29Si NMR can analyze the structure and formation mechanism of C-S-H from tricalcium silicate. When the hydration proceeds some polymerized silica appears, starting with Q^0 during the induction time and this polymerization is increased during the cement hydration process (Mackenzie and Smith, 2002 and Lin et al., 1997).

Hanna et al. [1995] also studied, by ^{29}Si and ^{27}Al NMR spectroscopy, the stabilization/solidification process of industrial wastes containing metals and organic material using Portland cement. According to the authors, metal presence retards the silicate polymerization.

The main factors of compressive strength behavior are: water/cement ratio; incorporated air; type of cement, aggregates and additives. There are some researches referring to incorporation of wastes in cement such as Flores-Velez and Dominguez (2002) who studied the incorporation of zinc-iron oxide nanoparticles from steel smelting industry in Portland cement.

According to Stegeman and Buenfeld [2002] the compressive strength presents important information about the products formed by S/S process. Structure effects, durability and setting are some parameters that may be observed in this analysis.

The unconfined compressive strength analysis presents important information about the products from cement hydration in S/S of wastes with heavy metals, referred to structural mechanical resistance, durability and the retarding or accelerating effects on cement hydration [Ortego et al., 1991].

Research presented in this chapter consists of the study of the stabilization/solidification process when tannery waste containing chromium is used in different contents, referring to Portland cement mass. Different additives were also used: sodium bentonite and organoclay, both used as commercial products and submitted to some changes in the laboratory. The experimental research performed for this study consisted of waste characterization by solid-state nuclear magnetic resonance spectroscopy, and X ray diffraction. Analyses of the solidified samples and of the solidification process were done by thermogravimetry and conventional and non-conventional differential thermal analysis, leaching and compressive strength tests.

Part of this work has been published in some journals and presented in international and Brazilian Conferences at Pinto et al. [2007, 2005c, 2005d, 2005e, 2004c, 2003, 2001].

2. METHODOLOGY

Commercial Portland cement Type II E-32 (identified as C) was used. Sodium bentonite and organoclays were also applied as solidification additives: Wyoming sodium bentonite (S) and a modified sodium bentonite (Na), obtained by cation exchange with sodium carbonate of verde-claro polycationic bentonite (VC-n) from Paraíba State, Brazil [Valenzuela Diaz, 2001].

Commercial organoclay (identified as Ti) and modified Organoclay (identified as O) was also studied. The last one was obtained from the modified sodium bentonite by cation exchange reaction with a quaternarium ammonium salt, the same used for the commercial organoclay, alquil-benzil-dimetil-amonium chloride, using a process developed by Valenzuela-Diaz [2001].

Solid tannery waste from preliminary treatment of wastewater was collected after its industrial drying process, in tannery industry located at City of Franca, São Paulo State, Brazil. In the laboratory the material was dried at 60°C and ground using mortar and pestle until 100% passing through ASTM #100 sieve.

Inductively Coupled Plasma – Mass Spectrophotometer (ICP-MS), Perkin-Elmer Elan 9000, performed chemical analysis of the raw material. The studied compositions are shown in Table 1.

Table 1. Composition of the solidified samples referred to respective cement (C) mass

Composition	Water (%)	Tannery waste (%)	Modified sodium bentonite (%)	Modified organoclay (%)	Commerci al Sodium bentonite (%)	Commercial organoclay (%)
CW	60	-	-	-	-	-
CNa	60	-	5	-	-	-
CO	60	-	-	5	-	-
CS	60	-	-	-	5	-
Cti	60	-	-	-	-	5
CT	60	10	-	-	-	-
CTNa	60	10	5	-	-	-
CTO	60	10	-	5	-	-
CTS	60	10	-	-	5	-
CTTi	60	10	-	-	-	5
CTNaO	60	10	5	5	-	-
CTSTi	60	10	-	-	5	5
CT15	60	15	-	-	-	-
CT15NaO	60	15	5	5	-	-
CT15STi	60	15	-	-	5	5
CT20	70	20	-	-	-	-
CT20NaO	70	20	5	5	-	-
CT20STi	70	20	-	-	5	5

The mass percentages refer to the unhydrated cement mass. To have a good workability the water/cement (w/c) mass ratio was different according to the quantity of waste used. For 10 and 15% in mass of tannery waste samples used, the w/c ratio was 0.6; for 20% in mass of tannery waste sample used, the w/c ratio was 0.7. Each composition was weighted and the dry material mixed initially. When the dry compounds were well-mixed, proportional water mass was added and the system was mixed again manually until a homogeneous paste was formed.

The samples were kept in cylindrical plastic molds, with 5 cm diameter and 10cm height. Each mold with the sample was tapped on the ground to eliminate possible air bubbles. The molds were capped. The samples had a curing time of 28 and 90 days in a wet chamber at 25°C with 100% of humidity. For compressive strength analysis the curing time was 28 days.

2.1. NMR Analysis

Paramagnetic species interfere with nuclear magnetic resonance experiments, and ordinary Portland cement contains enough iron to represent a serious interference [Nestlé, 2000]. Thus for NMR analysis white Portland cement (identified as WC) with low iron content was used. The composition of the white cement was similar to that of Type I Portland, except for iron content.

^{29}Si MAS NMR, using a Bruker BZH 400/89, 9.39Tesla, Oxford Instruments, analyzed the samples. The samples were ground and compacted in zircon rotors. A solid state Chemagnetics CMX Infinity probe was used to rotate the sample at 3.5kHz. The spectrum frequency was 79.4793MHz, pulse width of 5.50μs, the relaxation delay between scans was 2s, spectrum width of 50kHz and line broad factor of 50Hz. Chemical shifts were reported in ppm. The software Mestre-C, supplied by *Mestre C* was used to analyze the results.

2.2. XRD Analysis

Tannery waste was characterized by X ray powder diffraction using a Siemens D5000 diffractometer. The radiation was K-α of copper, with steps of 0.040° (2θ) and time for steps of 0.500 s. The compounds were analyzed according to the d-spacing values, with 2θ between 2° and 70°.

After this period the samples were removed from the molds and ground in agata mortar and pestle to a particle size passing through sieve #100.

The raw samples used in this research were analyzed also by this technique. The samples with exception of organoclays were micronised in ethanol to obtain a particle size up to 5μm using a Micronising Mill, Mc Crone. The samples then were centrifuged for 20minutes at 2000rpm and the liquid phase was removed. The raw samples were dried at 60°C for 24hours before placed in the round molds.

2.3. NCDTA Analysis

Non-conventional Differential Thermal Analysis (NCDTA) was used to study the initial stages of cement hydration. It presents the same basic principle as the conventional Differential Thermal Analysis (DTA), but differs in the fact that there is not an externally controlled heating or cooling process as in DTA. The difference between the sample and reference temperatures (Ts-Tr), which is measured as a function of time, is due to the exothermic reactions that occur within the sample and the reference is at ambient temperature. The higher is the inert content in total mass, the lower will be the instantaneous value of (Ts-Tr) for a same total generated heat. Thus, the total heat evolved up to a time t, on cement mass basis, was estimated by multiplying the integral of the NCDTA curve from t=0 to t=t by the rate between the total mass of the paste and its cement mass.

As shown in Figure 1, each measuring device consists of a polystyrene cup inserted in a Styrofoam foam vessel. Both cup and vessel have their respective leads. The thermocouples are inserted into the samples trough these leads. There are 3 units for samples and one for the reference. Thermocouples are inserted into the sample and reference. A Vernier Lab Pro interface connects the thermocouples to the computer, which acquires and processes the temperature data.

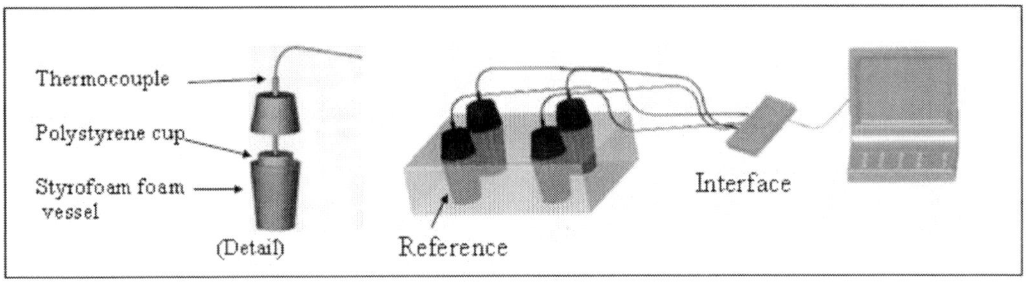

Figure 1. Schematics of the NCDTA system.

Reprinted from Journal of Thermal Analysis and Calorimetry Vol 80, C.A.Pinto, J.Dweck, J.J.Sansallone, F.K.Cartledge, M.E.Tittlebaum and P.M. Büchler. Early stages of solidification/stabilization of stormwater runoff solid residuals in cement, pp. 715-720, Copyright (2005), with permission from Springer.

2.4. DTA/DTG Analysis

Simultaneous thermogravimetry and differential thermal analysis (TG/DTA) analysis were performed from 35°C to 1000°C, after a previous drying step at 35°C, by using 100mL/min of air, as the purge gas, with a heating rate of 10°C/min, in a simultaneous TG/DTA, TA Instruments equipment, model SDT2960. The reference sample was alpha-alumina and a sample mass of about 15 mg was used in each analysis. All results are shown on cement calcined basis. When this basis is applied to thermogravimetric analysis (TG) data of the hydrated cement pastes, one may calculate and compare on the same basis, the mass losses due to dehydration or decomposition steps of the main phases of paste samples with different cement hydration degrees. The final calcined mass of cement paste samples has the same oxide composition than that of the non-hydrated calcined cement [Dweck et al., 2000]. When other aggregates are added, from the cement/aggregate mass ratio for each aggregate and from the respective calcined oxide masses of each raw material, all thermal analysis data was also calculated on a cement calcined basis for the same reason. Details of calculations on this basis are shown by Dweck et al. [2009].

2.5. UCS Analysis

The Standard Test Method for Unconfined Compressive Strength (UCS) of Cylindrical Concrete Specimens used was a Brazilian method identified as NBR 7215/96: Portland cement: Determination of compressive strength. The testing machine was operated with a rate of loading of 0.25 ±0.05 MPa/s. The compressive strength was calculated by dividing the maximum load carried by the specimen during the test by the average cross-sectional area. The stress σ (MPa) can be determined by the formula:

$$\sigma = \frac{F}{A} \tag{3}$$

where: F- force (lbf); A-area (in^2)

2.6. TCLP Analysis

Each composition was weighted and the dry material was mixed during the initial stages. When the dry compounds were well-mixed, proportional water mass was added and the system was mixed again manually until a homogeneous paste was formed.

The samples were kept in cylindrical plastic molds, with 5 cm diameter and 10cm height. Each mold with the sample was tapped on the ground to eliminate possible air bubbles. The molds were capped and kept in a wet chamber at 25°C with 100% of humidity. The method used in the TCLP tests was Method 1311 - EPA- SW-846. The samples were prepared in two series of cure time: 28 and 90 days.

After this period the samples were removed from the molds and a small subsample was used to determine the extraction fluid appropriated to the experiment. It was weighted with 5g of sample and mixed with 96.5mL of water- stirring for 5 minutes. The solution used a magnetic stir bar and plate. The pH measured was 11, so 3.5mL of HCl was added and the system was heated at 50°C. When the temperature was achieved the heat was held at this temperature for 10 minutes. At room temperature the pH was measured and it was above 5. The extraction fluid was selected as number 2, containing acetic acid solution at pH 2.88±0.05. The pH was measured before the experiment started.

All samples were broken to a particle size smaller than 1cm as required by the TCLP test. The solidified samples were prepared in triplicate and 100g of each sample was weighed. The extract fluid mass used was calculated according to the equation 1

$$W_E = \frac{20 \cdot P_S \cdot W_W}{100} \tag{1}$$

W_E: Weight of the extraction fluid (g)
P_S: Percent solids of the specimen (%)
W_W: Weight of the specimen (g)

Considering that all samples were 100% solids, the extract fluid mass used was 2000g. The system containing the solidified samples and the extract fluid were kept in polypropylene bottles. Rotary agitation was performed every 20 hours, rotated at approximately 30rpm at room temperature. After this period the rotation was stopped and the extract solutions were filtrated. The pH was recorded. The filters used were 0.7 m diameter glass fiber filters. Before each filtration, nitric acid solution was used to wash the system. Then, 150mL of each extract solution was collected. The pH was measured and nitric acid solution was added to the samples until pH was below 2. All samples were stored under refrigeration, at 4ºC, until chemical analysis.

The extract solutions were acid digested to analyze metals following the SW-846 Method 3015. Nitric acid and hydrochloric acid were used in 9 and 3mL respectively and mixed with 50mL of extract solution into an Erlenmeyer capped with an appropriate watchglass. The system was heated in a hot plate at 150ºC for 1.5 hour and at 175ºC for 30 minutes. After adding 10mL of dionized water in each flask, the samples were filtrated, using filter paper # 4 and washed with nitric acid. 10mL of each sample was collected in centrifuge tubes and analyzed in Inductively Coupled Plasma - Mass Spectrophotometer (ICP-MS), Perkin-Elmer Elan 9000. Dilutions with nitric acid in some samples were necessary.

The metal concentration can be calculated according to the equation:

$$Cm = Ce \times 10^{-3} \times Vf \times Vs^{-1} \times DF \tag{2}$$

where: *Cm* – metal concentration (mg/L); *Ce* – measured metal concentration (ppb); *Vf* – volume of filtrate (L); *Vs* – sample volume (mL); *DF* – dilution factor (%).

3. Experimental

The elemental composition of the raw materials is presented in table 2.

Table 2. Elemental composition of the raw materials in mg/g

Samples *Elements*	**C**	**T**	**Na**	**S**	**O**	**Ti**
Na	0.39	28.50	12.80	7.08	0.10	1.28
Mg	0.01	8.52	5.26	1.80	4.26	1.34
Al	19.80	0.51	14.60	5.82	12.70	3.96
Ca	315.00	51.80	2.35	5.90	1.66	3.25
Cr	0.04	200.00	0.32	0.03	0.06	0.01

Mn	2.02	0.06	0.05	0.04	0.04	0.09
Fe	16.50	1.67	21.50	3.15	16.50	2.45
Ni	0.11	0.02	0.16	0.03	0.12	0.03
Cu	0.01	0.03	0.02	0.00	0.02	0.01
Zn	0.04	0.24	0.06	0.04	0.05	0.04
As	0.029	0.002	0.001	0.011	0.001	0.002
Cd	0.000	0.002	0.001	0.002	0.001	0.002
Pb	0.01	0.01	0.01	0.03	0.01	0.01

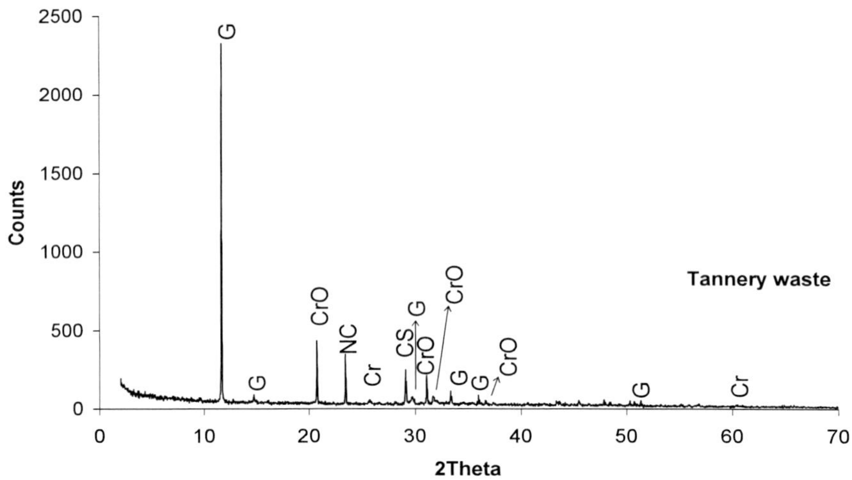

Reprinted from Materials Science Forum, vol. 530-531, Pinto, C. A. ; Diaz, F R V ; Sansalone, J J ; Dweck, J ; Cartledge, F K ; Büchler, P M . X-Ray Diffraction study of particulate tannery waste solidified in cement. pp. 478-484 Copyright (2006), with permission from Trans Tech Publications.

Figure 2. X Ray Diffraction of dried tannery waste.

3.1. NMR

NMR analysis was studied considering the chemical shifts (ppm) according to Table 3.

Table 3. Tetrahedral silicates

Notation	Chemical Shifts (ppm)	Structure
Q^0	-66 to –74	SiO_4
Q^1	-78 to –83	$SiO_3(OSi)$
Q^2	-86 to –88	$SiO_2(OSi)_2$
Q^3	-91 to –99	$SiO(OSi)_3$
Q^4	-107 to -110	$SiO(OSi)_4$

Fitzgerald [1999].

The XRD for the tannery waste, represented in Figure 2, presents, as main peaks, calcium sulfate (abbreviated as G), chromium oxide (abbreviated as CrO), sodium carbonate (abbreviated as NC) and chromium hydroxide (abbreviated as Cr) [Pinto et al., 2006].

The NMR spectra analysis is represented in the figures and the intensities due to the silica peaks are shown in Table 4 and Figures 3 to 17.

Table 4. ^{29}SI NMR chemical shifts and intensities for the raw samples as received

Raw samples	Chemical shift (ppm)	Intensities (%)
PC II-unhydrated	Q^0	100
VC-n	Q^3	100
Na	Q^3	100
O	Q^3	100
S	Q^3	76.7
	Q^4	23.3
Ti	Q^3	100

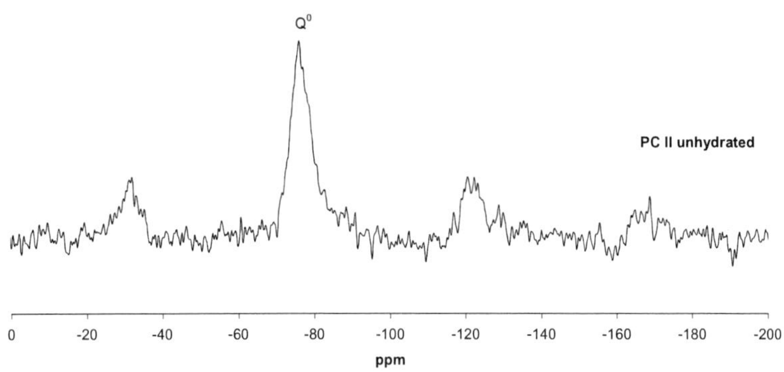

Figure 3. ^{29}Si NMR spectra of Portland cement type II unhydrated.

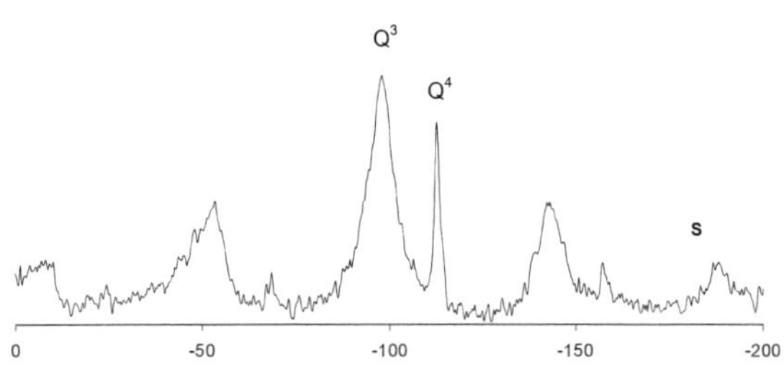

Figure 4. ^{29}Si NMR spectra of commercial sodium bentonite.

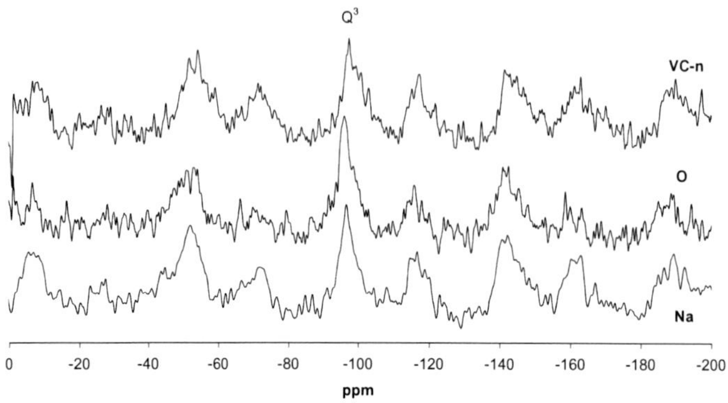

Figure 5. ^{29}Si NMR spectra of Verde-clara clay (VC-n) and modified clays: sodium verde-clara (Na) and organoclay(O).

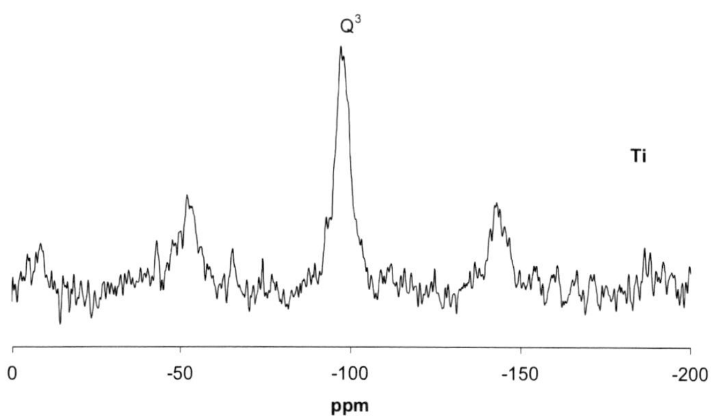

Figure 6. ^{29}Si NMR spectra of commercial organoclay.

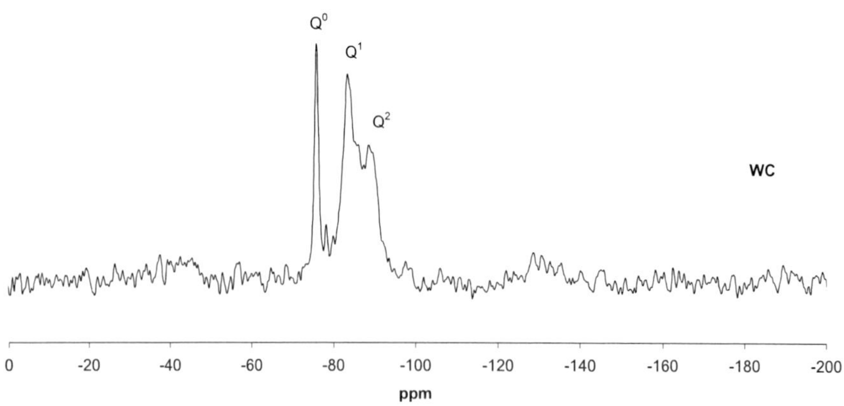

Reprinted from *Sixth International Latin American Conference on Powder Technology* ,vol. 1, Pinto, C. A. ; Dweck, J ; Sansalone, J J ; Cartledge, F K ; Albanez, N. ; Hamassaki, L T ; Büchler, P M . Evaluation of S/S Process of Tannery Waste in Cement by Structural and Mechanical Analysis. pp. 207 Copyright (2007), with permission from Brazilian Ceramic Association – ABC.

Figure 7. ^{29}Si NMR spectra of white cement hydrated after 28 days of curing time.

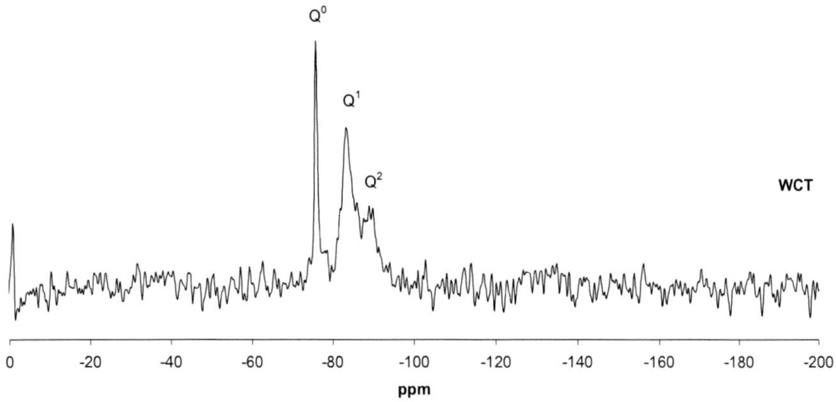

Figure 8. ^{29}Si NMR spectra of cement sample with 10% of tannery waste containing chromium.

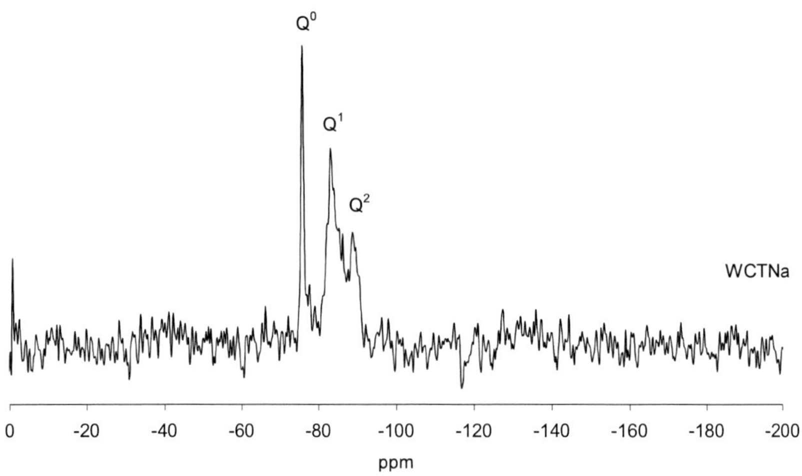

Figure 9. ^{29}Si NMR spectra of cement sample with 10% of tannery waste containing chromium and modified sodium bentonite.

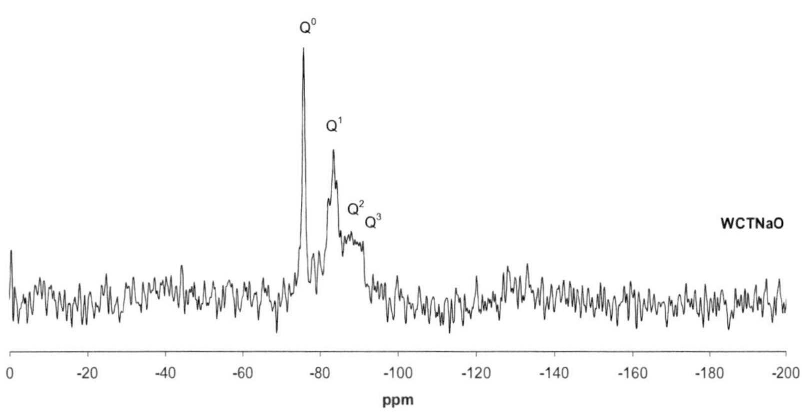

Figure 10. ^{29}Si NMR spectra of cement sample with 10% of tannery waste containing chromium and modified sodium bentonite and organoclay.

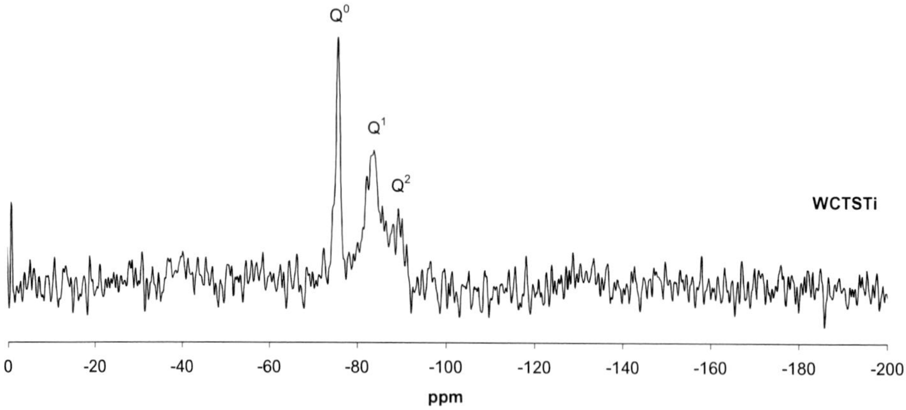

Figure 11. ^{29}Si NMR spectra of cement sample with 10% of tannery waste containing chromium and commercial sodium bentonite and organoclay.

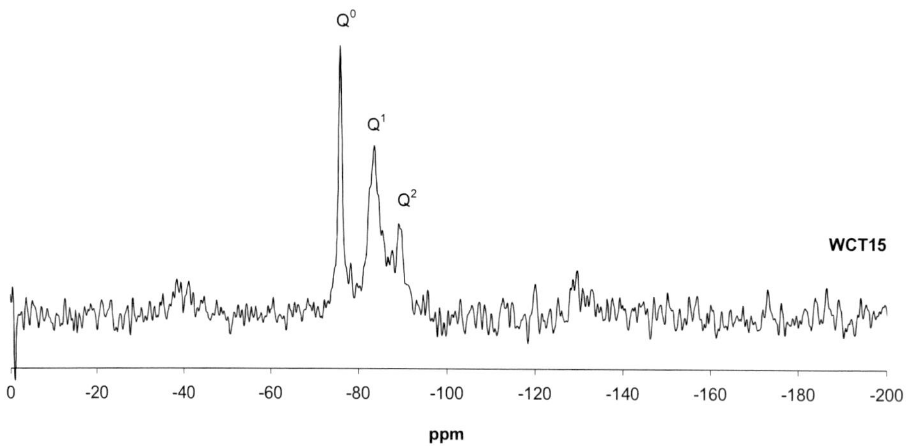

Figure 12. ^{29}Si NMR spectra of cement sample with 15% of tannery waste.

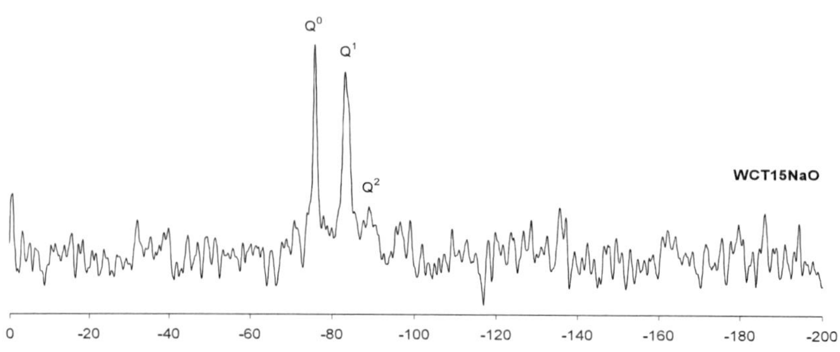

Figure 13. ^{29}Si NMR spectra of cement sample with 15% of tannery waste containing chromium and modified sodium bentonite and organoclay.

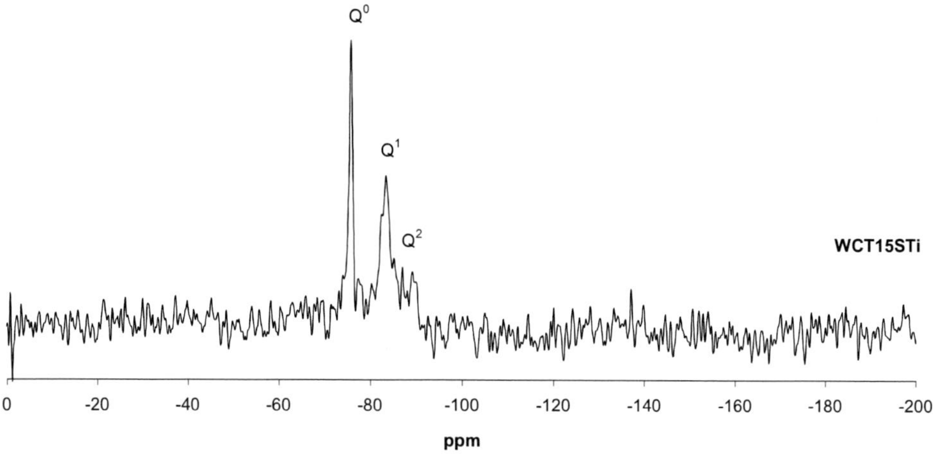

Figure 14. ^{29}Si NMR spectra of cement sample with 15% of tannery waste containing chromium and commercial sodium bentonite and organoclay.

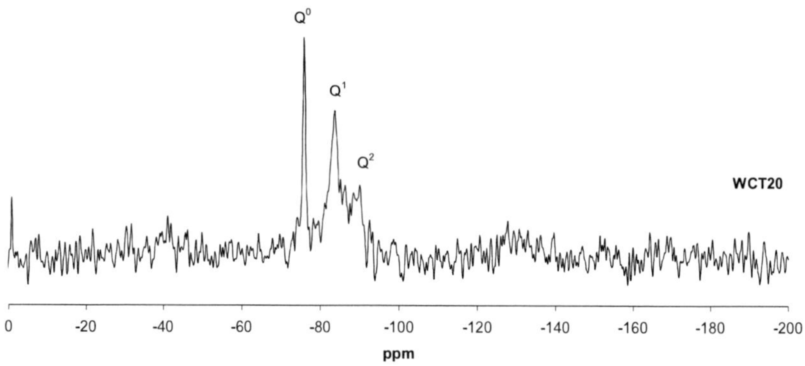

Figure 15. ^{29}Si NMR spectra of cement sample with 20% of tannery waste.

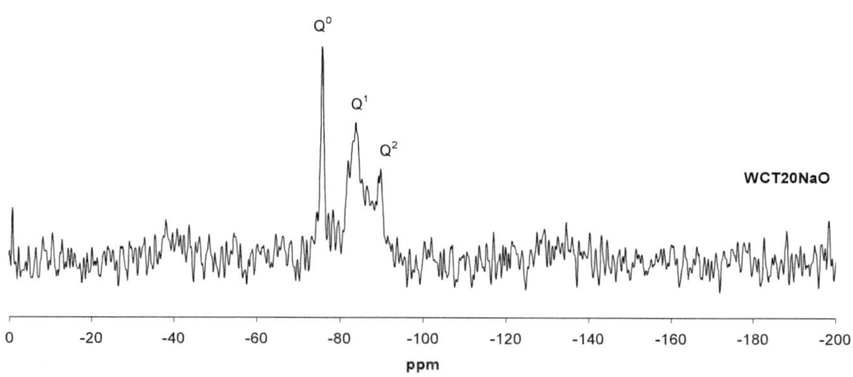

Figure 16. ^{29}Si NMR spectra of cement sample with 20% of tannery waste containing chromium and modified sodium bentonite and organoclay.

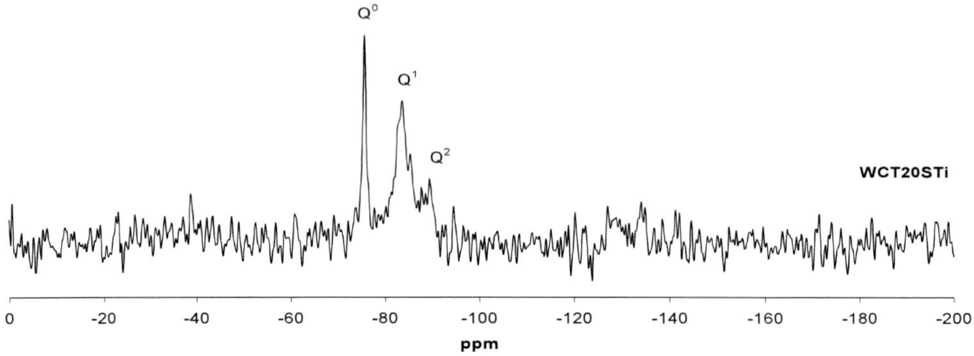

Reprinted from *Sixth International Latin American Conference on Powder Technology* ,vol. 1, Pinto, C. A. ; Dweck, J ; Sansalone, J J ; Cartledge, F K ; Albanez, N. ; Hamassaki, L T ; Büchler, P M . Evaluation of S/S Process of Tannery Waste in Cement by Structural and Mechanical Analysis. pp. 207 Copyright (2007), with permission from Brazilian Ceramic Association – ABC.

Figure 17. ^{29}Si NMR spectra of cement sample with 20% of tannery waste containing chromium and commercial sodium bentonite and organoclay.

Portland cement unhydrated presented SiO_4^{4-} peak, referred to C_3S and C_2S unhydrated, only for silica. The verde-clara clay, natural, presented 100% of intensity for $SiO(OSi)_3$, Q^3, and the same intensity was observed for this clay after laboratory cation exchange for sodium (Na) and the cation exchange for ammonium quaternarium (O). The commercial sodium bentonite presented different intensity due to a Q^4 peak, in a lower intensity compared to Q^3. The commercial organoclay presented the Q^3, only. The tannery waste was also analyzed and did not present any silica peak.

The control samples presented the same chemical shifts, Q^0, $Q^1$and Q^2. The samples containing the clays do not present the same peaks when analyzed separately, due to the quantity added of clay in each sample that it was 5% referred to cement mass. The cement hydration can be analyzed by the decrease of Q^0 units and increase of $SiO_3(OSi)$ (Q^1) and $SiO_3(OSi)_2$ (Q^2).

The low intensity of Q^0 for the hydrated white cement hydrated shows that after 28 days of curing time the initial hydration process is finished, presenting higher peak for $SiO_3(OSi)$ followed by $SiO_2(OSi)_2$.

The appearance of Q^0 in the control samples with clays is higher than the cement only hydrated and, on the other hand, the $SiO_2(OSi)_2$ (Q^2) peak presents lower intensity.

The samples with tannery waste were analyzed according to the quantity of waste added to the cement matrix, 10, 15 and 20%. The samples containing 10% of tannery waste presented the peaks Q^0, Q^1 and Q^2 with different intensities. The cement hydration is observed in all samples by the presence of the peaks $Q^1$and Q^2 and the reduction of the peak of silica monosilicate (SiO_4), Q^0.

The presence of waste retards the cement reactions, observed by the intensity of 24.5% for Q^0, higher than the white cement only. Previous research showed that some heavy metals, such as zinc, can react with some cement compounds and cadmium that forms calcium hydroxide and provides nucleation sites for precipitation of CSH and calcium hydroxide [Ortego et al. 1991, Cartledge et al., 1990].

When the samples containing 10% of waste and clays are analyzed, it can be observed that the presence of each clay separately presents a high intensity of Q0, with the exception of Na and O clays.

Samples using commercial clays showed a lower intensity for $SiO_2(OSi)_2$, caused by the interference of clay in the system. Sodium bentonite can compete with cement for the water, considering that it is hydrophilic clay. Organoclay can encapsulate some cement compounds as calcium hydroxide and obstruct the complete cement hydration [Nzengung et al, 1997].

The compositions containing 15 and 20% of waste presented the same chemical shifts as the previous samples. In all cases, the intensity of Q^0, which was 28% approximately, was high- showing that the increase of waste mass in the cement system reduces the cement hydration process compared to samples containing 10%, which presented a silicate (SiO_4) intensity of 24.5%.

When the samples containing 15 and 20% of waste are compared, apparently there was no difference in intensities for the chemical shifts. This difference can be observed only when clays are applied. In this case, the Q^0 is increased, with exception for WCT20STi, which was lower, possibly due to an experimental inaccuracy.

This behavior has been previously studied by Pinto et al. [2004a] where cement hydration was retarded by the waste presence, but this effect was reduced when sodium bentonite and organoclay were added to the system. Organoclay can encapsulate some cement compounds such as calcium hydroxide and obstruct complete cement hydration [Nzengung et al., 1997].

3.2. NCDTA

The early cement hydration stages in the several pastes with the waste and additives were studied by nonconventional differential thermal analysis as follows.

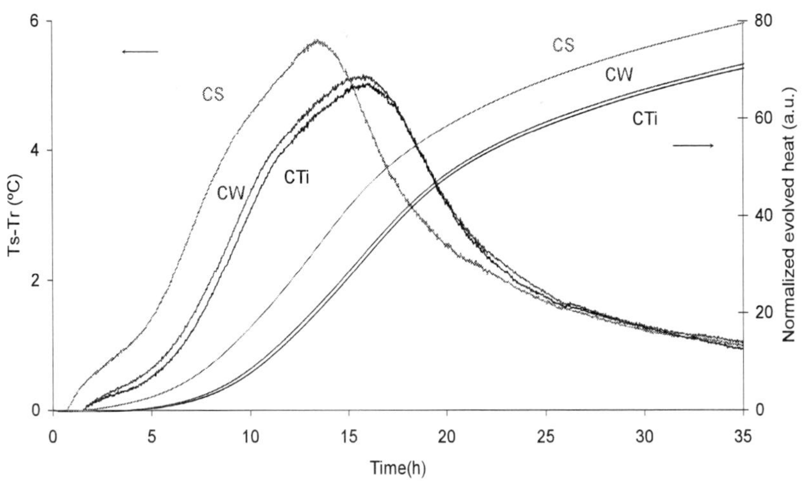

Figure 18. NCDTA curves and accumulated heat in the reaction as function of time normalized by cement mass from cement pastes with water (CW) and cement with commercial clays, sodium bentonite (CS) and organoclay(CTi).

Portland cement hydration occurs with a series of chemical reactions due to water in the system. In the first 10 minutes of hydration, there is an exothermic peak by formation of C-S-H and ettringite, especially from this last product that promotes the not hydrated cement particles covering. After a few hours, there is a fast hydration of tricalcium silicate, with a maximum peak after 10 hours. After that, the heat rate generated in the process starts being reduced gradually [Swaddiwudhipong, Chen and Zhang, 2002].

Figure 18 shows the NCDTA curves as measured and the accumulated heat evolved during first 35 hours of hydration. Commercial sodium bentonite presents a higher influence compared to organoclay. It accelerates the cement hydration reactions presenting a maximum peak of temperature difference after 14 hours of hydration approximately, considering that hydrated cement presents the same peak after 16 hours of curing time. The presence of organoclay retards the matrix hydration in about 30 minutes compared to the CW sample. This effect was also observed in the study of Dweck et al. [2003]. The results of normalized accumulated heat show that the heat generated during CS sample hydration is higher compared to the other samples because this composition has a higher temperature difference between the sample and the reference sample, of almost 1°C of difference when compared to CW and CTi samples for the same period of hydration.

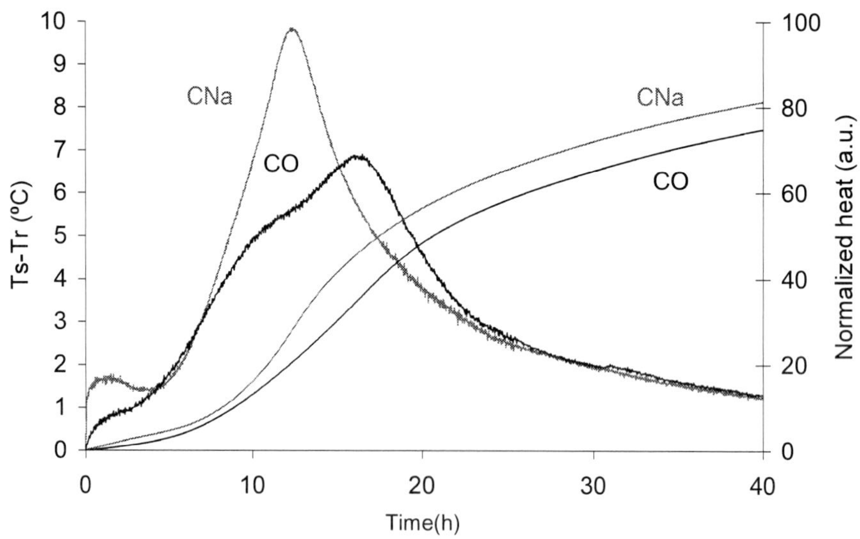

Figure 19. NCDTA curves and accumulated heat evolved in the reaction as function of time, normalized by cement mass of solidified sample containing modified clays, CNa and CO.

In the case of modified clays the same behavior was observed. Sodium bentonite consumes part of the water from cement hydration due to hydrophilic characteristics. Once this absorption evolves liberation heat the increase of temperature accelerates the cement hydration reactions. The higher the water at the composition, the better the capillary porous action and the easier is for the heat to transfer in the solidified matrix.

Thus, when this occurs, the generated heat becomes more intensive and liberates faster, according to Spence [1992].

According to Andriolo [1984] the cement hydration liberates heat immediately, after the water makes contact with it. The tricalcium aluminate hydration should also occur this way, but, due to the presence of calcium sulfate in the cement, the reaction is retarded.

Spence [1992] showed that the heat released from hydration, due to reactions of C_3A and of CaO is up to 200cal/g while C_3S and C_2S hydration release 12 and 62 cal/g. The released heat can be observed in the first hours of hydration. The small peak appears in the first hour of hydration that corresponds to initial reactions of C_3A to form ettringite. This heat is then reduced by the presence of sulfate and the second peak occurs due to reactions of C_3S to form tobermorite and aluminates forming more ettringite [Mehta and Monteiro, 1994].

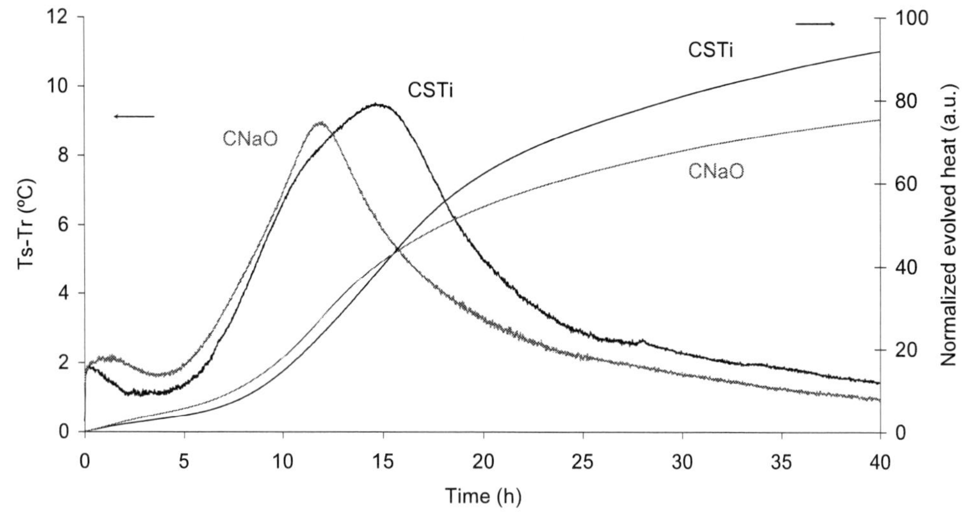

Figure 20. NCDTA Curves of accumulated evolved heat in the reaction as function of time normalized by cement mass of solidified samples containing commercial clays (CSTi) and modified clays (CNaO).

From Figure 20 it can be observed that all samples accelerate the cement hydration in the beginning and after this the maximum peak of CNaO occurs after 11 hours. The sample CSTi does not present the same initial acceleration, presenting a maximum peak after 15.5 hours.

The heat released in the reactions with samples containing commercial clays until 15.5 hours is lower than that for modified clays. After this period the behavior is changed and the opposite occurs.

In Figure 21, the influence of tannery waste in cement hyrdration can be observed. The waste promotes a higher reaction temperature in the beginning. The higher the quantity of waste, the higher the initial peak, due to hydrated calcium sulfate present in the waste [Dweck et al 2003], increasing the formation of ettringite by hydration reaction of tricalcium aluminate and calcium sulfate in the first hours of hydration. The second peak, due to calcium silicate hydration, occurs in the opposite way. A higher peak is observed for samples containing less waste, 10%, probably due to higher quantity of water used initially in the hydration reaction of C_3A in the formulations containing more waste causing the C_3S hydration reaction to be reduced by the lower quantity of water present.

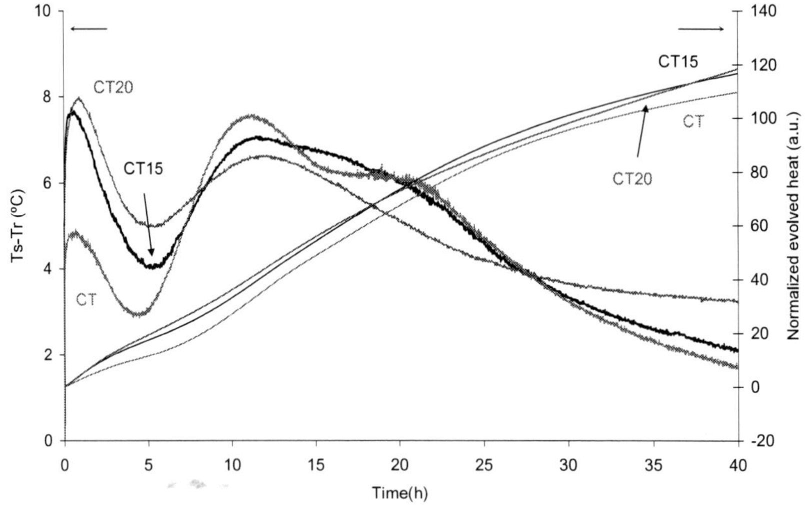

Figure 21. NCDTA curves and accumulated evolved heat in the reaction as function of time, normalized by cement mass of solidified samples CT, CT15 and CT20 containing 10, 15 and 20% of tannery waste respectively.

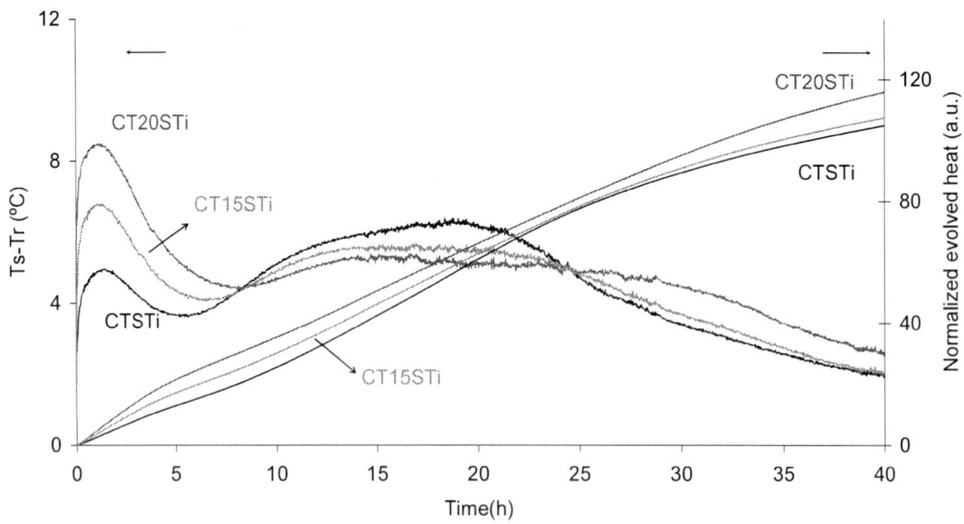

Figure 22. NCDTA curves and accumulated heat evolved in the reaction as function of time, normalized by cement mass of solidified samples containing commercial clays and tannery waste, in, 10, 15 and 20%.

In Figure 22 the influence of waste is noticed. The higher the quantity added, the higher the released heat. The second peak that starts after 5 hours of samples mixture, occurred initially for samples with 10% of waste, interfering in the C_3S and C_3A reaction due to calcium sulfate presence in the waste. Figure 23 shows the 15 and 20% waste cases.

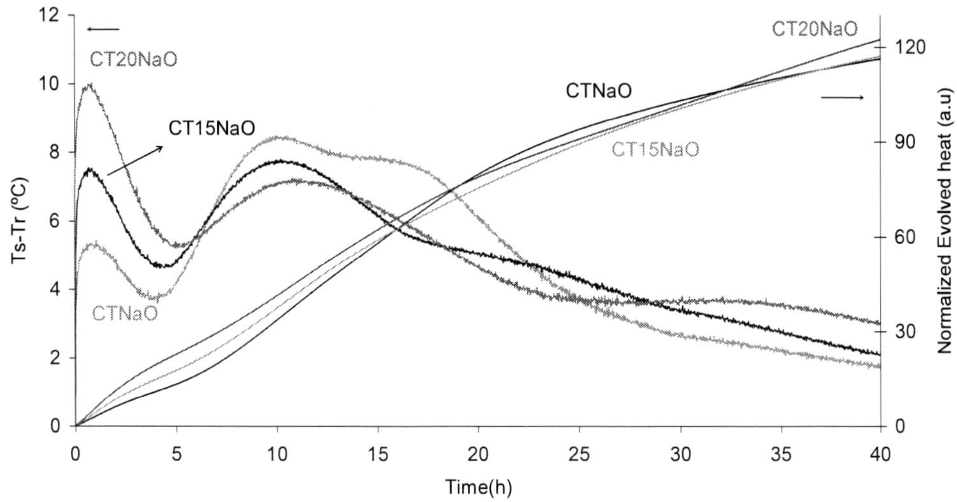

Figure 23. NCDTA curves and accumulated heat evolved in the reaction as function of time, normalized by cement mass of solidified samples containing modified clays and 10, 15 and 20% of waste.

The NCDTA curves show that the behavior of cement hydration was the same for samples containing commercial clays. The initial increase of temperature difference is due to waste presence. After 6 hours of hydration this effect is inverted by the lower amount of water available, that was previously consumed. The total heat generated after 40 hours of hydration is similar, independently of the waste content present in the paste.

3.3. DTA/DTG

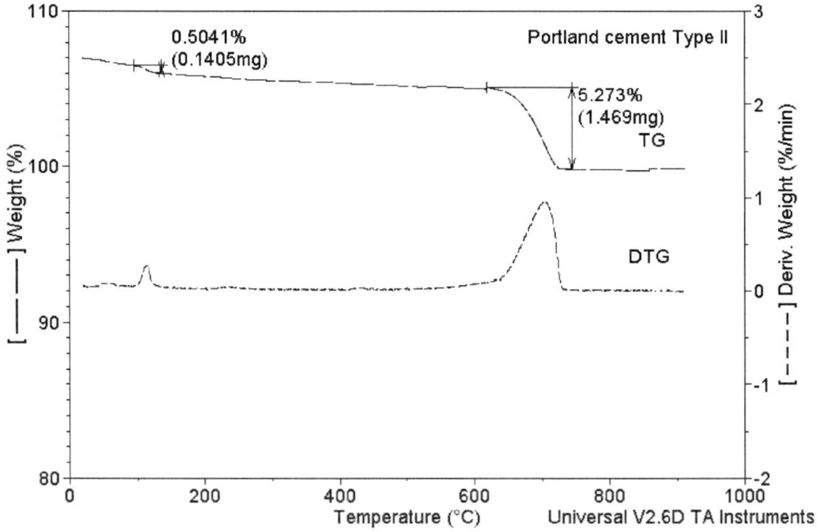

Figure 24. TG/DTG curves of unhydrated Portland cement.

Figure 24 shows Portland cement TG/DTG curves. From 100 to 120°C there is a mass loss due to cement dehydration, which is 0.5% of total mass lost approximately. It corresponds to the dehydration of hydrated calcium sulfate present in cement composition. The DTG peak between 600 and 700°C represents mass loss by decarbonation, around 5%, which occurs via CO_2 loss from the $CaCO_3$ present in cement composition.

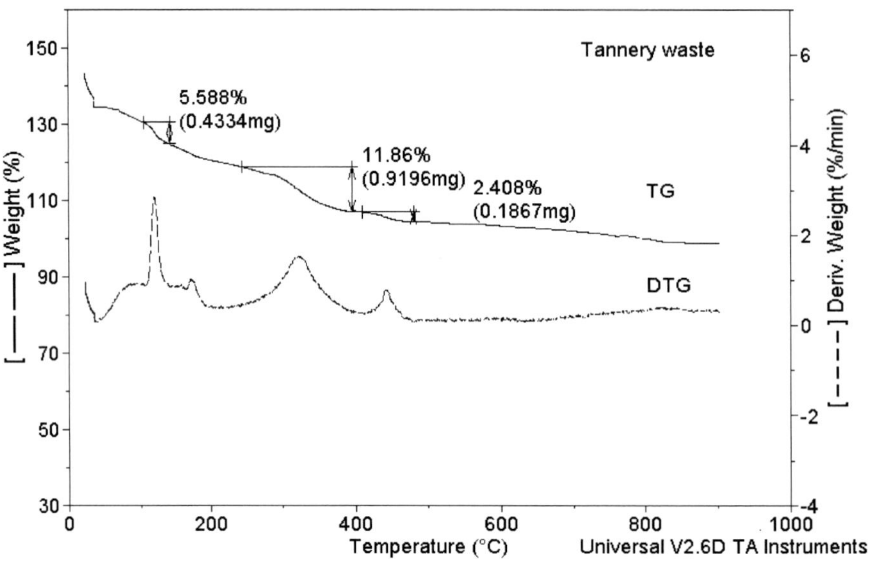

Figure 25. TG/DTG curves of the tannery waste.

Figure 25 shows the TG/DTG curves of the tannery waste as used in the compositions. From TG and DTG curves it can be seen that during the first isothermal step at 35°C and up to 100°C, during a heating rate of 10°C/min, there is a mass loss due to the residual water present in the waste.

Between 120 and 160°C, there is water loss from the decomposition of the dehydrated calcium sulfate present in the waste Dweck et al. [2000b]. From 200° to 500°C, there is a weight loss caused by organic material decomposition and to chromium hydroxide dehydroxilation seen in the respective DTG peak.

When the commercial organoclay is analyzed, there is a decomposition and volatilization of organic materials present in the sample due to the carbonic chain of ammonium quaternary between the layers. This decomposition occurs between 200 and 400°C. The dehydroxilation can be observed by the curve from 500 to 600°C and the decarbonation is shown between 600 and 700°C.

The commercial bentonite presents a water loss from 35 to 100°C. This clay, even as a powder material, presents humidity. Between 500 and 700°C there is a dehydroxilation of the sample.

The humidity present at verde clara clay can be observed at TG curve as well as by the mass loss from 35 and 130°C. The peak at DTG curve and the endothermic effect analyzed by DTA can also observe this loss. The dehydroxilation of the sample occurs between 350 and 500°C. A small endothermic peak is observed at this temperature range.

After cation exchange of the original clay to sodium there were not observed significant changes related to thermal analysis, showing a water loss and dehydroxylation peak between 400 and 600ºC, as observed in the original clay.

The organoclay obtained by cation exchange of the clay by organic radical presented a water loss between 35 and 100ºC. From 200 and 400ºC there is a mass reduction due to organic material loss by decomposition, volatilization and burn out. This loss occurs in several steps, depending of the organic material present. The dehydroxilation is observed between 400 and 500ºC, which is followed by a burnout phase of organic residues from 500 to 600ºC characterized by an exothermic effect. Similar steps were observed during commercial organoclay thermal analysis.

The solidified samples analyses are shown in next figures.

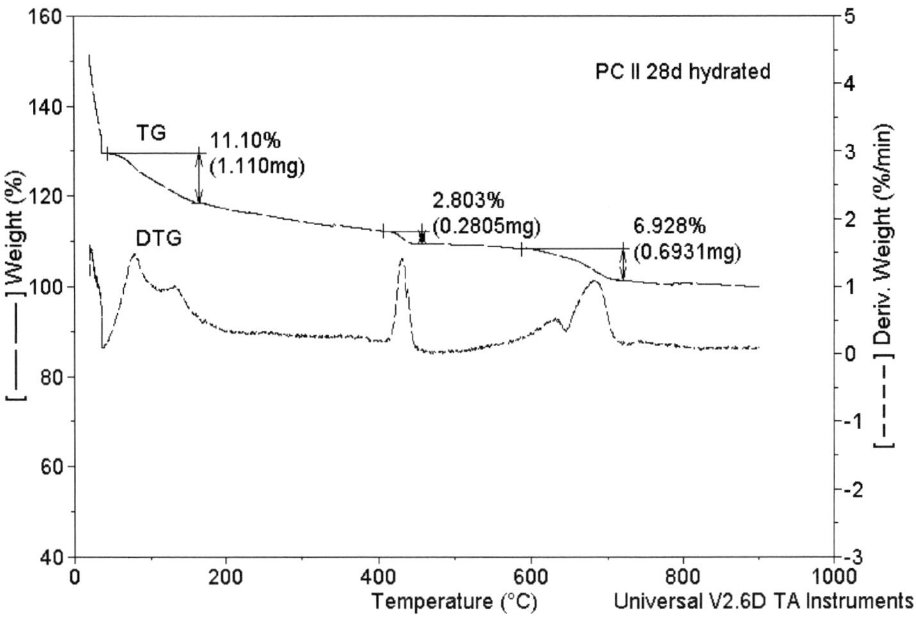

Figure 26. TG/DTG curves of Portland cement after 28 days of hydration.

Figure 26 shows that, after the drying step at 35°C, the hydrated cement 1 mass loss between 60 and 200ºC is due to dehydration of tobermorite ($3CaO.2SiO_2.xH_2O$, formed by calcium silicate hydration) and ettringite, ($3CaO.Al_2O_3.3CaSO_4$. $32H_2O$ formed by tricalcium aluminate hydration), which occur in this order, according to Dweck et al. [2002; 2000b]. Reaching 400ºC starts the dehydroxilation of calcium hydroxide, which ends at 450ºC. The decarbonation is represented by the TG/DTG curves between 550 and 730ºC, and is due to the decomposition of calcium carbonate [Dweck et al. 2000a].

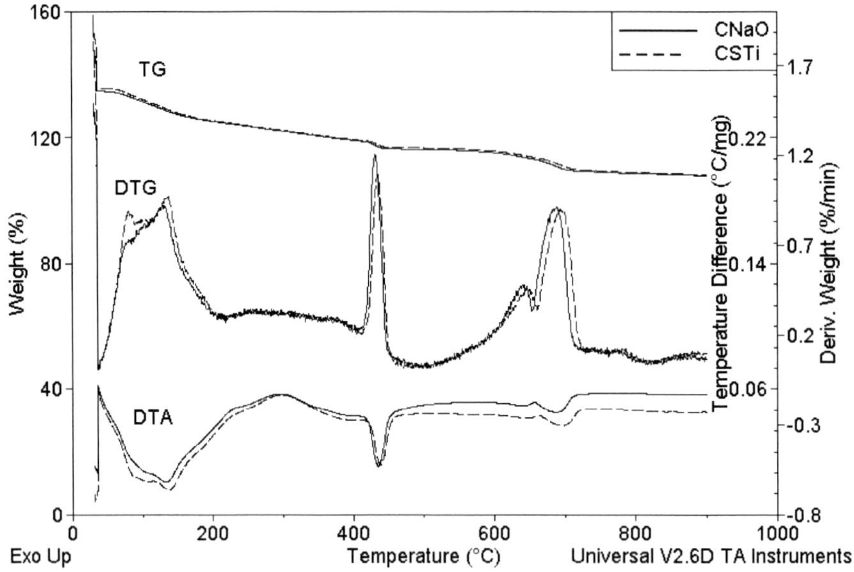

Figure 27. TG/DTG/DTA curves of Portland cement solidified with commercial and modified clays after 28 days of hydration.

The effect of the commercial and modified clay samples in cement is compared in Figure 27. The initial mass loss occurs by samples dehydration between 50 and 200ºC. An endothermic peak at respective DTA curves represents this effect. A small mass loss is verified from 200 and 400ºC by organic material decomposition present in organoclays. DTG peaks between 400 and 450ºC, with corresponding endothermic DTA peaks, represent the dehydroxilation. The last phase of mass loss, from 550 and 700ºC corresponds to the decarbonation of calcium carbonate not-well and well-crystallized respectively (Dweck 2000). In this case endothermic peaks are observed in corresponding DTA curves.

In the study of Pinto et al. [2004a] where sodium bentonite and organoclay are analyzed separately from the waste, it was observed that, when the organoclay, which is hydrophobic, is added to the cement+waste system, it allows the water added to be available for cement hydration, differently from the case of sodium bentonite, which is hydrophilic and absorbs part of the water present in the paste. By this way the samples containing cement and organoclay increase the formation of the hydration compounds, like tobermorite and ettringite even with waste presence. Another fact is that even after 9 weeks of hydration, on cement calcined mass basis, the thermal analysis curves show no changes. This indicates that in solidified samples containing waste, the main hydration reactions of cement occur in the first 4 weeks of solidification.

The waste presence in solidified samples with modified bentonite and organoclay affect cement hydration. As shown in Figure 28, it is observed by the difference between the mass losses at TG curves. From 35 and 200ºC occurs the highest mass losses by sample's dehydration. The tobermorite dehydration in the sample CTNaO is higher probably due to reaction of the waste with cement, forming silicate-hydrated compounds.

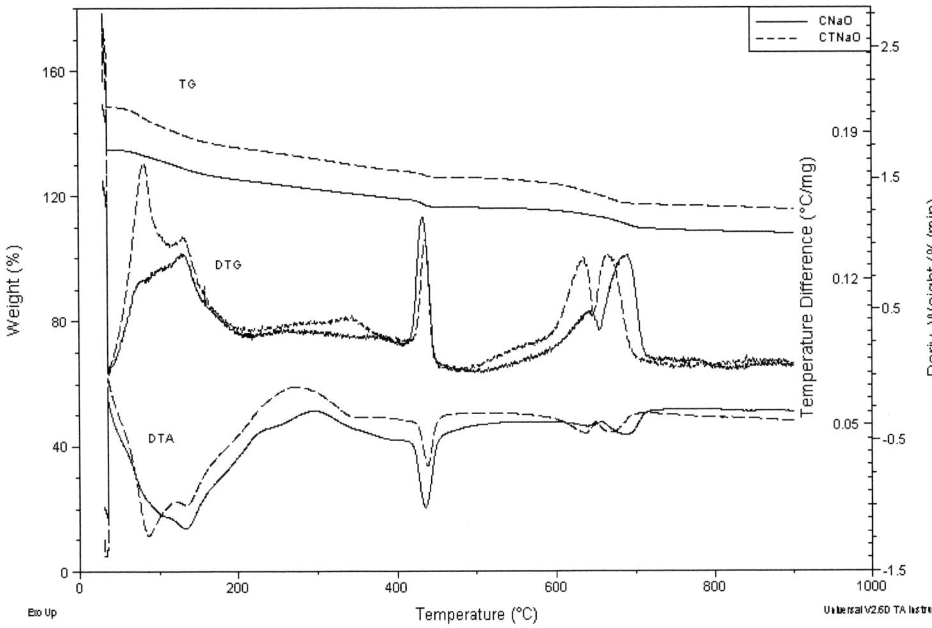

Figure 28. TG/DTG/DTA curves of Portland cement solidified with modified clays and 10% of waste respectively after 28 days of hydration.

The second peak around 150ºC corresponds to ettringite dehydration which, in this case, presents a difference between the samples curves. Between 200 and 400ºC there is a decomposition and burnout of organic material as shown by DTG curves. This peak is higher for samples containing waste. For CNaO sample, in the DTG curve there occurred an organic material mass loss that is present in the organoclay, but in lower intensity.

At 400ºC the dehydroxylation starts. The sample without waste started the dehydroxylation first and at a higher intensity than with sample CTNaO. The dehydroxylation and decarbonation reactions that occur during the thermal analysis according to Dweck et al. (2000b) are respectively:

$$Ca(OH)_2 \rightleftharpoons CaO + H_2O \quad \text{(R-1)}$$

$$Ca(OH)_2 + CO_2 \longrightarrow CaCO_3 + H_2O \quad \text{(R-2)}$$

The calcium carbonate decomposition is shown at R-3:

$$CaCO_3 \rightleftharpoons CaO + CO_2 \quad \text{(R-3)}$$

The sample containing waste (CTNaO) has a higher organic material due to its presence in the waste, the released CO_2 is also higher and reaction R-2 occurs. As a consequence, the dehydroxylation peak is reduced as can be observed by DTG curves of CTNaO sample.

Volzone [1995] studied, by DTA, the behavior of chromium hydroxide solution when mixed with smectite. According to the author, the peak at 130ºC approximately corresponds to sample dehydration and the dehydroxilation occurs between 400 and 680ºC.

The decarbonation is observed between 500 and 700°C. In this phase, the sample with waste starts the decarbonation in a lower temperature and presents a higher peak.

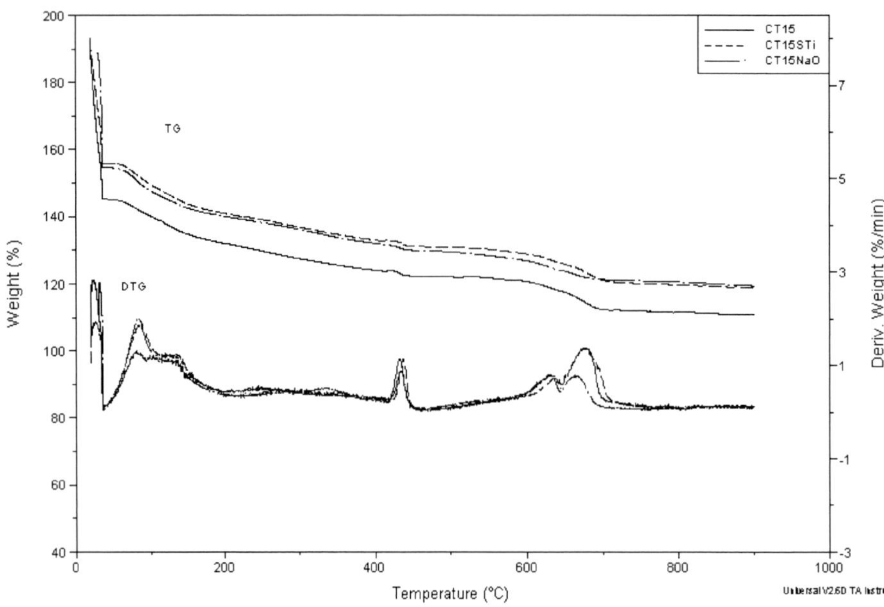

Figure 29. TG/DTG curves of Portland cement solidified with clays and 15% of waste after 28 days of hydration.

Figure 29 compares samples containing 15% of tannery waste with solidified samples with commercial clays (CT15STi) and with modified clays (CT15NaO). The differences represented in the horizontal at TG curves are due to clays presence in the samples. The behavior of mass loss is similar in all cases. Initially there is a water loss of tobermorite and ettringite compounds, between 50 and 200°C. Among the DTG curves at this phase there is a higher peak for CT15NaO and after, CT15STi.

A mass loss is continuous from 200 to 400°C, when the organic material of the waste and the carbonic chain from the organoclays both decompose. At DTG curve for CT15NaO at this phase between 300 and 350°C a higher peak is observed probably due to an excess of organic added at cation exchange. Between 400 and 450°C occurs a dehydroxilation mainly of calcium hydroxide and from 580 to 700°C occurs the decarbonation of the samples.

The analysis of the samples with 20% in mass of tannery waste shown in Figure 30 presents some differences among the TG/DTG curves. The dehydration of the main hydrated cement compounds, tobermorite and ettringite occurs between 50 and 200°C, where the first peak corresponds to tobermorite dehydration. In this phase no changes among the peaks were observed.

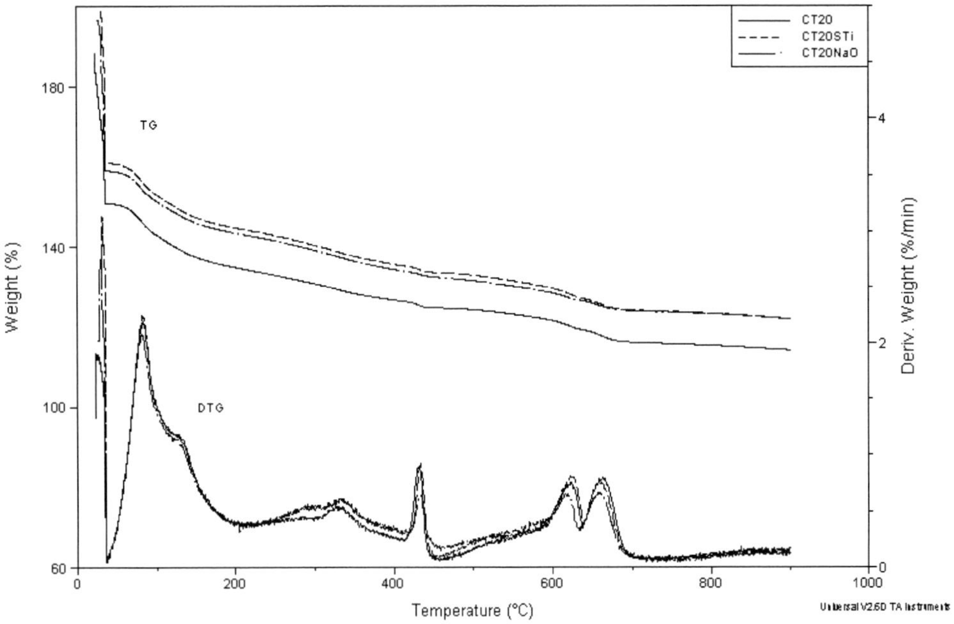

Figure 30. TG/DTG curves of Portland cement solidified with clays and 20% of waste after 28 days of hydration.

The mass loss by decomposition of organic material occurs between 200 and 400°C. It was noticed that the mass loss is higher compared to samples with 15% of tannery waste. Among the samples with 20% the compositions containing organoclays presented higher mass loss of organic material. From 400 to 450°C there is dehydroxilation and at this phase the peak is higher for CT20.

Decarbonation occurs in two phases. The first, between 580 e 630°C according to Dweck et al. (2002), corresponds to calcium carbonate not well crystallized and the second peak, between 630 and 700°C is due to decarbonation of well crystallized $CaCO_3$. In this step some differences are observed and a CT20NaO sample has a lower peak for both dehydroxilation and decarbonation.

The dehydration of samples containing 20% in mass of tannery waste is shown in Figure 31 by endothermic peaks between 50 and 200°C. The organic material decomposition has a more representative exothermic effect occurred from 250 and 350°C. Between 400 and 430°C there is an endothermic peak due to dehydroxilation and after 600°C the decarbonation can be observed in all samples.

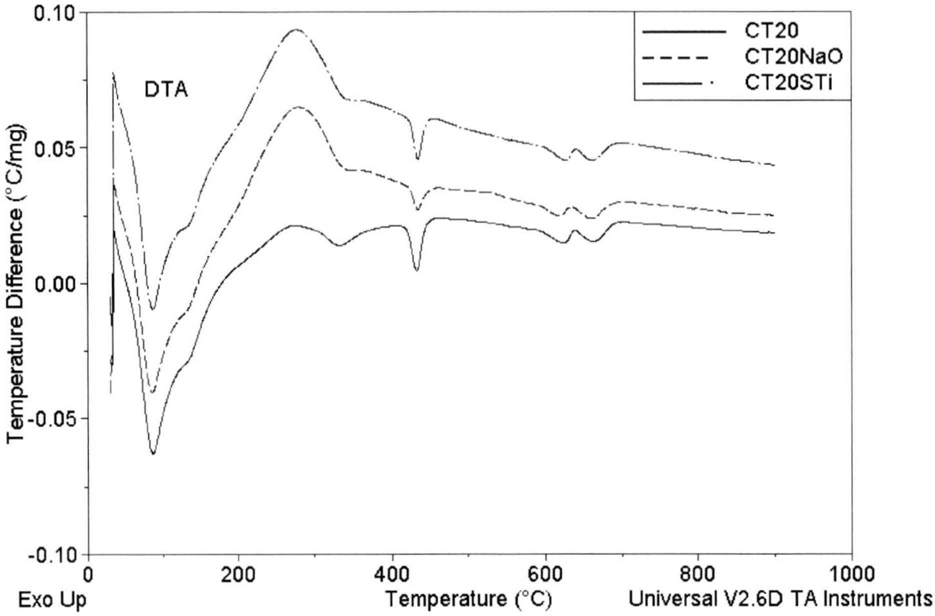

Figure 31. DTA curves of Portland cement solidified with clays and 20% of waste after 28 days of hydration.

3.4. UCS

The compressive strength results after 28 days of cement hydration, represented in Figure 32, were from 21.5 to 27.3 MPa. The clay presence increased the compressive strength.

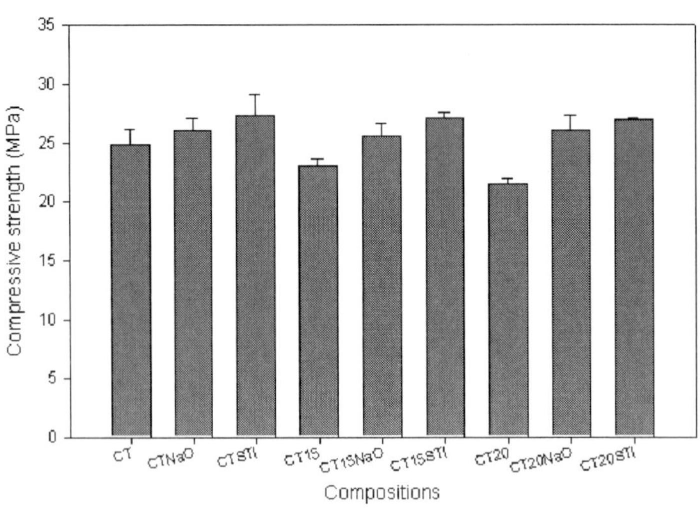

Figure 32. Compressive strength results of solidified samples after 28 days of cement hydration.

The modified clays increased the strength from 5% to 19% approximately and the commercial clays increased it from 10% to 20%. All the results were above the US EPA values, that according to Barth and Percin, [1990] the lowest acceptable value for stabilization/solidification process of wastes in cement is 0.34MPa.

The modified clays increased the strength from 5% to 19% approximately and the commercial clays increased it from 10% to 20%. This behavior is due to the competitive system bentonite-cement that reacts with water, reducing its quantity available for cement hydration and consequently higher compressive strength is observed. The bentonite's particle size similar to cement improves this behavior. The results were above the US EPA values, that according to Barth and Percin, [1990] the lowest value for stabilization/solidification process of wastes in cement is 0.34MPa.

The presence of waste interferes on the cement reactions [Pinto et al., 2005b; Dweck et al., 2003, Dweck, et al., 2002]. Previous research showed that some heavy metals, such as zinc react with some cement compounds and cadmium providing nucleation sites for precipitation of CSH and calcium hydroxide [Li and Wenquan Liang, 2002; Lin et al., 1997]. This behavior was previously studied by Pinto et al. [2004a] where cement hydration was retarded by the waste presence, but this effect was reduced when sodium bentonite and organoclay were added to the system.

3.5. TCLP

Organoclays can encapsulate some cement compounds such as calcium hydroxide and obstruct the complete cement hydration [Nzengung et al., 1997].

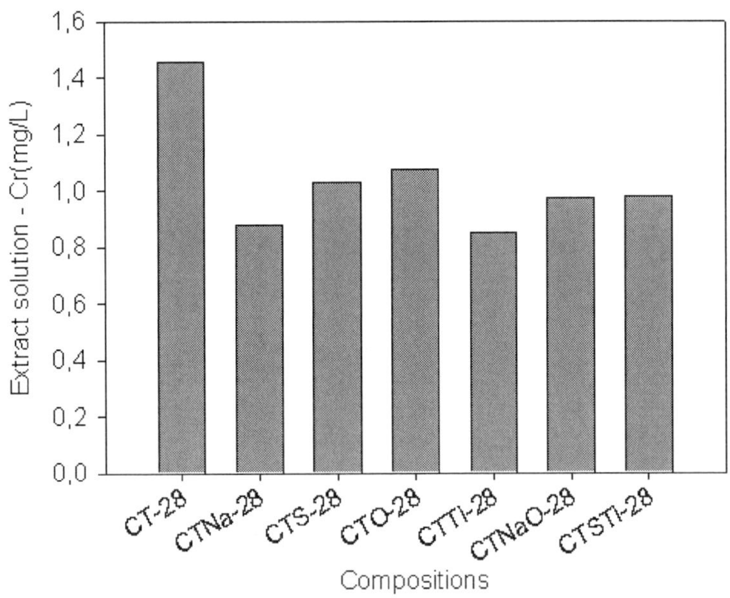

Figure 33. Extract solution concentration of Cr in samples containing 10% of tannery waste from TCLP solidified after 28 days of hydration.

The results of chromium leached from the solidified matrix showed that among the samples after 28 days of hydration containing 10% of tannery waste, the sample that presented the highest concentration was CT with 1.45mg/L of Cr. The solidified sample containing modified sodium bentonite, reduced in 39% the leached chromium and the original commercial bentonite reduced in 29%. The organoclays reduced the chromium leached in 26% for the modified organoclay and 41% for the commercial one. When both clays are added to the compositions the results where 0.97 mg/L for CTNaO and 0,98mg/L for CTSTi, which corresponds to 33 and 32% of chromium reduction respectively.

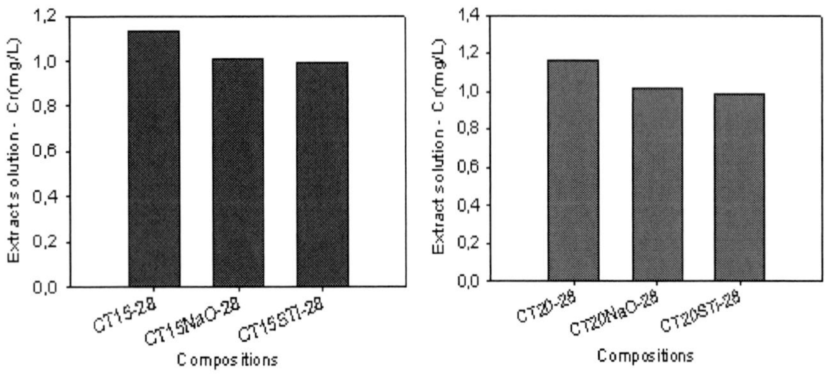

Figure 34. Extract solution concentrations of Cr in solidified samples containing 15 and 20% of tannery waste after 28 days of hydration respectively.

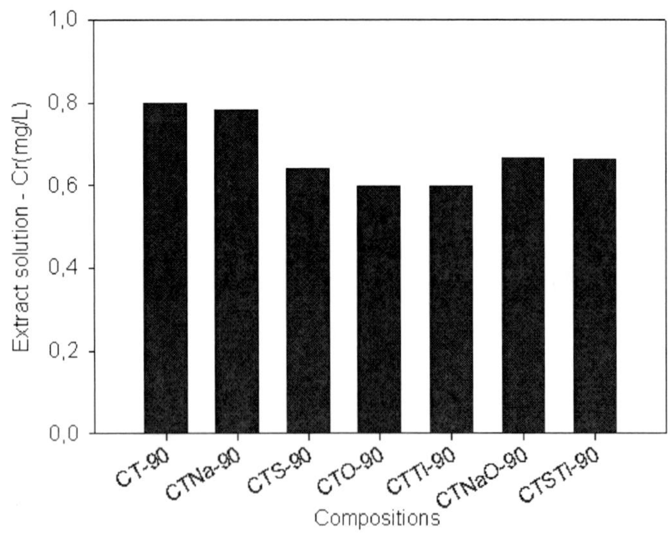

Reprinted from Materials Science Forum, vol. 498-499, Pinto, C.A.; Sansalone, J.J.; Dweck, J.; Cartledge, F.K.; Diaz, F. R. V.; Büchler, P. M. Hazardous wastes disposal: Stabilization/ Solidification of tannery residue in the presence of chromium, pp. 697-703 Copyright (2005), with permission from Trans Tech Publications.

Figure 35 Chromium concentrations in extract solutions from solidified samples containing 10% of tannery waste, after 90 days of hydration.

The samples with 15% of tannery waste and both clays presented 11 and 12.5% of chromium reduction leached from CT15NaO and CT15STi samples respectively. The samples with 20% presented a similar behavior with chromium reduction of 13 and 15% in CT20NaO and CT20STi cases respectively.

The 90 days curing time samples are shown in figures 4 and 5. Figure 4 presents the results for the samples containing 10% of tannery waste. The decrease of Cr compared to 28 days samples was observed. Modified bentonite (CTNa-90) apparently did not present significant difference compared to the control cement paste sample without any addition (CT-90), with a reduction of 2% in Cr. Commercial bentonite (CTS-90) decreased the Cr concentration by 20%. Modified and commercial organoclays presented the same value of Cr in the solution, 0.596mg/L, which corresponds to a reduction of 25%, approximately. When both clays were applied to the cement matrix in both cases, CTNaO-90 and CTSTi-90 presented the same values, 0.66mg/L, with a decrease of 17% of chromium in the extract solution.

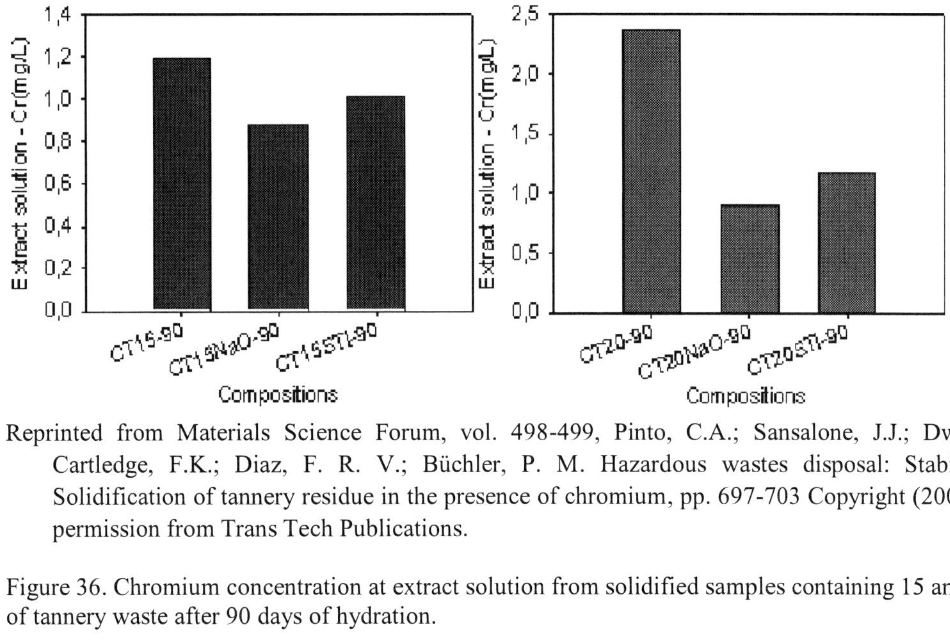

Reprinted from Materials Science Forum, vol. 498-499, Pinto, C.A.; Sansalone, J.J.; Dweck, J.; Cartledge, F.K.; Diaz, F. R. V.; Büchler, P. M. Hazardous wastes disposal: Stabilization/ Solidification of tannery residue in the presence of chromium, pp. 697-703 Copyright (2005), with permission from Trans Tech Publications.

Figure 36. Chromium concentration at extract solution from solidified samples containing 15 and 20% of tannery waste after 90 days of hydration.

Figure 36 shows the results for 15 and 20% in mass of residuals in the system. An increase in the Cr leached from solidified bodies can be observed compared to samples with 10% of waste. The increase of residuals added to the system increased the chromium concentration in the lixiviate. CT15NaO-90 presented 0.878mg/L of Cr, which corresponds to 26.7% lower than CT15-90, 1.2mg/L. The commercial clays (CT15STi-90) presence reduced by approximately 16% the chromium concentration in the leachate.

The samples containing 20% of tannery waste presented a similar behavior. Modified clays (CT20NaO-90) presented a reduction of 61% compared to CT20-90. The commercial clays reduced the chromium leached in 50% compared to cement sample only.

When only chromium is added to the cement, according to Gilliam and Wiles, (1992) it can react with calcium sulfate making it less available for the reaction with calcium aluminate to form ettringite.

The results showed that in all cases the chromium concentration leached was under the limit concentration of 5mg/L according to the environmental legislation. The clays presence had better results compared to the cement system only. The highest quantity of residuals used was 20% in mass, but due to the results, higher concentrations can be used, in terms of leaching tests.

Zhang, Akhter e Cartledge (1997) studied the behavior of some metals, including chromium on different types of cement matrixes. According to the results, chromium is immobilized when mixed with cement matrix. This could be observed by the leaching results.

The particle size of waste and clays interferes in cement hydration due to an interaction among the materials, modifying the products formed during the process's curing time. The compressive strength analysis was improved when organoclays and sodium bentonites were added to the cement matrix containing waste, due to the competition sites between sodium bentonite and cement for water and also due to the adsorption properties of clays which reduce the capacity of waste on interference on cement hydration reactions. The leaching test confirmed the same behavior observed in the previous analysis, showing the reduction of chromium leached at the system by the cement stabilization/solidification capacity and the improvement of these results when organoclays and sodium bentonite were used.

CONCLUSION

The silica polimerization analysis by NMR performed the study of the stabilization by solidification of the tannery waste process, which indicated that the waste and clay additives reduce the polymerization process that occurs during cement hydration.

By NCDTA used to study the early stages of cement hydration it was seen that, the tannery waste, due to its calcium sulfate content, accelerates the initial formation of ettringuite and that commercial sodium bentonite presents a higher initial accelerating effect when compared to organoclay.

Thermogravimetry and Derivative thermogravimetry allow one to compare quantitatively the main phases formed during the solidification process and to see the influence of the waste and the used additives presence, which was noticed in all studied cases.

It was seen that tannery waste reduces the cement hydration process that occurs during the first 28 days. When 10% is added to the cement matrix, there are reactions with cement compounds to form chromium hydroxide. A higher reduction is observed when 15% is applied. The samples containing 20% of waste did not modify significantly this effect when compared to the 15% cases.

Commercial and modified sodium clays presented an effect of reduction in the tobermorite and ettringite formation, probably due to sodium bentonite presence, which absorbs part of the water available for cement hydration, reducing the water available for cement hydration reactions.

When the waste content was increased, an increase of total final ettringite formation occurred, due to its calcium sulfate content. The consequent increase of organic material loss was also observed, which influenced the higher formation of carbonate compounds during thermal analysis. The presence of each clay reduced also the cement hydration, which was also indicated by NMR analysis.

The leaching tests showed that the clays presence reduces the chromium concentration in the lixiviate. Modified sodium bentonite, which was submitted to a cation exchange in the laboratory, presents with higher performance compared to commercial bentonite. The organoclays present better results for the commercial clay.

Samples after 90 days of cement hydration present lower Cr contents in the lixiviate in all cases, when compared to 28 days solidified samples, indicating a higher stabilization of the waste. When both clays are added to the system, no significant differences are observed in samples containing 10% tannery waste. The samples containing 15 and 20% of waste present the same behavior. The modified clays, sodium bentonite and organoclay, had a higher performance compared to the tested commercial clays.

The compressive strength results of all compositions presented similar performance and their values were above the minimum compressive strength necessary for the viability of the S/S process.

The chromium concentration leached in all samples was under the limit concentration of 5mg/L according to the environmental legislation, which shows the environmental safe conditions that can be achieved by using the Stabilization /Solidification process with cement to treat the tannery waste.

Acknowledgments

Our acknowledgments to the Brazilian Research Council (CNPq), Brazilian Ministry of Education (CAPES), UWMRC/USEPA and Louisiana Water Resources Research Institute (LWRRI) for financial support. The authors would like to acknowledge Dr. Dale Treleaven at Chemistry Department, Louisiana State University (LSU), USA for the use of the ^{29}Si NMR spectrometer. To Dr. Frank K Cartledge and Dr. John J.Sansalone for the use of laboratorial equipments and facilities at Chemistry and Civil and Environmental Engineering Departments at LSU.

References

Andriolo, F.R. Construções de concreto: manual de práticas para controle e execução. In Portuguese. São Paulo: Pini, 1984.

Ates, E.; Orhon,D.; Tünay,O. Characterization of tannery wastewaters for pretreatment-selected case studies. Water Science Technology, 1997vol.36, n.2-3, pp. 217-223.

Avudainayagam, S.; Naidu, R.; Kookana, R.S.; Alston, A.M.; McClure, S.; Smith, L.H. Effects of electrolyte composition on chromium desorption in soils contaminated by tannery waste. Australian Journal of Soil Research, 2001, 39, 1077-1089.

Balasubramanian, S; Pugalenthi,V. Determination of total chromium in tannery wastewater by inductively coupled plasma-atomic emission spectrometry, flame atomic absorption spectrometry and UV-visible spectrophotometric methods. Talanta, 1999 vol. 50, pp. 457-467.

Boyd, S.A.; Teppen, B.J.; Li, H.; Johnston, C.T. Sorption of nitroaromatic compounds (NAC) by smectite clay: Geochemical modulation of NAC bioavailability and toxicity.

Geochimica et Cosmochimica ACTA, 2004 v. 68 (11), pp. A451-A451 Suppl. S, jun 2004.

Calvanese,G.; Cioffi,R.; Santoro,L. Cement stabilization of tannery sludge using quaternary ammonium salt exchanged bentonite as pré-solidification adsorbent. Environmental Technology. 2002 Vol. 23, pp. 1051-1062.

Cartledge, F.K.; Butler, L.G.; Chalasani, D.; Eaton, H.C.; Frey, F.P.; Herrera, E.; Tittlebaum, M.E.; Yang, S-L.. Immobilization mechanisms in solidification/stabilization of Cd and Pb salts using Portland cement fixing agents. Environmental Science Technology, 1990, 24, 867-873.

Celik, A.; Yildiz, N.; Calimli, A. Adsorption of some organic compounds by hexadecyltrimethylammonium-bentonite. Reviews in Chemical Engineering, 2000 v.16 (3), pp. 301-309.

Cheeseman, C.R.; Asavapisit, S.; Knight, J. Effect of uniaxially pressing ordinary Portland cement pastes containing metal hydroxides on porosity, density and leaching. Cement and Concrete Research, 1998 v. 28, n. 11, p.1639-1653, nov.1998.

Chun Y., Sheng, G.Y., Boyd S.A. Sorptive characteristics of tetraalkylammonium-exchanged smectite clays. Clays and Clay Minerals, 2003 v. 51 (4), pp. 415-420, Aug 2003.

Conner, J.R.; Hoeffner,S.L. The history of stabilization/solidification technology. Critical reviews in Environmental Science and Technology, 1998, 28(4), pp. 325-396.

Dweck, J.; Cherem da Cunha, A.L.; Pinto, C.A.; Gonçalves, J.P.; Büchler, P.M., Thermogravimetry on calcined mass basis – hydrated cement phases and pozzolanic activity quantitative analysis Journal of Thermal Analysis and Calorimetry 2009 (On-line first) DOI: 10.1007/s10973-008-9761-0

Dweck, J ; Pinto, C. A. ; Vianna, M M G R ; Sansalone, J J ; Diaz, F R V ; Cartledge, F K ; Büchler, P M . Non Conventional Differential Thermal Analysis applied to Stabilization of waste containing high metals In: Proc. II Pan-American Thermal Analysis and Calorimetry Congress and Brazilian. Congress of Thermal Analysis and Calorimetry 2004c. (In Portuguese)

Dweck, J.; Ferreira da Silva, P. F.; Aderne, R.S.; Büchler, P. M.; Cartledge, F. K. Evaluating cement hydration by non-conventional DTA. An application to waste solidification. Journal of Thermal Analysis and Calorimetry 2003, 71(3), 821-827.

Dweck, J.; Ferreira da Silva, P. F.; Büchler, P. M.; Cartledge, F. K. Study by thermogravimetry of the evolution of ettringite phase during type II Portland cement hydration. J. of Ther. Anal. and Calor., 2002, 69(1), 179-186.

Dweck, J.; Büchler, P.M.; Cartledge, F.K. The effect of different bentonites on cement hydration during solidification/stabilization of tannery wastes. Journal of Thermal Analysis and Calorimetry, 2001, v. 64, n. 2, p.1011-1016, Jan. 2001.

Dweck, J., Büchler, P.M., Coelho, A.C.V.; Cartledge, F.K. Hydration of a Portland cement blended with calcium carbonate. Thermochimica Acta, 2000 v. 346, n. 1 2, p.105-113, mar. 2000a.

Dweck, J.; Büchler, P.M.; Coelho, A.C.V.; Cartledge, F.K. Solidification/stabilization of a tannery waste with blended cement and Wyoming bentonite. Journal of Environmental Science Health A , 2000b, 35 (5), 715-740.

Elkhadiri, I.; Diouri, A.; Boukhari, A.; Aride, J.; Puertas, F. Mechanical behavior of various mortars made by combined fly ash and limestone in Moroccan Portland cement. Cement and Concrete Research 2002, 32, 1597-1603.

Flores-Velez, A.M.A., Dominguez,O. Characterization and properties of Portland cement composites incorporating zinc-iron oxide nanoparticles. Journal of Materials Science. 2002 Vol. 37, pp. 983-988.

Gilliam, T.M.; Wiles, C.C. Stabilization and solidification of hazardous, radioactive and mixed wastes. ASTM Publication Code Number:04-012400-56. USA.1992.

Hanna, R.A.; Barrie, P.J.; Cheeseman, C.R.; Hills, C.D.; Büchler, P.M.; Perry, R. Solid State ^{29}Si and ^{27}Al NMR and FTIR study of cement pastes containing industrial wastes and organics. Cement and Concrete Research. 1995, 25 (7), 1435-1444.

Hills, C.; Pollard, S.J.T. The influence of interference effects on the mechanical, microstructural and fixation characteristics of cement-solidified hazardous waste forms. Journal of Hazardous Materials 1997, 52 (2-3), 171-191.

Irassar,E.F.; Bonavetti,V.L.; González,M. Microstructural study of sulfate attack on ordinary and limestone Portland cements at ambient temperature. Cement and Concrete Research, 2003 Vol. 33, pp. 31-41.

Jauberthie,R.; Rendell,F. Physicochemical study of the alteration surface of concrete exposed to ammonium salts. Cement and Concrete Research, 2003 vol 33, pp. 85-91.

Jose, C.L.V; Diaz, F.R.V.; Buchler P.M. Characterization and adsorption of phenol by organophilic clays. Advanced Powder Technology III Materials Science Forum, 2003 v. 416-4, pp.550-554.

Lai, L.T.; Kasbohm, J.; Eidam, J.; Kramer, A.; Below, H.; Nam, P.T.; Qui, D.H. Anthropogenic impacts on heavy metals pollution in soil and sediments of craft-oriented villages in the Nam Dinh Province-Vietnam. Proceedings of 12th International Clay Conference, Bahia Blanca, Argentina, 2001.

Lambert, J.B.; Mazzola, E.P. Nuclear magnetic resonance spectroscopy. An introduction to principles, applications, and experimental methods. Pearson Education, Inc. USA, 2004.

Lea, F.M. The chemistry of cement and concrete. 3.ed. New York, Chemical Publishing Co., 1971.

Li, B.; Wenquan Liang, Z. H. Study on high-strength composite Portland cement with a larger amount of industrial wastes. Cement and Concrete Research. 2002, 32, 1341-1344.

Lin, C.K.; Chen, J.N.; Lin, C.C. An NMR, XRD and EDS study of solidification/stabilization of chromium with Portland cement and C_3S. Journal of Hazardous Materials 1997, 56, 21-34.

Lo, I.M.C. Solidification/stabilization of phenolic waste using organic-clay complex. Journal of Environmental Engineering. 1996 pp. 850-855, September.

Mackenzie, K.J.D.; Smith, M.E. Multinuclear solid-state NMR of inorganic materials. Pergamon Materials Series, UK, 2002.

Mehta, P.K., Monteiro, P.J.M. Concreto: estrutura, propriedades e materiais. In Portuguese. São Paulo: Pini, 1994.

Milacic, R.; Stupar, J. Fractionation and oxidation of chromium in tannery waste and sewage sludge-amended soils. Environmental Science Technology, 1995 v. 29, n. 2, p. 506-514, Feb. 1995.

Mulholland, K.L. Dyer,J.A. Pollution prevention: methodology, technologies and practices. American Institute of Chemical Engineers, New York. 214 pages, 1999.

Nennemann,A.; Kulbach,S.; Lagaly,G. Entrapping pesticides by coagulating smectites. Applied Clay Science. 2001 Vol. 18, pp.285-298.

Nzengung, V.A; Voudrias, E.A; Nkedikizza, P.; Wampler, J.M; Weaver, C.E. Organic cosolvent effects on sorption kinetics of hydrophobic organic chemicals by organoclays. Environmental Science and Technology, 1997, 31 (5), 1470-1475.

Ortego,J.D.; Barroeta,Y.; Cartledge,F.K.; Akhter,H. Leaching effects on silicate polymerization. An FTIR and 29Si NMR study of lead and zinc in Portland cement. Environmental Science and Technology. 1991 Vol 25, pp. 1171-1174.

Pera,J.; Thevenin,G.; Chabnnet,M. Design of a novel system allowing the selection of an adequate binder for solidification/stabilization of wastes. Cement and Concrete Research, 1997 vol. 27(10), pp. 1533-1542.

Pinto, C. A. ; Dweck, J ; Sansalone, J J ; Cartledge, F K ; Albanez, N. ; Hamassaki, L T ; Büchler, P M . Evaluation of S/S Process of Tannery Waste in Cement by Structural and Mechanical Analysis. In: Sixth International Latin American Conference on Powder Technology, 2007, Rio de Janeiro, Brazil. Sixth International Latin American Conference on Powder Technology, 2007. v. 1. p. 207.

Pinto, C. A. ; Diaz, F R V ; Sansalone, J J ; Dweck, J ; Cartledge, F K ; Büchler, P M . X-Ray Diffraction study of particulate tannery waste solidified in cement. Materials Science Forum, 2006 v. 530-531, p. 478-484.

Pinto, C.A.; Sansalone, J.J.; Dweck, J.; Cartledge, F.K.; Diaz, F. R. V.; Büchler, P. M. Hazardous wastes disposal: Stabilization/Solidification of tannery residue in the presence of chromium. Materials Science Forum, 2005a, 498-499, 697-703.

Pinto, C.A.; Dweck, J.; Sansalone, J.J.; Cartledge, F.K.; Tittlebaum, M. E.; Büchler, P. M. Early stages of solidification/stabilization of storm water runoff solid residuals in cement. Journal of Thermal Analysis and Calorimetry., 2005b, 80, 715-720.

Pinto, C. A. ; Sansalone, J J ; Dweck, J ; Cartledge, F K ; Diaz, F R V ; Büchler, P M . Waste treatment: tannery waste performance in stabilization/solidification in cement process. In: XII International Conference on Heavy Metals in the Environment, 2005, Rio de Janeiro. Proceedings:XII International Conference on Heavy Metals in the Environment, 2005c.

Pinto, C. A. ; Diaz, F R V ; Sansalone, J J ; Dweck, J ; Cartledge, F K ; Büchler, P M . Study of particulate tannery waste solidified in cement by X-Ray Diffraction. In: V International Latin-American Conference on Powder Technology, 2005, Costa do Sauípe - BA. Proceedings V International Latin-American Conference on Powder Technology, 2005d.

Pinto, C. A. ; Sansalone, J J ; Cartledge, F K ; Hamassaki, L T ; Dweck, J ; Diaz, F R V ; Büchler, P M . Leaching and mechanical behavior of tannery waste stabilized/solidified in cement. In: 2nd Mercosur Congress on Chemical Engineering and 2nd Mercosur Congress System Engineering, ENPROMER, 2005, Village Rio das Pedras-RJ. Proceedings of 2nd Mercosur Congress on Chemical Engineering and 2nd Mercosur Congress System Engineering, 2005e.

Pinto, C. A. ; Hamassaki, L T ; Diaz, F R V ; Dweck, J ; Büchler, P. M. . Thermal and Mechanical Characterization of Tannery Waste Solidification and Stabilization. Journal of Thermal Analysis and Calorimetry, 2004a v. 77, n. 3, p. 777-787.

Pinto, C. A. ; Vianna, M M G R ; Hamassaki, L T ; Dweck, J ; Diaz, F R V ; Büchler, P. M. Caracterização térmica e mecânica do resíduo de curtume solidificado em cimento . In: Congresso Brasileiro de Ciência e Tecnologia em Resíduos e Desenvolvimento Sustentável, 2004, Florianópolis-SC. Anais Congresso Brasileiro de Ciência e Tecnologia em Resíduos e Desenvolvimento Sustentável, 2004b. In Portuguese.

Pinto, C.A.; Hamassaki, L.T.; Valenzuela-Diaz, F.R.; Dweck, J.; Buchler P.M. Tannery waste solidification and stabilization - Thermal and mechanical characterization. Journal of Thermal Analysis and Calorimetry, 2004c v. 77 (3), pp. 777-787.

Pinto, C. A. ; Sansalone, J J ; Dweck, J ; Cartledge, F K ; Diaz, F R V ; Büchler, P M . Hazardous wastes disposal: Stabilization/Solidification of tannery residue in the presence of chromium. In: IV International Latin-American Conference on Powder Technology, 2003, Guarujá- SP. Proceedings IV International Latin-American Conference on Powder Technology, 2003.

Pinto, C. A. ; Toffoli, S. M. ; Hamassaki, L T ; Diaz, F R V ; Büchler, P M . Application of bentonite and organophilic clay in stabilization/solidification of tannery waste.. In: XII International Clay Conference, 2001, Bahia Blanca - Argentina. Proceedings of XII International Clay Conference, 2001.

Souza Santos, P. Science and technology of clays. In Portuguese. São Paulo, Brazil. Edgar Blucher, 1992.

Spence, R.D. (ed.) Chemistry and microstructure of solidified waste forms. Boca Raton: Lewis Publishers, 1992.

Stegeman, J.A.; Buenfeld,N.R. Prediction of unconfined compressive strength of cement paste with pure metal compound additions. Cement and Concrete Research, 2002 V.32, pp. 903-913.

Stepniewska, Z.; Bucior, K.. Chromium contamination of soils, waters and plants in the vicinity of a tannery waste lagoon. Environmental Geochemistry and Health, 2001, 23, 241-245.

Swaddiwudhipong, S.; Chen, D.; Zhang, M.H. Simulation of the exothermic hydration process of Portland cement. Advances in Cement Research, 2002, v. 14, n. 2, p. 61-69, Apr. 2002.

The Merck index: an encyclopedia of chemicals and drugs. 8.ed. Rahway: Merck, 2001.

Tikalsky,P.J.; Roy,D.; Scheetz, B.; Krize, T. Redefining cement characteristics for sulfate-resistant Portland cement. Cement and Concrete Research. 2002 Vol. 32,pp. 1239-1246.

Tillman, F.D., Bartelt-Hunt, S.L., Smith, J.A., Alther G.R. Evaluation of an organoclay, an organoclay-anthracite blend, clinoptilolite, and hydroxy-apatite as sorbents for heavy metal removal from water. Bulletin of Environmental Contamination and Toxicology, 2004, v. 72 (6), pp. 1134-1141 Jun 2004.

Tsivilis,S.; Batis,G.; Chaniotakis,E.; Grigoriadis,G.; Theodossis,D. Properties and behavior of limestone cement concrete and mortar. Cement and Concrete Research. 2000, Vol. 30, pp. 1679-1683.

Tsivilis,S.; Chaniotakis,E.; Badogiannis,E.; Pahoulas,G.; Hias,A. A study on the parameters affecting the properties of Portland limestone cements. Cement and Concrete Composites. 1999 Vol. 21, pp. 107-116.

United States Environmental Protection Agency. Test methods for evaluating solid waste, physical/chemical methods. 3.ed. Washington, 1992. (SW-846).

Valenzuela-Diaz, F.R. Preparation of organophilic clays from Brazilian smectitic clay. Advanced Powder Technology II Key Engineering Materials. 2001, 189-191, 203-207.

Vianna, M..M.G.R.; Franco, J.H.R.; Pinto, C.A.; Diaz, F.R.V.; Buchler, P.M. Sorption of oil pollution by organoclays and a coal/mineral complex. Brazilian Journal of Chemical Engineering, 2004, v.21 (2), pp. 239-245, Apr-Jun 2004.

Vieira, M.F.A.; Brito, A.L.F.; Campos, A.M.B.; Santos, D.L.; Goldfarb, J.; Costa, M.A.S.M.; Bastos, S.A. Reutilization in the tanning industry of the residual liquid from chrome tanning. Proceedings of ENPROMER 2001, 2001, 1, 511-516.

Vlyssides, A.G.; Israilides,J. Detoxification of tannery waste liquors with an electrolysis system. Environmental Pollution, 1997, Vol. 97, nº1-2, p.147-152,.

Volzone, C. Hydroxy-chromium smectite: influence of Cr added. Clays and Clay Minerals, 1995, v. 43, n. 3, p. 377-382,.

Volzone, C. Changes in kaolinite, vermiculite and smectite clays after adsorption by hydroxy-chromium species. Australian Journal of Soil Research. 1998, Vol. 36, pp. 423-428,.

Volzone, C. OH-Cr (III) in dioctaedral and trioctahedral smectites: texture and structure changes. Materials Chemistry and Physics. 1997, Vol. 47, pp. 13-16.

Volzone, C.; Cesio,A.M. Influence of different chromium species on smectite clay. Anales de la Asociación Química Argentina 1999,Vol. 87, nº 1/2, pp. 59-63.

Walsh, A.R.; O'Halloran, J. Accumulation of chromium by a population of mussels exposed to leather tannery effluent. Environmental Toxicology and Chemistry, 1998, 17 (7), 1429-1438.

Weng, C.H.; Huang, C.P.; Allen, H.E.; Leavens, P.B.; Sanders, P.F. Chemical interactions between Cr(VI) and hydrous concrete particles. Environmental Science & Technology, 1996, vol.30, nº2, 371-376.

In: Management of Hazardous Residues Containing Cr(VI) ISBN: 978-1-61470-637-3
Ed: Maria José Balart Murria

Chapter 14

AN OVERVIEW OF THE PROPERTIES AND TREATMENT OF CHROMITE ORE PROCESSING RESIDUE (COPR)

Maria Chrysochoou[1] and Dimitris Dermatas[2]

[1] Department of Civil and Environmental Engineering
University of Connecticut Storrs CT 06269, USA
E-mail: maria.chrysochoou@uconn.edu
[2] Waste Management Authority Eastern Macedonia-Thrace, Komotini, GREECE

ABSTRACT

Chromite Ore Processing Residue was produced in large quantities in various countries and was deposited as backfill material in residential and commercial areas. There are two problems with COPR: its content in carcinogenic hexavalent chromium and its volume expansion and heave that leads to structural failures. Heave drew attention to two COPR deposition sites, SA7 and DMT; SA7 is currently under court order to remove and dispose all COPR in a hazardous waste landfill. Investigations were thus initiated to understand the mechanism of COPR heave and find a solution to mitigate it while also applying a reductive treatment for Cr(VI).

Based on the site observations and the background information on COPR, it was considered that the driver for manifestation of heave was mineralogical transformations that led to volume expansion. The understanding of a) qualitative mineralogy and reaction pathways and b) quantitative mineralogy and distribution in the sites was therefore essential to delineate the heaving mechanism. The same analyses would also yield Cr(VI) speciation and the geochemical pathway to treat it. The analyses were performed separately for three types of COPR with different physical and geotechnical properties: grey black (GB) granular and hard brown (HB) COPR found in both sites, and fine clayey COPR (C) found only at SA7.

The chemical analyses showed that the major elements were essentially the same, including calcium, aluminum, iron, magnesium and chromium. These were the metals that would comprise the bulk of COPR mineralogy and drive the geochemistry of the system. The distribution of these metals in the different types of COPR was somewhat different; the most important observation for heave was the lower concentration of Al in C COPR and the higher in DMT COPR; and for reductive treatment, the accumulation of Cr(VI) in HB and C COPR.

The mineralogical analyses showed that COPR resembles iron-rich cement, with hydration and carbonation as major reactions, the latter limited by the presence of CO_2, the former limited kinetically by the rate of parent material reaction (brownmillerite and periclase). GB COPR contained high amounts of parent minerals, exhibiting severe kinetic inhibition of hydration reactions after more than 50 years of deposition. HB COPR contained almost exclusively hydrated phases, was thus more reactive, and was found as distinct layers in both sites. A plausible explanation for a difference in hydration kinetics is a difference in temperature between the two materials, which emerges as a significant factor for manipulation of the geochemistry. C COPR resembled a diluted GB by high amounts of portlandite and calcite; it is considered a site-specific occurrence. The mineralogy also showed that Cr(VI) was mainly bound in a cementitious phase, Calcium Aluminum Chromium Oxide Hydrates (CACs); hydrotalcites and hydrogarnets also have potential to bind Cr(VI) but the extent of it remains unresolved.

Based on the identified mineralogy, the possible heaving mechanisms were pure hydration of parent phases to katoite and brucite, formation and hydration/dehydration of ettringite, formation and hydration of CACs, and formation of calcite. Swell tests showed that pure hydration is too slow to cause pronounced heave; carbonation led to consolidation. CAC formation was not achieved, so that this mechanism remains unresolved. Ettringite is known to cause swell in cements and lime-treated soils and caused swell in the lab. The DMT investigation showed ettringite present in high amounts around the heaved areas. The SA7 pilot treatment using ferrous sulfate led to the rapid formation of high amounts of ettringite and pronounced heave. . This finding may reverse to a benefit, because sulfate may accelerate brownmillerite hydration and also releases chromate from the solid; the addition of enough sulfate can therefore dissolve all brownmillerite to form a geotechnically stable ettringite-dominated matrix and render Cr(VI) available for any reductant, preferably a slow-release one. However, this is not the only heave mechanism, as heave is also observed in the absence of ettringite. The association of swell with zones of lithification points to an accelerated GB-to-HB transformation as a viable heave mechanism.

1. Introduction

Chromite Ore Processing Residue (COPR) is generated as a by-product of chromite ore processing to isolate and extract chromium. Chromite ore, which may be expressed with the chemical formula $(Mg,Fe)(Cr,Al,Fe)_2O_4$, is roasted at a high-temperature (ca. 1200 ^oC) to remove metal impurities, mainly Fe, Mg and Al. Addition of lime (CaO) during the roasting process leads to the formation of insoluble phases that are removed as solid waste, while chromium is oxidized to its hexavalent form and recovered as soluble sodium chromate.

COPR as it comes out of the roasting process is a granular material that resembles a sandy soil (Figure 1), but develops cementing properties over time; it was therefore considered to be a suitable foundation material for construction purposes. As a result, numerous COPR deposition sites exist in areas near chromite ore processing plants.

Figure 1. Representative image of black unreacted COPR from the SA7 site.

Such plants are found throughout the world, including England, Japan, Germany and the United States (McKee, 1988), as well as India, Pakistan, China and the former Soviet Union (Darrie, 2001). Three plants were located in Hudson County, New Jersey and deposited 2.75 million tons of COPR in the period 1905-1976 (Weng et al., 1996). Because COPR was considered a suitable foundation material, deposition sites are often prime real estate located in residential or commercially developed areas. With the beginning of the environmental movement in the 1970s, local population became gradually aware of the hazard posed by the significant amounts of residual hexavalent chromium (Cr(VI)) in COPR. The observation of Cr salt crystals in the walls of homes, schools, and workplaces, and in encrustations of the soil surface, quickly raised concerns). Investigations at 150 sites led to their characterization as hazardous and to the initiation of extensive risk assessment studies conducted by the New Jersey Department of Environmental Protection (NJDEP) (Burke et al., 1991, Lioy et al., 1992). The total Cr content of COPR was found as high as 46,000 mg/kg, with Cr(VI) being up to 30% of the total Cr in some cases (Lioy et al., 1992). The current regulatory limit for Cr(VI) for residential soil use in the state of New Jersey was 240 mg/kg (NJDEP, 1998) and was reduced to 20 mg/kg for both residential and non-residential sites (NJDEP 2007).

Apart from the environmental threat posed by the presence of Cr(VI), the use of COPR for construction purposes proved to be problematic from the engineering aspect as well, due to delayed volumetric expansion that manifested as heave and was documented at some sites. Study Area 7 (SA7) is a 34-acre site located in Hudson County, N.J., between Route 440 and the Hackensack river. About 1.5 million tons of COPR that was generated by an adjacent processing plant, owned by Mutual Chemical, were deposited at SA7 in the period 1905-1954 (Dermatas et al., 2006). Land grants permitting fill placement were issued from 1900 to 1925. In 1954, Mutual Chemical stopped processing chromium ore. After hydraulic deposition, the COPR was spread mechanically across the site, graded, and covered with soil for the purpose of commercial site development. A drive-in theater was constructed at the site and then an adjacent large department store (1966). A furniture depot and a car wash facility were also constructed. All structures but the furniture depot ultimately had to be demolished due to the pronounced surface heave. Figure 2 shows the site at the beginning of the investigation in 2003, with a prominent heave feature. The SA7 investigation was initiated due to litigation pursued by a local environmental activist group against the site owner. The court ultimately ordered the excavation and final disposal of all deposited COPR in a hazardous waste landfill, which was completed in 2009.

Figure 2. Image of the SA7 site in 2003 (heaved pavement is visible in the center).

Heaving phenomena were also observed at Dundalk Marine Terminal (DMT), a 350-acre site located in Baltimore, Maryland. The southern portion of the terminal was constructed on marshlands by means of fill placement. The fill material at DMT included mixtures of COPR, sand dredging, and other locally available fill materials such as clay and borrow sand (Tinjum et al. 2008a). The need for constant reworking activities due to sporadic heave at the site triggered an investigation that aimed at identifying the volume expansion mechanism, in search of a more permanent and economic solution.

Both the treatment of hexavalent chromium in COPR and the mitigation of heave phenomena are largely dependent on its geochemistry; i.e. understanding the chemistry and mineralogy and the overall geochemical reaction pathway of COPR is essential to the design of successful treatment options that tackle both the environmental and geotechnical aspects. Accordingly, this overview is divided into three sub-chapters: a) COPR geochemistry; b) COPR environmental treatment; and, c) COPR heave and its mitigation. It should be noted that this overview refers to pure COPR derived from the high-lime process, and not to COPR-soil mixtures or waste derived from different processing technologies, which may have different geochemistry and behavior.

2. COPR Chemistry and Mineralogy

Chrysochoou et al. (2010) provides a detailed description of the properties of different COPR types encountered at SA7 and DMT. Tinjum et al. (2008a) also provided an overview of the physical characteristics of different COPR types at DMT. Briefly, two main COPR types with different physicochemical characteristics were observed: a grey-black granular material (Figure1) and a hard brown material that was encountered as bulk layers at DMT and as apparently random layers at SA7. Figure 3 shows an example of GB-HB stratification at DMT.

Figure 3. GB-HB-GB stratification at DMT.

Another type of COPR found only at SA7 was a fine-grained material that consisted primarily of calcium-bearing compounds (portlandite $Ca(OH)_2$ and calcite $CaCO_3$) and lower amounts of typical COPR minerals. The study concluded that the presence of this layer was an artifact of the depositional process at SA7, as it was not found at any other site, and will thus no longer be discussed here.

COPR chemistry is dominated by calcium due to the lime addition, and by the metals that complex with Cr in chromite ore: iron, aluminum and magnesium. Figure 4 shows the concentrations of these metals for three sites, SA7, DMT and Glasgow (data obtained from Chrysochoou et al. (2010) and Geelhoed et al. (2002)). COPR from the Glasgow site was Mg-rich, while SA7 and DMT COPR had elevated iron content, which most probably reflects differences in the composition of the original chromite ore. Glasgow COPR also had a higher Cr content and more residual Cr(VI) compared to SA7 and DMT GB COPR. It should be noted that the presence of hard brown COPR in Glasgow was only reported in one early publication (Manzur et al. 2001), while all subsequent publications referred to grey-black type of COPR. Thus it, the extent of HB COPR presence and its properties in Glasgow sites, remains unknown.

The addition of lime during the roasting process created a highly alkaline material at sites where COPR was not mixed with native soil; COPR pH was in the range 12-12.5, which corresponds to the pH of lime-saturated water (Hausmann 1990). COPR also has a very high Acid Neutralization Capacity (ANC). Moon et al. (2008) reported a total ANC of 37 eq/kg grey-black COPR from the SA7 site to reduce COPR pH to below 1, with an acid-COPR equilibration time of 2 months. Chrysochoou et al. (2010) reported a slightly lower ANC of 32 eq/kg for SA7 GB COPR and 30 eq/kg for both GB and HB COPR from the DMT site; the equilibration time used in the ANC tests was one week.

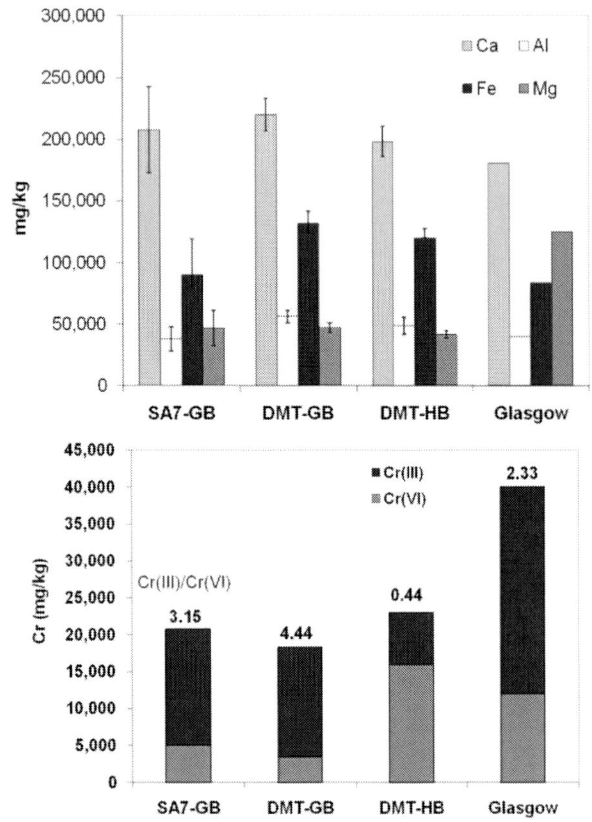

Data obtained from Chrysochoou et al. (2010) and Geelhoed et al. (2002).

Figure 4. Concentrations of major metals (left) and Cr species (right) in COPR from three depositions sites.

Tinjum et al. (2008c) also performed ANC analyses on GB COPR obtained from an unidentified COPR deposition site on the mid-Atlantic U.S. coast and reported a total ANC of 3.5 eq/kg at 20 min equilibration and ~23 eq/kg at 48 h equilibration time. The difference in the results is attributed to the kinetic inhibition of mineral dissolution and specifically brownmillerite, which is the COPR parent mineral and stores the majority of the material alkalinity. This will be discussed in detail in a later section.

COPR mineralogy is cementitious. The most common mineral phases observed both in U.S. and Glasgow COPR are listed in Table 1.

Brownmillerite and periclase are the COPR "parent" phases, formed during the roasting process at 1200 °C. Brownmillerite belongs to the general group of ferrites that are one of the major clinker phases of Portland and oilwell cements (EBH, 1998). Periclase may also be encountered as a minor component in cements. Both hydrate into different products depending on the pore water chemistry, i.e., the availability of other compounds (silicate, carbonate, sulfate, and, in the case of COPR, chromate).

Table 1. Common crystalline phases identified in most SA7 and DMT COPR samples

	SA7	DMT	Chemical formula	Mineral group	Gs
Afwillite	x		$Ca_3(HSiO_4)_2(H_2O)_2$	Nesosilicates	2.62
Brownmillerite	x	x	Ca_2FeAlO_5	Oxides	3.76
Brucite	x	x	$Mg(OH)_2$	Hydroxides	2.39
Calcite	x	x	$CaCO_3$	Carbonates	2.71
CAC[1]-9	x	x	$Ca_4Al_2O_6(CrO_4)\cdot 9H_2O$	n.a.	1.38
CAC[1]-12	x	x	$Ca_4Al_2O_6(CrO_4)\cdot 12H_2O$	n.a.	1.27
CAC[1]-14	x	x	$Ca_4Al_2O_6(CrO_4)\cdot 14H_2O$	n.a.	1.28
CAC[1]-Na	x	x	$NaCa_4Al_2O_6(SO_4)_{1.5}\cdot 15H_2O$	n.a.	1.33
Ettringite	x	x	$Ca_6Al_2(SO_4)_3(OH)_{12}\cdot 26H_2O$	Sulfates	1.80
Katoite	x	x	$Ca_3Al_2(OH)_{12}$	Hydrogarnets	2.76
Hydroandradite[2]	x	x	$Ca_3Fe_2(SiO_4)_{1.15}(OH)_{7.4}$	Hydrogarnets	3.16
Periclase	x	x	MgO	Oxides	3.78
Portlandite	x	x	$Ca(OH)_2$	Hydroxides	2.23
Quintinite	x	x	$Al_2Mg_4(OH)_{12}(CO_3)(H_2O)_3$	Hydrotalcites	2.14
Sjögrenite	x	x	$(Mg_6Fe_2(OH)_{16}(CO_3)(H_2O)_4$	Hydrotalcites	2.11

[1] Calcium Aluminum Chromium Oxide Hydrate
[2] Only crystal structure file

Hydrogarnet is the most stable hydration product of brownmillerite at ambient temperature in the absence of anions (Rogers and Aldridge, 1977). It may occur as a solid solution with variable Al and Fe contents, while silicate (SiO_4^{2-}) may substitute for hydroxyls. Crystalline FeOOH (either goethite or lepidocrocite) may also be a product of brownmillerite hydration and was identified in only a few patterns from hard brown SA7 and DMT COPR (Chrysochoou 2006). Calcium-iron hydrates have also been reported as ferrite hydration products, but could not be detected by XRD due to their disordered nature (Emanuelson and Hansen, 1997). The direct hydration product of periclase is brucite. Portlandite (hydrated lime), apart from being a product of brownmillerite hydration, may also be considered as the direct hydration product of excess unhydrated lime (quicklime) that is not bound in brownmillerite during the roasting process. Quicklime was not identified in any of the COPR XRD patterns, as its hydration is a fast exothermic reaction that occurs even in contact with atmospheric moisture (Hausmann, 1990). Portlandite is therefore also considered to be a COPR "parent" phase, being present in COPR upon deposition. Finally, residual chromite may also be added to the COPR parent phases; however, it is not expected to react at ambient conditions; it was present in very low or non-detectable quantities and is therefore not considered to be a major component for the COPR reaction pathways.

Brownmillerite and periclase form more complex hydration products in the presence of anions, such as chromate, sulfate, silicate, and carbonate. The hydration of brownmillerite in the presence of anions leads to the formation of hexagonal calcium aluminate hydrates (CAH), which have a layered structure and accommodate anions and water in the crystal interlayer; they are commonly encountered in the cement literature as monophases, because of their capacity to accommodate one (divalent) anion in their unit cell. Calcium Aluminum

Chromium Oxide Hydrates (CACs) and $NaCa_4Al_2O_6(SO_4)_{1.5}{\bullet}12H_2O$ (Table 1), hydrotalcites (quintinite and sjoegrenite), monosulfate ($Ca_4Al_2O_6(SO_4){\bullet}12H_2O$), and hydrocalumite ($Ca_4Al_2O_6Cl_2{\bullet}10H_2O$) are examples of such monophases. Ettringite is described as a triphase in cement literature, because it accommodates three sulfate groups in its unit cell (Moore and Taylor, 1970). It is considered the most common hydration product of aluminoferrite in the presence of calcium hydroxide (EBH, 1998). Silicate, apart from being substituted in hydrogarnet, also formed Calcium Silicate Hydrates (CSH), either amorphous, or crystalline as afwillite (Table 1). Afwillite was only identified in SA7 COPR, in which quartz and other soil minerals mixed with COPR provided a source of soluble silica at the strongly alkaline COPR pH. Finally, CO_2, coming in contact with COPR, forms primarily calcite, progressively consuming calcium from portlandite and other cementitious phases and decreasing COPR pH. Figure 5 illustrates the reaction pathway of COPR minerals.

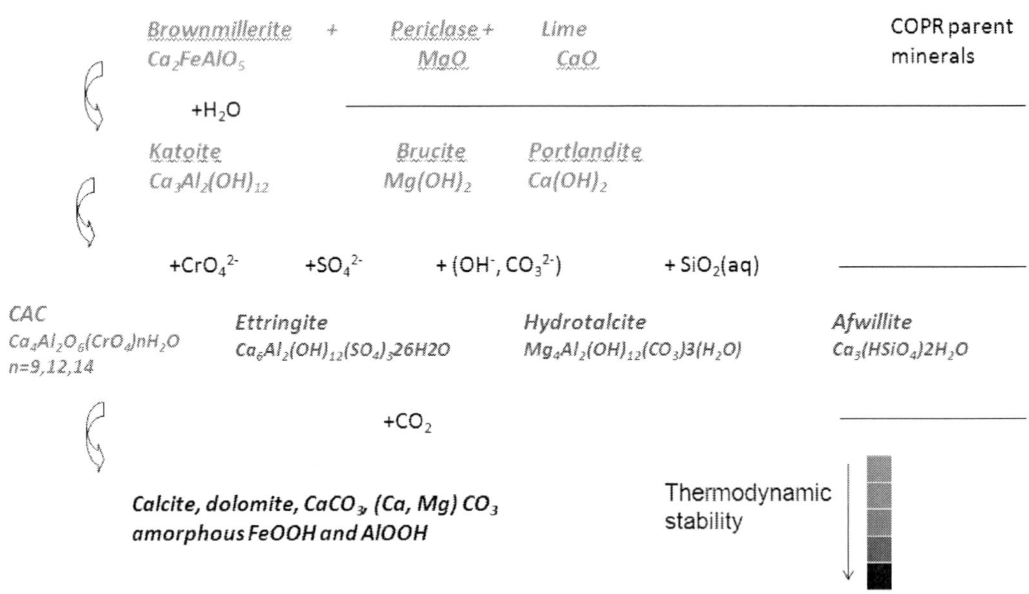

Figure 5. COPR reaction pathway – color scheme illustrates relative thermodynamic stability of different minerals.

GB and HB COPR have qualitatively the same mineralogy, but are quantitatively very different (Chrysochoou et al. 2009, Chrysochoou et al. 2010). COPR mineralogy was quantified using the Rietveld method (Hillier et al., 2003, Chrysochoou and Dermatas, 2007). The Rietveld analysis provides a relative weight distribution of the crystalline phases that are used as input, based on their properties and the observed reflections. This means that the amorphous material and the crystalline phases that are below the detection limit cannot be directly quantified. The quantification of the amorphous material (including trace crystalline phases) is performed indirectly through the addition of an internal standard (a crystalline phase of known structure) at a known weight ratio. Average concentrations for the different layers at the two sites (SA7 and DMT) are shown in Table 2 (data reproduced from Chrysochoou 2006).

Table 2. Average mineral concentrations in %wt. of the total crystalline phases in the SA7 and DMT COPR zones

	SA7-GB (unsat)	SA7-GB (sat)	DMT-GB	DMT-HB (sat)
Brownmillerite	28	30	59	2
Periclase	2	2	4	0
Brucite	7	9	4	8
Portlandite	3	5	2	1
$CaCO_3$	15	17	6	7
Hydrotalcite	8	6	6	18
Hydrogarnet	14	10	15	31
Ettringite	6	5	1	0
CAC	8	6	4	32
Afwillite	5	6	0	0
Soil minerals	3	4	0	0

Overall, grey-black COPR encountered both at SA7 and DMT yielded high concentrations of brownmillerite (30% and 60% of the crystalline phases on average). Katoite was the most abundant hydration product (10-15%), with lower concentrations of other phases (4-8% CACs, 6-8% hydrotalcites, 1-6% ettringite). DMT GB COPR contained more brownmillerite and less portlandite and calcite compared to SA7 B1 and B2 COPR. DMT HB COPR was, on the other hand, dominated by hydration products (31% hydrogarnet, 32% CACs, 18% hydrotalcites), with low or non-detectable concentrations of brownmillerite and periclase. The amorphous content of COPR was observed to be between 30% and 40% (Chrysochoou and Dermatas, 2007, Dermatas et al., 2006, 2007), which may be used as a correction factor to obtain absolute mineral concentrations from Table 2.

The mineralogical analyses correlated well with the physical properties of the respective sample groups. The high amount of parent phases in sandy GB COPR was consistent both with the color and the granular nature of the material (synthesized brownmillerite is a black sandy material). Conversely, lithification of hard brown COPR can be attributed to the high amount of hydration products, similar to the cementing action produced by Calcium Aluminate Hydrates (CAH) and Calcium Silicate Hydrates (CSH) in cements (EBH, 1998). Accordingly, the higher amount of brownmillerite hydration products explained the cementation observed occasionally in the unsaturated grey-black COPR compared to DMT GB, which was loose material in all obtained samples.

The chemical analyses were also, in general, consistent with the mineralogical analyses. The higher calcium concentration in SA7 samples compared to DMT was reflected in the higher concentrations of Ca-bearing minerals (calcite and its polymorphs, portlandite, afwillite). At DMT, the mineralogy confirmed that the metal concentrations in the HB COPR were lower compared to GB because of hydration-induced dilution; in the case of HB, dilution was additionally caused by the presence of soil admixtures and calcite. The high Cr(VI) concentration in HB COPR also agreed with the high concentration of CACs.

Mineralogy also affects other chemical parameters, such as the ANC and Cr leaching properties. A high amount of hydration products results in a high ANC at the alkaline pH range (10-12) for the HB COPR (Chrysochoou et al., 2010); this, along with the presence of

Cr(VI)-bearing minerals, favors the immobilization of Cr(VI) in the matrix even upon acid attack (e.g., TCLP conditions). Conversely, the high amounts of brownmillerite contribute to a higher ANC at neutral to acidic pH values in GB COPR; in this case, alkalinity is quickly consumed at high pH and Cr(VI) dissolves more easily into solution upon acid addition. Consequently, mineralogy plays a crucial role in the design and implementation of a reductive treatment, which primarily aims at releasing Cr(VI) into a solution to be available for redox reactions.

The understanding of Cr(VI) speciation and leaching patterns is crucial to the success of COPR reductive treatment and will thus be discussed separately.

3. Cr(VI) Speciation and Leaching in COPR

Cr(VI) occurs as oxyanionic $HCrO_4^-$ or CrO_4^{2-}, the latter form being prevalent at the strongly alkaline COPR pH. Sorption mechanisms cannot effectively immobilize oxyanions at neutral and alkaline pH, because common sorptive surfaces (clays, Fe- and Al-oxyhydroxides) become negatively charged at slightly acidic to neutral pH (Sposito, 1989). Furthermore, there are few solubility-controlling phases for oxyanions; Ca, Ba, and Pb are known to form chromate precipitates, with solubility products that vary from $10^{-2.26}$ ($CaCrO_4$) to $10^{-12.6}$ (PbCrO4). The existence of solubility-controlling phases for Cr(VI) may be desirable to restrict in situ leaching, but becomes a problem when reductive treatment is pursued, as it requires the dissolution of chromate-binding phases to render chromate available in solution.

Chrysochoou and Dermatas (2007) reported the presence of Calcium Aluminum Chromium Oxides Hydrates (abbreviated as CAC) in COPR with the chemical formula $Ca_4Al_2(CrO_4)(OH)_{12}{\cdot}nH_2O$ (n=3,6,8). These compounds were identified by bulk X-ray Diffraction (XRD) and quantification with the Rietveld method showed that they accounted for 50% on average (25-75% range) of the total Cr(VI) in the solid. Hillier et al. (2003, 2007) reported that hydrogarnet ($Ca_3Al_2(OH)_{12}$) was also an important host phase for chromate in COPR found in Glasgow, Scotland. Hillier et al. (2003) reported Cr(VI)-hydrocalumite as a Cr(VI) host instead of CAC to account for the observed peaks at 7.9 Å. Hydrocalumite ($Ca_4Al_2Cl_2(OH)_{12}{\cdot}6H_2O$) has identical structure with the CACs, with chlorine instead of chromate in the interlayer. This phase shares peaks with the hydrotalcite group ($Mg_6Al_2(CO_3)(OH)_{16}{\cdot}4H_2O$), which was reported instead of hydrocalumite in SA7 COPR. The presence of chromate instead of chlorine in the interlayer is thought to expand the basal d-spacing from 7.9 Å (PDF 42-0558) to 9.7 Å (PDF 41-0478) with the same waters of hydration (six).

All these phases belong to the general group of layered double hydroxides (LDH) or anionic clays. LDH have the same layered structure, which consists of positively charged brucite-like sheets, held together by anions and water molecules. These have the general formula $\left[M_{1-x}{}^{2+}M_x{}^{3+}(OH)_2\right]^{x+} A^{n-}{}_{x/n}$, where M stands for a metal and A for an anion. LDH are capable of multiple substitutions, both for the cations in the octahedral sheets and for the anions in the interlayer. Rousselot et al. (2002) report the incorporation of Al^{3+}, Fe^{3+}, Ga^{3+} and Sc^{+}, while Hou et al. (2003) investigated the expansion of LDH structures with a large variety of anions (sulfate, selenate, chloride, fluoride and others). Goswamme et al. (1998)

showed that a variety of LDH structures (Mg-Al, Ni-Al) could incorporate chromate in the interlayer. The actual chemistry of the observed LDH structure is significant in terms of the pH stability of the phases; Ca-Al phases tend to stable at higher pH values (>11), while Mg-Al phases are stable down to pH 8 or lower. Chrysochoou et al. (2009) utilized micro-XANES analyses to demonstrate that both Ca-Al and Mg- and Fe-rich LDH phases bound Cr(VI) in COPR. A Na-rich analog of CACs with the approximate chemical formula $Na_{0.6}Ca_4Al_2O_6(CrO_4)_{1.3}(OH)_{12}{\cdot}xH_2O$, with Na varying between 0.5 and 1 and chromate between 1.25 and 1.5, was found to bind high amounts of chromate and account for ~30% of the total Cr(VI). This Na-CAC phase is thought to have similar properties as Na-free CACs and dissolve when the pH drops below 10.5. Chrysochoou et al. (2009) also observed that 50% of the Cr(VI) was bound in Mg-rich LDH and potentially in brownmillerite, in physically remote locations that could not be easily accessed by acid and reductants. Geelhoed et al. *(2002)*, Chrysochoou *(2006)* and Tinjum et al. *(2008c)* showed hydrotalcite and brucite peaks surviving down to pH 7 upon acid addition to COPR. This fraction is thus considered harder to effectively treat by the addition of acid and reductants.

Wazne et al. (2009) studied the Cr(VI) leaching behavior of GB COPR from the unsaturated zone of the SA7 site. They found that 93% of the total Cr(VI) as measured by alkaline digestion was leachable at pH 8, with 9.2 eq H^+ /kg required to attain pH 8 after 1 week. The leaching tests were performed on COPR pulverized to <150 μm and at a Liquid-to-Solid (L:S) ratio of 20, so that Cr(VI) leaching from the matrix was not diffusion limited.

Tinjum et al. (2008c) found that only 30% of the total Cr(VI) was mobilized from <2mm solid COPR at pH 8 after 48h of equilibration. Geelhoed et al. (2002) found that Cr(VI) leaching from field-moist COPR increased between 4 and 26 days of equilibration and that maximum leaching occurred at pH 8, the overall mobilization rate of Cr(VI) remained low at ~16% of the total Cr(VI) concentration at a L:S of 5.

Based on these studies, it can be concluded that if COPR pH can be lowered to 8, then Cr(VI) leaching will be thermodynamically favored and that the rate of leaching will be controlled by the rate of Cr(VI) diffusion from the solid. The latter will be controlled by the COPR particle size, which will have to be reduced to <150 μm to facilitate fast release. Given the difficulty associated with fine pulverization of large COPR amounts, this is likely the limiting step for effective reductive treatment of Cr(VI) in COPR. There have been several efforts to treat COPR in the past few years, which are summarized below.

4. Overview of COPR Treatment Efforts

Several methods for COPR treatment have been proposed:

Chemical reduction using a variety of reductants (ferrous sulfate, calcium polysulfide, sodium dithionite, nano ZeroValent Iron etc.) (Geelhoed et al. 2003, Su and Ludwig 2005, Dermatas et al. 2006, Graham et al. 2006, Cao and Zhang 2006, Wazne et al. 2007, Moon et al. 2007, 2008, 2009, Tinjum et al. 2008c);

Biologically-mediated reduction (Higgins et al. 1998, Stewart et al. 2007);

Vitrification (Meegoda et al. 1999, 2000);

Asphalt treatment (Moon et al. 2009);

Pyrolysis using sucrose, starch, flour and sewage sludge (Wang et al. 2007, Zhang et al. 2009).

The main focus of remedial efforts has been on chemical reduction. Biologically-mediated approaches have the disadvantage that they are most effective at pH values below 10. Higgins et al. (1998) showed that reduction could occur at pH values below 10 following acid addition, while Stewart et al. (2007) found that the increase in pH from 7.7 to 9.3 considerably slowed bacterial activity. Stewart et al. (2009) showed that Cr(VI) reduction was possible in a clay layer underlying a COPR site at pH 10.5and that Cr(VI) reduction was occurring indirectly through microbially-mediated Fe(II) production in an alkaline soil layer, not through direct enzymatic reduction. Pure COPR has high amounts of iron and it is not inconceivable that this mechanism can take place, especially following the addition of acid that could release and reprecipitate iron as amorphous hydroxide (Tinjum et al., 2008b). However, this approach would need to be established and it still requires high amounts of acid to facilitate sustained pH reduction.

Pyrolysis and vitrification of COPR have a common disadvantage: the need to excavate COPR, treat it ex situ with an energy-intensive and costly process and presumably to reuse it as fill material. Especially vitrification at 1,600 °C (Meegoda et al., 1999) is a cost-prohibitive approach of treating large amounts of deposited COPR. Pyrolysis studies were conducted by two groups of researchers in China, where COPR production is still ongoing at ~1 million tons per year (Wang et al. 2007). Even though these researchers suggested that pyrolysis technology may also be used for the remediation of COPR deposition sites, it is again doubtful if the cost of excavating, treating and re-depositing COPR is lower than excavation and disposal. Furthermore, the regulatory complications and public perception problems of excavating and redepositing COPR would likely be enormous. The application of this technology is considered more appropriate as post-treatment immediately following COPR production.

Cost-efficiency issues aside, pyrolysis can be an effective reduction method. Zhang et al. (2009) showed that Cr(VI) was reduced to concentrations below 100 mg/kg at temperatures above 400 °C in an exposure time of 10 min. Particle size played a major role in efficiency at lower temperatures (300 °C), but higher temperatures apparently facilitated the diffusion of volatile organics into COPR aggregates, eliminating the need for pulverization. The Cr(VI) concentration was reduced to 24 mg/kg at 600 °C, which approached the current NJDEP limit of 20 mg/kg. Thus, if the economics work and a beneficial use solution is found for the treated material, pyrolysis with waste organics appears to be an effective solution for COPR treatment.

Considerably more effort has been invested into devising an in situ treatment approach for COPR sites, which translates into in situ chemical reduction. Geelhoed et al. (2002) reported the first trials of ferrous sulfate injection into a COPR column, and showed that iron was quickly oxidized and precipitated at the top of column, while the infiltrating sulfate caused mobilization of chromium(VI). Su and Ludwig (2005) reported that the addition of sodium dithionite to ferrous sulfate could sustain reducing conditions long enough to facilitate Cr(VI) reduction in a field trial.

An array of reductacts were screened in bench scale treatability studies at the beginning of the SA7 COPR investigation in 2003; elemental iron (Fe), ferrous sulfate ($FeSO_4{\cdot}7H_2O$), pyrite (FeS2), calcium polysulfide (CaSx), sodium bisulfide (NaSH), and sodium dithionite

($Na_2S_2O_4$). Because of the multitude of factors involved in the testing (dosage, pH, particle size, two COPR zones), two matrices (a screening and an optimization matrix) were tested based on fractional factorial design. The testing criteria were the EPA regulatory method for total Cr(VI), alkaline digestion (AD), and the Toxicity Characteristic Leaching Procedure (TCLP). The results showed that most of the commonly used reductants (ZVI, pyrite, sodium dithionite) were not successful in reducing Cr(VI) (Wazne et al. 2005). The main reason was the competing oxidation of the active ingredients (elemental iron, ferrous iron) by oxygen, which proceeded quickly at high pH. Conversely, decreasing the pH to 9 yielded more satisfactory results for all reductants. Ferrous sulfate and calcium polysulfide showed improved performance, and were chosen as the most promising reductants to perform a pilot scale treatability study.

The pilot study involved the screening of three different types of mixing equipment: in situ Horizontal Rotary Mixers (HRM), in situ Vertical Augers (VA) and ex situ pugmills. Three different stoichiometries of ferrous sulfate were tested, amounting to 3X, 5X and 8X (19%, 32% and 51% wt. of $FeSO_4{\cdot}7H_2O$) the necessary amount to reduce the average Cr(VI) content of GB COPR (4,000 mg/kg). Those were combined with variable amounts (0, 10, 20 and 30%) sulfuric acid (H_2SO_4). Particle size reduction prior to pugmilling was also attempted, and the target water content was varied between 30 and 60%. All in all, five HRM, 3 VA and 23 pugmill plots were tested for ferrous sulfate. Additionally, two pugmill plots were tested for calcium polysulfide (2X and 2.25X stoichiometries at natural pH). All plots were monitored for pH, water content, total Cr(VI) and TCLP Cr over 600 days following treatment. Selected samples were also tested for Cr(VI) using X-ray Absorption Near Edge Spectroscopy (XANES) at 240 days curing. Details of the experimental methods and selected results can be found in Dermatas et al. (2006), Wazne et al. (2007), Moon et al. (2007).

Table 3. Overview of the pilot treatability study at SA7

	HRM	VA	Pugmill
FS 3X	No acid, 23% H_2SO_4[1], 25% $NaHSO_4$	No acid, 22% H_2SO_4	No acid, 20%, 30% H_2SO_4
FS 5X	No acid	No acid	No acid, 10%, 20%, 30% H_2SO_4
FS 8X	No acid	-	No acid, 10%, 20% H_2SO_4
CPS 2X, 2.25X	-	-	No acid

[1]35° Baume (41%) solution

The effect of particle size reduction and of the target water content were also addressed. However, particle size distribution analyses on ground material showed that the attempt to substantially reduce the particle size of GB COPR in large quantities had generally failed. While grinding did reduce the size of coarser grained particles (coarse to medium sand), it did not increase the amount of fine particles. The reason for the difficulty to grind GB COPR to fine powder is that it consists primarily of brownmillerite, which is a relatively hard mineral (its hardness is 5.5 out of 10 in the Moh's scale), while the cementing matrix formed around the brownmillerite particles has also to be broken up for effective particle size reduction. Thus, particle size effects could not be effectively addressed. The effect of water content was tested by varying the target moisture content of several pugmill FS-treated samples between 30% and 60%. However, no significant differences in the Cr(VI) reduction performance were observed, so that this issue will not be further discussed.

The HRM and VA results are presented in Dermatas et al. (2006), while Wazne et al. (2007) discusses CPS results from the pilot study. Figure 6 shows the Cr(VI) concentrations in the pugmill samples, as measured by the alkaline digestion and colorimetric analysis (EPA method 3060A and 7196).

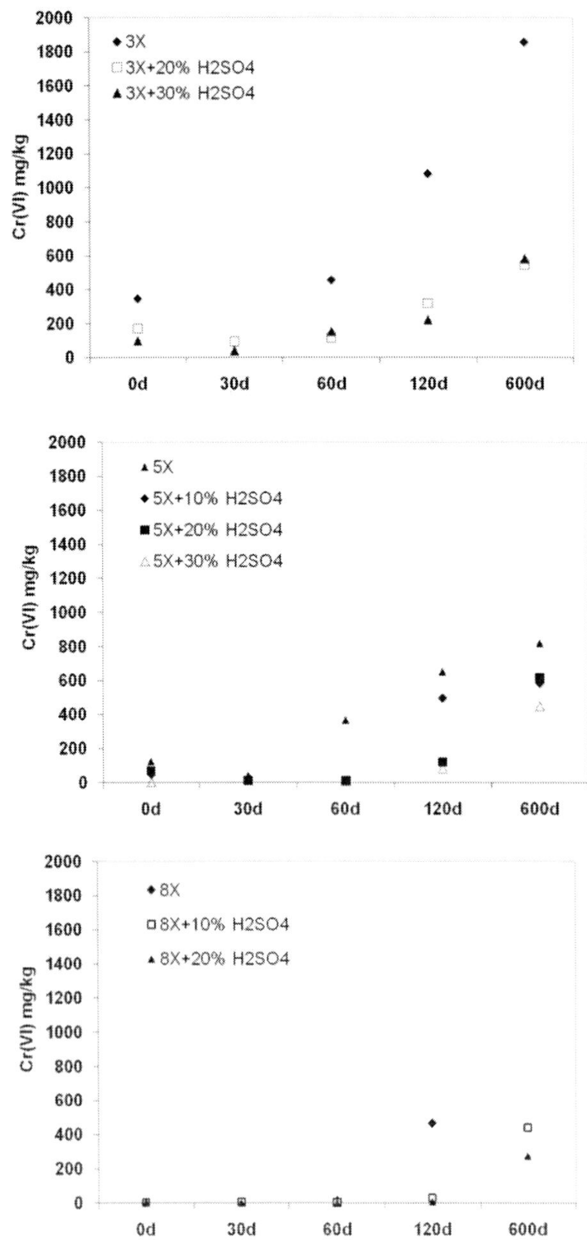

Figure 6. Cr(VI) results for SA7 FS pugmill treated samples.

The following common observations are made for all samples:

AD results increased over time.
All results exceeded 240 mg/kg at 600 days curing time.

There are two potential explanations for the apparent Cr(VI) rebound over time:

a) The reduction was initially successful, and the Cr(III) was re-oxidized to Cr(VI) over time.
b) The reduction was an artifact of the regulatory test (AD).

The only potential oxidizing agents in treated COPR would be atmospheric oxygen and pre-existing manganese oxides. Previous studies (Schroeder and Lee, 1975, Eary and Rai, 1987) have shown that re-oxidation of Cr^{3+} by dissolved oxygen is very limited or does not occur at all. Cr^{3+} following reduction by Fe^{2+} precipitates with Fe^{3+} in a α-(Fe,Cr)OOH structure that is extremely insoluble under alkaline conditions (He et al., 2004) and thus unavailable for further reactions.

Manganese oxides have the ability to re-oxidize Cr^{3+} through diffusion of cations in the solid phase (Manceau and Charlet, 1992); however, the total Mn concentration in COPR is low (<1,000 mg/kg) compared to the Cr^{6+} concentration. Additionally, Mn occurs mostly as an isomorphous substitution in COPR minerals rather than as pure oxide (data obtained by electron probe microanalysis). Jointly, these factors suggest that Cr^{6+} concentration rebounding was not due to re-oxidation of Cr^{3+}.

The second potential explanation addresses the influence of AD itself on Cr(VI) speciation. The EPA method 3060A (AD) states: "To quantify total Cr(VI) in a solid matrix, three criteria must be satisfied: (1) the extracting solution must solubilize all forms of Cr(VI), (2) the conditions of the extraction must not induce reduction of native Cr(VI) to Cr(III), and (3) the method must not cause oxidation of native Cr(III) contained in the sample to Cr(VI). Method 3060 meets these criteria for a wide spectrum of solid matrices. Under the alkaline conditions of the extraction, minimal reduction of Cr(VI) or oxidation of native Cr(III) occurs." However, the method was developed to quantify Cr(VI) in untreated solid matrices that did not contain appreciable amounts of reductants and assumed that alkaline conditions would suppress any reduction reactions. He et al. (2004) demonstrated that the latter assumption was false, showing that ferrous iron could reduce Cr(VI) in alkaline environments, if sufficient amounts in excess of the stoichiometric requirement were available in solution.

Based on the above, it was concluded that the initial low Cr(VI) concentrations obtained by AD were a result of COPR reaction with the available ferrous iron during the test. Progressively, ferrous was converted to ferric through competing oxidation by oxygen and was no longer available to react with Cr(VI) during AD performed in later time points. The rebound seizes and AD results stabilize in one value when all ferrous has converted to ferric. The rate of ferrous oxidation is a function of sample exposure of the atmosphere, which is again a function of the mixing equipment used and the water content. The rate of ferrous oxidation and the total amount of ferrous available determine the time frame for stabilization of the AD results.

These results were also supported by the observed pH regime in the different plots (Figure 7).

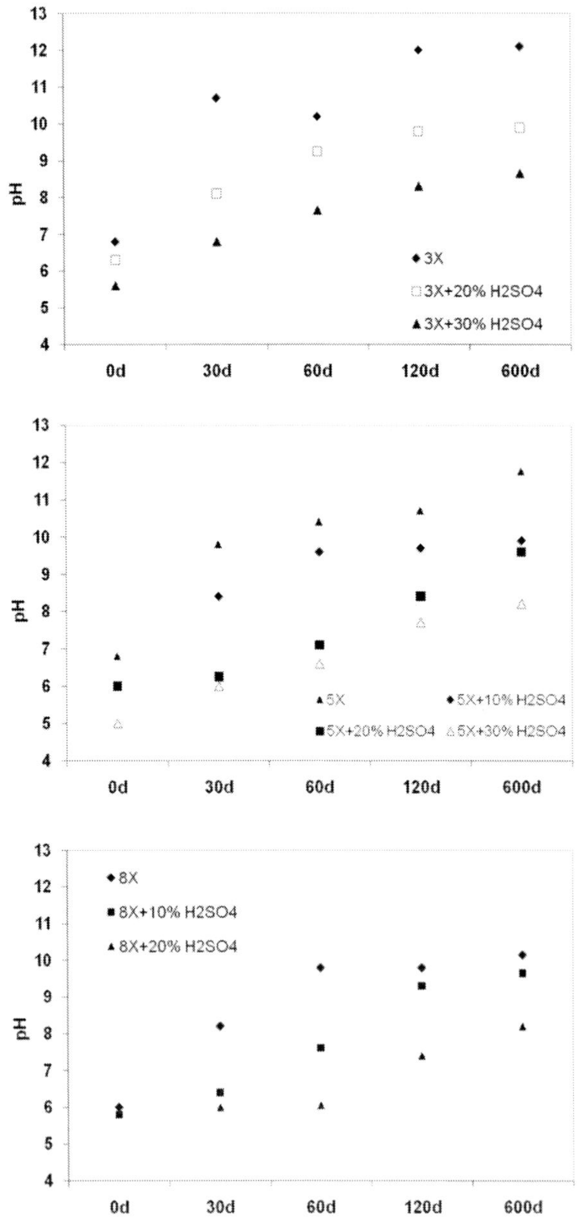

Figure 7. pH results for SA7 FS pugmill treated samples.

The pH regime followed similar trends in all FS treated samples, initially dropping abruptly immediately following mixing, and then progressively increasing. However, there were differences observed between the different types of equipment, as well as mixing dosages. To explain those, the mechanisms of pH changes need first be understood.

The addition of FS to COPR imparts acidity based on the oxidation of ferrous iron, either by Cr(VI) or by oxygen, according to the reactions

$$3Fe^{2+} + CrO_4^{\ 2-} + 8H_2O = 4Fe_{0.75}Cr_{0.25}(OH)_3 + 4H^+ \ (1)$$

and

$$Fe^{2+} + \frac{1}{4}O_2(g) + \frac{5}{2}H_2O = Fe(OH)_3 + 2H^+ \ (2).$$

Ferrous iron may also precipitate as ferrous hydroxide ($Fe(OH)_2$) or ferrous carbonate ($FeCO_3$), reactions that also consume OH^- and decrease the pH.

The released acidity caused the initial decrease of pH from strongly alkaline (>12) values to near-neutral and even acidic values in most treated samples. The decrease in pH then changed the thermodynamic equilibria in the solid phase, causing COPR mineralogy to change (see also subsequent section for a detailed discussion). The cementitious COPR minerals, which are stable at high pH (hydrogarnets, hydrotalcites, brucite, CACs) dissolve to form amorphous hydroxides. Furthermore, the lower pH favors the gradual dissolution of brownmillerite, which is extremely slow at alkaline conditions. Both processes release alkalinity (OH^-), gradually increasing the pH. However, the reactions that release alkalinity proceed at a slower rate compared to the initial release of acidity, which explains the observed pH trends.

As illustrated previously, pH shifts due to the addition of acidity are predicted by Acid Neutralization Capacity (ANC) curves. Because the different FS and acid dosages in the pugmills correspond to variable acid equivalents ranging from ~1 up to 8 eq/kg, an apparent ANC curve was generated by plotting the added equivalents of each pugmill against the respective 600-day data pH data. This ANC curve was then compared against the experimental ANC curve in Figure 8.

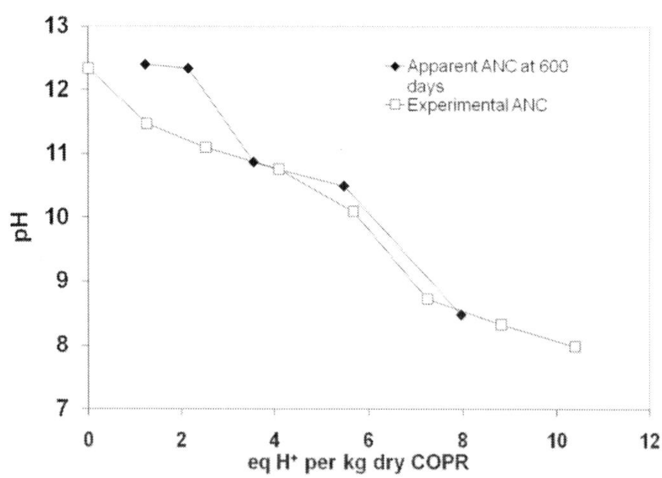

Figure 8. Experimental (lab-based ANC) vs. ANC generated by treatability field data.

The comparison between the two curves shows that the experimental ANC (taken after two months of equilibration with HCl) failed to predict COPR pH for low acid equivalents (<3 eq/kg). More acid is required to decrease COPR pH to 11 than predicted by the theoretical curve. The reason for this may be that more brownmillerite dissolved in the pugmill samples compared to the samples used to produce the ANC curve. The agreement between the two curves is very good at higher acid equivalents, indicating that 8 eq/kg acid or 30% H_2SO_4 (93% purity) can permanently decrease COPR pH to 8.5.

The results of the calcium polysulfide (CPS) mixes were also consistent with the FS observations. Cr(VI) reduction was found complete by AD, yielding non-detectable Cr(VI) (<2 mg/kg) up to 120 days (Wazne et al. 2007) and 3 mg/kg at 600 days for both the 2X and the 2.25X stoichiometries. Thus, CPS is the only treatment that conforms with the current NJDEP regulatory standard of 20 mg/kg for Cr(VI), as measured by AD. These observations agree with the results of Graham et al. (2006), who found that CPS was highly effective for treatment of Glasgow COPR. However, XANES analyses for total Cr(VI) showed that CaSx performance was again an artifact of AD and yielded a Cr(VI) concentration of 2,910 mg/kg (Wazne et al. 2007) that was even higher than the 1,848 mg/kg obtained for the ferrous sulfate HRM 8X mix (Dermatas et al. 2006). The reason for the inconsistencies between the XANES and AD results for the two reductants are again to be sought in Cr(VI) speciation.

CaSx treatment occurred at natural pH and no pH decrease was caused through its addition to the matrix, contrary to ferrous sulfate, which acidified COPR upon mixing. This means that no mechanism for Cr(VI) dissolution was provided upon CPS addition, other than equilibration between the solid phase and the solution upon depletion of dissolved Cr(VI) by sulfide. This mechanism is, however, slower compared to pH-induced dissolution of Cr(VI)-bearing minerals. Furthermore, oxygen-mediated oxidation of sulfide is slow at the strongly alkaline COPR conditions that do not favor or even allow microbial activity. Thus, the staying power of sulfide ions in solution is much higher compared to ferrous iron, so that it is available over long periods of time to reduce Cr(VI) during AD. In the case of SA7 COPR, enough sulfide was available to reduce residual Cr(VI) after 600 days of curing. Apparently, treatment would continue to be successful (in terms of AD) as long as sulfide was available. If sulfide is still available, when all Cr(VI) diffuses out of the solid and is reduced, then treatment will be permanently successful, no matter which method is used to assess it. Thus, the rates of Cr(VI) dissolution, sulfide (or any reductant) oxidation, and the quantities of both are necessary knowledge to design a Cr(VI) treatment that meets the regulatory criteria in the long term.

5. COPR Heaving Mechanisms

No precise history of heaving phenomena at SA7 or DMT has been recorded, so that a limited set of data was available that could help delineate the nature of the underlying expansion mechanism. The presented information is mostly anecdotal; however, common characteristics were identified for both sites.

The main feature of COPR heaving is that it is sporadic, i.e., it appears as heave ridge or mount and pavement buckling at isolated locations, rather than as uniform expansion of the site. Figure 9 shows an example of a heave ridge observed at SA7 in 2003. The largest heave feature at SA7 was similar in nature, with the top of the ridge reaching ~1 m above ground surface; it was located on the slab directly above the foundation of the department store that remained in place after its demolition.

Figure 9. Heave ridge observed at SA7 (9/22/03).

The first appearance of heave at SA7 following the reworking and pavement activities in the mid 1950s has been anecdotally placed at about 7 years following the construction of the drive-in theatre. No information is available for subsequent heave observations, except that reworking activities were undertaken several times to eliminate heave features prior to complete demolition of the buildings in the late 1970s.

Heave was also manifested in the form of ridges at DMT, but no systematic report of the heave features are known to the authors.

Based on the observations at both sites it was postulated that:

a) Volume expansion was induced by mineralogical changes in COPR; mineral formation induced swell pressures that sporadically overcame confining pressures and was thus ultimately manifested as heave.
b) The presence of physical obstructions played a crucial role in the manifestation of volume expansion as surface heave; it was hypothesized that concrete obstructions could not absorb swell pressures generated by adjacent COPR and redirected them towards the ground surface.
c) Either lithification or compaction of COPR are required for the manifestation of heave. A granular material of high void ratio (volume of voids to volume of solids) can dissipate swell pressures, while these are propagated within a lithified material and manifested as heave when they overcome the confining pressures of the rigid matrix.

The first step to verify these hypotheses was to identify potential mineralogical transformations that would lead to volume expansion. These mechanisms were then tested with laboratory swell tests and examined in the light of field data collected from both sites.

The main mineralogical transformations occurring in COPR are hydration reactions of parent minerals, brownmillerite and periclase, and carbonation. Mineralogically-induced volume changes can be described in terms of changes of specific gravity (Gs). Both brownmillerite and periclase are compact oxide structures of high density; when they hydrate, they produce phases of lower density; this implies that hydration reactions lead to an overall volume increase, the extent of which depends on the Gs and the amount of the newly formed phase; apparently, kinetics of mineral formation affect the rate of volume expansion and thus the rate of heaving phenomena development. Carbonation reactions, on the other hand, lead to increase of mass due to CO_2 uptake, rather than an overall volume increase, as calcite density is higher compared to portlandite and most hydration products. Based on all this, the following heaving mechanisms were formulated:

1) Brownmillerite hydration to form hydrogarnets
2) Periclase hydration to form brucite
3) CAC and hydrotalcite formation
4) Ettringite formation
5) Calcite formation

The first step in assessing the contribution of each reaction to COPR swell was to conduct a literature review on similar heaving phenomena observed in cements and pozzolanically stabilized soils, as they closely resemble COPR.

Brownmillerite is encountered in larger quantities in oilwell cements and sulfate-resisting Portland cements. Although its hydration behavior has been well studied, there is no mention of expansive behavior of ferrite-rich cements. The lack of volume expansion phenomena in ferrite-rich cements can be attributed to the reported brownmillerite hydration rate being extremely slow. This apparent kinetic inhibition was traced to the early formation of pozzolanic product and ferric oxihydroxide passivation layers around COPR brownmillerite nodules (Chrysochoou et al., 2005). However, as it can be seen in Table 1 the qualitative mineralogy for GB versus HB is quite similar but quantitatively it is at the opposite ends of the brownmillerite reaction pathway; DMT-GB resembles almost unreacted COPR whereas HB2-DMT resembles almost fully hydrated COPR, with SA7-GB and DMT-HB being somewhere in between the COPR reaction pathway. Moreover, the fact that HB layers are found interbedded with GB layers, predominately at DMT but also at SA7 and Glasgow, UK, clearly points to a post-deposition, in-situ transformation of GB to HB. This seems to be the case, as there is no reported modification in the roasting process that could have resulted in two streams of COPR produced in short time intervals coupled by the fact that the raw COPR coming out of the roasting process and then deposited on-site closely resembled, by all historical accounts, DMT-GB COPR. Now if one looks at the Gs values for the mineral phases present in GB versus HB COPR (Table 1), it follows that GB to HB (brownmillerite to hydrogarnet) transformation may be a viable heaving mechanism, provided that the rate of transformation is somehow dramatically increased over the literature-reported extremely slow brownmillerite hydration rate. The question is then what does it take for the GB to HB transformation rate to substantially increase so that heave can take place, a question that will be dealt with in a following paragraph.

The hydration of periclase to brucite has been reported to create expansion in Mg-rich cements (Chatterji, 1995). Considering the relatively low amount of periclase in COPR (Table 2), the slow rate of hydration and the difference in porosity between cements and COPR, this mechanism was not considered a leading candidate as the main heaving mechanism. Calcite formation has been reported in the literature as a cause of cementation and consolidation of marine sediments (Nakamura et al., 1993, Fukue et al., 1996). No report on volume expansion of cements and soils due to calcite formation was found in the literature. The formation of monophases (CAC/hydrotalcites) is analogous to the formation of monosulfate in cements. Although monosulfate formation is a very common reaction in cements (EBH, 1998), it has not been considered to cause expansion. Even though CAC and hydrotalcite formation is accompanied by significant volume increase, their platy morphology is generally not conducive to expansion in high porosity systems as their crystals can be accommodated in the existing voids.

Sulfate attack is considered one of the major mechanisms of concrete deterioration and is related to the formation of ettringite (Ogawa and Roy, 1981, 1982a, 1982b). Taylor et al. (2001) present a comprehensive overview of the issues related to Delayed Ettringite Formation (DEF), which is the prevalent term for ettringite-induced concrete expansion. Because of the acicular habit of ettringite and its low specific gravity, it has the potential to exert high swell pressures and cause cracking and concrete failure. Hydration and dehydration of ettringite crystals was also proposed as a viable expansion mechanism (Mehta and Wang, 1982). Ettringite has also been reported in the literature as the cause of swell in lime-stabilized clays (Mitchell, 1984, Dermatas, 1995); lime provided the necessary Ca source and the high pH, while clay minerals provided alumina; potential sulfate sources were gypsum, sodium sulfate or sulfate carried by ground- and seepage water.

Interestingly, mineralogical analyses of the SA7 soil boring samples showed that ettringite presence at the site was significant; ettringite was present in 78 out of 120 XRPD patterns (65%), its concentration (without considering amorphous phases) ranged from 1% to 33%. Although these are relative concentrations that are reduced when the amorphous content is taken into account, it seems that high ettringite concentrations were sporadically present at the site. However, no correlation with site-specific heave features could be safely made.

As ettringite was detected in COPR and its thermodynamic stability over other COPR phases was established, its formation and hydration were considered as viable causes for COPR expansion. It was thus further investigated, along with the other potential heave mechanisms, by means of laboratory swell tests. Two types of swell tests were employed to study COPR volume expansion behavior, unconfined and confined. The experimental conditions and the results of the two testing rounds are presented in Moon et al. (2005), and Moon et al. (2006a), respectively. The major findings of the swell tests can be summarized as follows:

a) As pertinent literature suggested, no detectable conversion of brownmillerite to hydrogarnets or periclase to brucite could be observed after 12 months of subjecting GB COPR to wet and dry cycles, both at ambient temperature and 50 ^{o}C. No change in volume behavior was observed in the samples. These mechanisms could therefore not be conclusively assessed, as transformation was not achieved; however, the extremely slow rate of hydration is considered inconsistent with the relatively rapid phenomena observed in the field. The periclase to brucite mechanism was readily ruled out though, due to their consistently low concentrations throughout all COPR layers at both sites.
b) Calcite formation led to consolidation, as also reported in the literature. This mechanism was therefore ruled out.

c) CAC formation may contribute to swell; ongoing unpublished research performed at the University of Wisconsin indicates that high activity of chromate in the unsaturated zone and in the presence of wet and dry cycles may play a significant role in heave phenomena. However, dehydration/hydration of CACs was ruled out as a heave mechanism owing to the observed stability of the identified CAC hydrate phases against drying or wetting.
d) Ettringite formation resulted in volume expansion both in unconfined and in confined swell tests. The observed expansion was related with the quantity of ettringite formed and its particle size.
e) Lithification was confirmed to play an important role in swell development. Gradual cementation was observed in all samples that eventually swelled.
f) GB COPR tested under ambient to 50 ^{o}C temperature saturated conditions did not swell. This may be an indication that GB material kept constantly saturated may not expand. This is attributed to its inherently high porosity and low degree of lithification, where swell pressure development can be accommodated within the existing void space.

Even though, based on the laboratory swell tests, ettringite formation appears to be the most viable heave mechanism and while site investigation revealed extensive ettringite presence (Dermatas et al., 2005a; Chrysochoou, 2006), ettringite presence could not be correlated to the observed heave features at SA7. Conversely, at DMT ettringite was indeed correlated with the observed heave features in the location of the trenches (Chrysochoou, 2006). Although this does not become apparent from the average mineral concentrations reported in Table 2, the presence of isolated areas with high ettringite concentration in the upper lithified zone coincided with the presence of heave features. However, the positive identification of the mechanism was no possible and the role of other mechanisms including the brownmilerite to hydrogarnet (GB to HB) mechanism and the role of CACs could not established in a conclusive manner. Moreover, during field pilot testing of COPR treatment with ferrous sulfate, ettringite formation was clearly identified as the observed heave culprit (Dermatas et al., 2006).

The fact that even though ettringite is a viable mechanism does not necessarily mean that it is the predominant mechanism for COPR heave in all locations, especially because of the limited presence of sulfate in most COPR areas. For example, despite the sporadic high ettringite concentrations at SA7, no correlation with heave features could be established. Further investigation is required on the role of other mechanisms and especially the role of GB hydration and lithification in the vadose zone.

At SA7, a massive layer of HB-like COPR was linked to a major heave feature at the site, pointing again to the significance of lithification in the manifestation of heave. It may also hint that GB-to-HB is a potential mineralogical transformation that can generate adequate volume expansion. As noted previously though, one cannot easily account for the field conditions that would make this possible under realistic time frames, commensurate with observed heave development at either SA7 or DMT. At DMT for instance, where massive layers and not thin interbeds of HB COPR were encountered, the GB to HB transformation seems to have taken place in-situ sometime in the past, and for all practical purposes it may still be going on to some degree. The lack of any well documented historical data rules out making any reliable assessments to this end. However, by all accounts, at DMT, GB COPR was initially (starting at 1967) brought in and deposited, in a dry state, first in stockpiles and then spread laterally in order to reclaim unused waterlogged marshland. Thus, the GB to HB transformation had to take place in situ over the course of approximately 30 years or so. This time frame is by all accounts very short when one considers that the GB to HB transformation is extremely slow at ambient temperature, as evidenced both by the mineralogy of the DMT GB COPR material, as well as by the mineralogy of COPR deposited at SA7. SA7 COPR still contains significant amounts of brownmillerite, even though it has been deposited for a considerably longer time than the DMT COPR. A viable mechanism for a preferred acceleration of the hydration reactions in the HB material could have been its exposure to a high temperature environment. The presence of hydrothermal conditions could explain the rapid transformation of GB to HB/HB2 material. The hypothesis of a hydrothermal environment is further supported by several properties of the HB/HB2 COPR, as described below.

The XRPD analyses of DMT HB COPR yielded patterns that presented high noise and amorphous humps, indicating poor crystallinity and significant presence of amorphous material, which are distinctive characteristics of materials subjected to hydrothermal conditions (Chrysochoou, 2006).

The micromorphological analyses (Chrysochoou, 2006) showed that the HB matrix consists of nodules that are weathered internally, as opposed to GB nodules that are mostly compact. HB nodules mostly comprise a red matrix that corresponds to the amorphous iron oxides and hydroxides that give HB COPR its red color. The nodules in HB are cemented together by a cementitious matrix, which consists of small, disordered crystals. The internal weathering of the HB nodules and the poor crystallinity of the surrounding matrix also point to hydrothermal conditions.

The underlying cause which could have brought about hydrothermal conditions is not clear; the most probable hypothesis is that brownmillerite hydration, which is in itself an exothermic reaction, could have increased the temperature in the capillary zone, where excess water was not available to dissipate the heat. However, there is no field evidence that such as mechanism is indeed possible.

To further investigate GB to HB transformation slurry tests (Moon et al., 2006a) and high temperature experiments (Dermatas et al., 2006b; Sanchez, 2006) were conducted. Slurry tests using both ultrafine (10 μm) fresh synthetic brownmillerite and actual GB COPR showed that by using ultrafine synthetic brownmillerite the transformation of brownmillerite to hydrogarnet was very fast (less than 1 hour) and was complete within 7 days (Moon et al., 2006b). CAC formation was also thermodynamically favored upon chromate influx, as it was observed in the XRPD pattern within one hour of reaction. However, employing actual COPR as opposed to synthetic brownmillerite, hydrogarnet and CAC formation were not observed at all; no major phase transformation was observed in these samples (Moon et al., 2006b). This confirmed that the dissolution of COPR brownmillerite is indeed very slow, thus GB to HB transformations appear to be kinetically inhibited. Based on these results it can be hypothesized that COPR particle size reduction could potentially overcome this apparent kinetic inhibition. During high temperature (200°C) testing, brownmillerite dissolution was clearly observed in all GB COPR samples tested following 24 hours and for most of the samples, the mineral dissolved completely after 14 days (Dermatas et al., 2006b; Sanchez,

2006). The latter was also confirmed in the field when during pilot COPR treatment brownmillerite dissolution of up to 80% where high temperatures were recorded for up to 30 days (Dermatas et al., 2006a). Consequently, both lab and field observations clearly point to high temperature being a definite way to overcome the GB COPR apparent hydration kinetic inhibition. However, there is not concrete evidence that this mechanism is viable under field conditions.

An alternative hypothesis to explain rapid brownmillerite hydration was presented by Sass et al. (2008), who hypothesized that the reaction of divalent anions (chromate, carbonate) with hydrated monophases that are in turn produced by initial hydration of brownmillerite are responsible for the GB-to-HB transformation. In other words, Sass et al. (2008) proposed that chromate acts as a catalyst that accelerates brownmillerite hydration and showed that pure brownmillerite hydration proceeded fast to produce lithified pellets in the presence of chromate. However, pure brownmillerite will also react fast in the presence of pure water and this mechanism has not been demonstrated for actual GB COPR.

It is clear from the preceding discussion that the research undertaken to understand heave mechanisms has been inconclusive up to the date of publications referenced herein. Ongoing research undertaken elsewhere is focusing on the fundamental mechanism considered to be responsible for lithification of COPR as well as the role of chromate both in COPR lithification and the manifestation of heave phenomena. A second aspect of ongoing research is to better understand the hydrogeologic and thermodynamic conditions under which COPR will transform and the differences between vadose (unsaturated) zone and saturated zone systems. These conditions include the presence of water, the chemical activity of the chromate in the solution, the Gibbs free energy of the mineralogical transformations and factors that promote (catalyze) or restrict (passivate) the transformation reactions. The effect of negative pore pressures in the vadose zone and their contribution in the attainment of higher chromate activities than possible in saturated COPR systems is a key focus of ongoing research.

ACKNOWLEDGMENTS

The authors wish to thank Honeywell Inc. for the financial support of the COPR investigation at the SA7 and DMT sites. All analyses were conducted at Stevens Institute of Technology and we are grateful to Christos Christodoulatos, Mahmoud Wazne and Deok Hyun Moon for their collaboration in this project. Geosyntec Consultants conducted the trenching investigation and the geotechnical analyses at DMT. Finally, we recognize posthumously Mr. Paul Dutko for the important efforts and guidance he provided in the early stages of the COPR project. We also thank Chris French for providing valuable input into the heaving mechanism discussion and James Tinjum for sharing the content of ongoing research efforts at the University of Wisconsin.

REFERENCES

Burke T., Fagliano J., Goldoft M., Hazen R.E., Iglewicz R. and McKee T., 1991. Chromite Ore Processing Residue in Hudson County, New Jersey, Environ. Health Persp., 92:131-137.

Cao J. and Zhang W.-X. (2006). Stabilization of chromium ore processing residue (COPR) with nanoscale iron particles, *Journal of Hazardous Materials*, B132, 213-219.

Chatterji S., 1995. Mechanism of expansion of concrete due to the presence of dead-burnt CaO and MgO, Cement and Concrete Research, 25(1), 51-56.

Chrysochoou M., Dermatas D., Moon D.H., Grubb D.G. and Christodoulatos C. (2010). The importance of mineralogy in the geoenvironmental characterization of Chromite Ore Processing Residue, J. Geotech. Geoenv. Eng., in press.

Chrysochoou, M., Fakra, S., Marcus, M.A., Moon, D.H. and Dermatas, D., (2009). "Microstructural Analyses of Cr(VI) Speciation In Chromite Ore Processing Residue (COPR.", *Environ. Sci.Tech.*, 43(14), 5461-5466.

Chrysochoou M. and Dermatas D. (2007). "Application of the Rietveld method to assess Cr(VI) speciation in Chromite Ore Processing Residue." *J. Haz. Mat.* Vol. 141: 370-377.

Chrysochoou, M. (2006). "Investigation of environmental and geotechnical implications associated with deposition of Chromite Ore Processing Residue (COPR)." Ph.D. thesis, Stevens Institute of Technology, Hoboken, NJ.

Chrysochoou M., Moon D. H., Dermatas D., Wazne M., Christodoulatos C., Meng X., Kaouris M., Morris J., French C. and Sass B. (2005), "Mineralogical Analysis of Chromite Ore Processing Residue (COPR) by X-Ray Powder Diffraction (XRPD)", Proc. of the 8th Battelle In-situ and On-site Bioremediation International Symposium, Baltimore, Maryland, June 6-9, 2005.

Darrie R.G., 2001. Commercial extraction technology and process waste disposal in the manufacture of chromium chemicals from ore, Environmental Geochemistry and Health, 23:187-193.

Dermatas, D., Chrysochoou, M., Pardali, S., and Grubb, D.G., (2007). "Influence of X-Ray Powder Diffraction (XRPD) sample preparation on the Rietveld quantification of Chromite Ore Processing Residue (COPR)." *J. Environ. Qual.* 36(2): 487-497.

Dermatas, D., Chrysochoou, M., Moon, D.H., Grubb, D.G, Wazne, M. and Christodoulatos, C. (2006). "Ettringite-Induced Heave in Chromite Ore Processing Residue (COPR) upon Ferrous Sulfate Treatment." *Environ. Sci. Tech.* Vol.40: 5786-5792.

Dermatas D., Sanchez A., Moon D.H., Chrysochoou M., Christodoulatos C. and Grubb D.G. (2006b), Brownmillerite hydration in Chromite Ore Processing Residue (COPR) at elevated temperatures, Conference Proceedings Protection and Restoration of the Environment VIII, Chania, Greece.

Dermatas D., Chrysochoou M., Moon D. H., Pardali S., Christodoulatos C., Lazarte C. A., Pendleton C. H., Bonaparte R., Briggs R., Myers M., French C., Morris J. and Kaouris M. (2005a), "Mineralogical Characterization of Chromite Ore Processing Residue (COPR) at Dundalk Marine Terminal Area 1800", Proc. of the 8th Battelle In-situ and On-site Bioremediation International Symposium, Baltimore, Maryland, June 6-9, 2005.

Dermatas D., 1995. Ettringite-induced heaving in soils: State-of-the-art, *Appl. Mech. Rev.*, 38(10), 659-673.

Eary L.E., Rai D. 1987. Kinetics of Cr(III) oxidation by manganese dioxide, *Environmental Science and Technology*, 21, 1187-1193.

Elsevier, Butterworth, Heinemann, (EBH) (1998). *Lea's chemistry of cement and concrete*, 4th edition, Hewlett P.C. (Ed.), Elsevier, Butterworth, Heinemann, Oxford.

Emanuelson A. and Hansen S., 1997. Distribution of iron among ferrite hydrates, *Cement and Concrete Research*, 27(8), 1167-1177.

Fukue M., Nakamura T., Kato Y., Naoe K. (1996), Effects of carbonate on cementation of marine soils, *Marine Georesources and Geotechnology*, 36(1), 51-60.

Geelhoed J.S., Meussen C.L., Hillier S., Lumsdon D.G., Thomas R.P., Farmer J.G. and Paterson E., 2002. Identification and geochemical modeling of processes controlling leaching of Cr(VI) and other major elements from chromite ore processing residue, *Geochimica Cosmochimica Acta*, 66(22), 3927-3942.

Geelhoed J. S., Meussen J. C. L., Roe, M. J., Hillier S., Thomas, R. P., Farmer J. G. and Paterson E., 2003. Chromium remediation or release? Effect of iron(II) sulfate addition on chromium(VI) leaching from columns of chromite ore processing residue. *Environmental Science and Technology*, 37: 3206-3213.

Goswamee R.L., Sengupta P., Bhattacharyya K.G. and Dutta D.K., 1998. Adsorption of Cr(VI) in layered double hydroxides, *Applied Clay Science*, 13:21-34.

Graham, M.C., Farmer, J.G., Anderson, P., Paterson, E., Hillier, S., Lumsdon, D.G. and Bewley, R.J.F. (2006). "Calcium polysulfide remediation of hexavalent chromium contamination from chromite ore processing residue." *Sci. Tot. Environ.* Vol.364: 32-44.

Hausmann, M.R., 1990. Engineering Principles of Ground Modification, McGraw Hill, New York.

He, Y.T., Chen, C. and Traina, S.J. (2004). "Inhibited Cr(VI) reduction by aqueous Fe(II) under hyperalkaline conditions" *Environ. Sci. Tech.* Vol.38: 5535-5539.

Hillier, S., Lumsdon, D.G., Brydson, R. and Paterson, E. (2007). "Hydrogarnet: A host phase for Cr(VI) in Chromite Ore Processing Residue (COPR) and other high pH wastes." *Environ. Sci. Tech.* Vol. 41: 1921-1927.

Hillier S., Roe M.J., Geelhoed J.S., Fraser A.R., Farmer J.G. and Paterson E., 2003. Role of quantitative mineralogical analysis in the investigation of sites contaminated by chromite ore processing residues, *The Science of the Total Environment*, 308:195-210

Higgins T.E., Halloran A.R., Dobbins M.E. and Pittignano A.J., 1998. In situ reduction of hexavalent chromium in alkaline soils enriched with Chromite Ore Processing Residue, *J. Air and Waste Manage. Assoc.*, 48, 1100-1106.

Hou X., Bish D.L., Wang S., Johnston C.T. and Kirkpatrick R.J., 2003. Hydration, expansion, structure and dynamics of double-layered hydroxides, American Mineralogist, 88:167-179.

Lioy P.J., Freeman N.C.G., Wainman T., Stern A.H., Boesch R., Howell T., Shupack S.I., 1992. Microenvironmental analysis of residential exposure to chromium-laden wastes in and around New Jersey homes, Risk Analysis, 12:287-299.

Manceau, A and Charlet, L. (1992). "X-ray absorption spectroscopic study of the sorption of Cr(III) at the oxide-water interface I. Molecular mechanism of Cr(III) oxidation on Mn oxides." *J. Colloid Interface Sci.* Vol.148: 425-442.

Manzur L.C.G., Thomas R.P., Geelhoed J.S., Farmer J.G. and Campbell C.D., 2001. Variability in chemical properties of a COPR disposal site and assessment of potential toxicity using a biosensor, *Environmental Geochemistry and Health*; 23: 213-217

McKee T., 1988. Chromate Chemical Production Industry: Waste Treatment Past and Present, Report to the New Jersey Department of Environmental Protection, Trenton, NJ.

Meegoda J.N., Kamopornvijit, W., Vaccari D.A., Ezeldin A.S., Noval B.A., Mueller R.T. and Santora S., 1999. Remediation of chromium-contaminated soils: bench-scale investigation, Practice Periodical of Hazardous, *Toxic and Radioactive Waste Management*, 3(3), 124-131.

Meegoda J.N., Partymiller K., Tichards M.K., Kamopornvijit, W., Librizzi W., Tate T., Noval B.A., Mueller R.T. and Santora S., 2000. Remediation of chromium-contaminated soils: pilot-scale investigation, Practice Periodical of Hazardous, Toxic and Radioactive *Waste Management*, 4(1), 7-15.

Mehta P.K., Wang S, 1982. Expansion of ettringite by water adsorption, *Cement and Concrete Research*, 12, 121-122.

Mitchell J.K., 1984. Practical problems from surprising soil behavior, *J. of Geotech. Eng.*, 112(3), 250-289

Moon D.H., Wazne M., Koutsospyros A., Christodoulatos C., Gevgilili H., Malik M. and Kalyon D.M. (2009). Evaluation of the treatment of chromite ore processing residue by ferrous sulfate and asphalt, *Journal of Hazardous Materials*, 166, 27-32.

Moon, D.H., Wazne, M., Jagupilla, S.C., Christodoulatos, C., Kim, M.G. and Koutsospyros, A. (2008). “Particle size and pH effects on remediation of chromite ore processing residue (COPR) using calcium polysulfide (CaS5).” *Sci. Tot. Environ.*, 399, 2-10.

Moon, D.H., Wazne, M., Dermatas, D., Christodoulatos, C., Sanchez, A.M., Grubb, D.G., Chrysochoou, M., and Kim ,M.G. (2007).”Long-term treatment issues with chromite ore processing residue (COPR): Cr^{6+} reduction and heave.” *J. Haz. Mat.* 143(3): 629-635.

Moon D. H., Dermatas D., Chrysochoou M. and Shen G. (2006a), “An Investigation of the Heaving Mechanism Related to Chromite Ore Processing Residue”, *Journal of ASTM International*, ASTM, International, Vol. 3, No. 6, May 2006.

Moon D.H., Dermatas D., Chrysochoou M., Grubb D.G. (2006b), “An Investigation of the Phase Transformation of Brownmillerite in Chromite Ore Processing Residue, Proc. of the International Conference: Protection and Restoration of the Environment VIII, Chania, Greece, July 2006.

Moon, D.H., Dermatas, D., Sanchez A.M., Wazne, M., Chrysochoou M., Christodoulatos C., Grubb D.G., Kim, M.G. (2006c), “Investigation of heave and remediation of chromite ore processing residue (COPR) materials using ferrous sulfate”, Proc. of the Second International Conference on Environmental Science and Technology, American Academy of Science, Houston, Texas, USA, August 19-22, 2006.

Moon D. H., Chrysochoou M., Dermatas D. and Christodoulatos C., (2005). Investigation of Ettringite Formation in Chromite Ore Processing Residue, In-situ and On-site Bioremediation Symposium, Conference Proceedings, Battelle, Columbus, Ohio.

Moore A.E. and Taylor H.F.W., 1970. Crystal structure of ettringite, *Acta Crystallographica*, B26, 386-393.

Nakamura T., Fukue M.,Naoe K. (1993), Effects of calcium carbonate on geotechnical properties of sediments in Tokyo Bay, Proceedings of the Third (1993) International Offshore and Polar Engineering Conference, p. 647-651.

NJDEP (New Jersey Department of Environmental Protection). (2007). “Chromium Moratorium”, Memorandum, February 8.

New Jersey Department of Environmental Protection (NJDEP), 1998. Summary of the basis and background of the soil cleanup criteria for trivalent and hexavalent chromium, Site Remediation Program.

Ogawa K., Roy D.M. (1981), C_4A_3S hydration ettringite formation, and its expansion mechanism: I: Expansion: ettringite stability, *Cement and Concrete Research*, 11, 741-750.

Ogawa K., Roy D.M. (1982a), C_4A_3S hydration ettringite formation, and its expansion mechanism: II: Microstructural observation of expansion, *Cement and Concrete Research*, 12, 101-109.

Ogawa K., Roy D.M. (1982b), C_4A_3S hydration ettringite formation, and its expansion mechanism: III: Effect of CaO, NaOH and NaCl; Conclusions, *Cement and Concrete Research*, 12, 247-256.

Rogers D.E. and Aldridge L.P., 1977. Hydrates of calcium ferrites and calcium aluminoferrites, *Cement and Concrete Research,* 7(4), 399-409.

Rousselot I., Taviot-Gueho, Leroux F., Leone P., Palvadeau P. and Besse J.-P., (2002). "Insights on the structural chemistry of hydrocalumite and hydrotalcite-like materials: investigation of the series $Ca_2M^{3+}(OH)_6Cl_2{\cdot}H_2O$ (M^{3+}: Al^{3+}, Ga^{3+}, Fe^{3+} and Sc^{3+}) by X-ray Powder Diffraction." *J. Solid State Chem.* 167: 137-144

Sanchez A. M. (2006), Brownmillerite Hydration in COPR at Elevated Temperatures, MS Thesis, Stevens Institute of Technology, Hoboken, NJ, May 2006.

Sass B., Kremser D., Bhargava M and Lipps J. (2008). Diagenesis of buried Chrome Ore Processing Residue" GeoCongress 2008 conference Proceedings, New Orleans March 21-23rd 2008, pp.399-407.

Schroeder D.C., Lee G.F.,1975. Potential transformations of chromium in natural waters, *Water Air Soil Pollution*, 4, 355-365.

Sposito G., 1989. The chemistry of soils, Oxford University Press, New York.

Stewart D.I., Burke I.T., Hughes-Berry D.V. and Whittleston R.S. (2009). Microbially mediated chromate reduction in soil contaminated by highly alkaline leachate from chromium containing waste, Ecological Engineering, in press.

Stewart D.I., Burke I.T. and Mortimer R.J.G. (2007). Stimulation of Microbially Mediated Chromate Reduction in Alkaline Soil-Water Systems', *Geomicrobiology Journal*, 24:7, 655 – 669.

Su C.; Ludwig R.D. Treatment of hexavalent chromium in chromite ore processing solid waste using a mixed reductant solution of ferrous sulfate and sodium dithionite, *Environ. Sci. Technol.* 2005, *39*, 6208-6216.

Tinjum J., Houck C. and French C. (2008a). "Field Investigation Techniques for Characterization and Delineation of COPR." GeoCongress 2008 conference Proceedings, New Orleans March 21-23rd 2008, pp.407-415.

Tinjum, J.M., Benson, C.H. and Edil, T.B. (2008b). "Treatment of CrVI in COPR Using Ferrous Sulfate–Sulfuric Acid or Cationic Polysulfides." *J. Geotech. Geoenv.* Eng., 134(12), 1791-1803.

Tinjum, J.M., Benson, C.H. and Edil, T.B. (2008c). "Mobilization of Cr(VI) in chromite ore processing residue through acid treatment." *Sci. Tot. Environ.*, 391, 13-25.

Taylor H.F.W., Famy C. and Scrivener K.L. (2001), Review Delayed Ettringite Formation, Cement and Concrete Research, 31, 683-693.

Wang T., He M. and Pan Q. (2007). A new method for the treatment of chromite ore processing residues, *Journal of Hazardous Materials*, 149, 440-444.

Wazne, M., Papazoglou, P., Papastavrou, M., Meng, X., Dermatas, D. and Christodoulatos, C. (2005). "Remediation of Chromite Ore Processing Residue by Chemical Reductants" *Conference Proceedings In situ and on site bioremediation Symposium* Battelle, Columbus, Ohio.

Wazne, M. Jagupilla, S.C., Moon, D.H., Jagupilla, S.C., Christodoulatos, C. and Kim, M.G. (2007). "Assessment of calcium polysulfide for the remediation of hexavalent chromium in chromite ore processing residue (COPR)." *J. Haz. Mat.* doi:10.1016/ j.jhazmat.2007.01.012.

Wazne M., Jagupilla S.C., Moon D.H., Christodoulatos C. and Koutsospyros A. (2009). Leaching mechanisms of Cr(VI) from Chromite Ore Processing residue, *Journal of Environmental Quality*, 37, 2125-2134.

Weng C.H., Huang C.P., Allen H.E., Cheng A. H.D. and Sanders P.F., 1994. Chromium leaching behavior in soil derived from chromite ore processing waste, *The Science of the Total Environment*, 154, 71-86.

Weng C.H., Huang C.P., Allen H.E., Leavens P.B. and Sanders P.F., 1996. Chemical interactions between Cr(VI) and hydrous concrete, *Environmental Science and Technology*, 30:371-376.

Whitlock A.R. and Galitz C.L. (1997), Effects of Chromite Ore Processing Residue on concrete structures, *Journal of Soil Contamination*, 6(6), 751-765.

Zhang D., Kong H., Wu D., He S., Hu Z. and Hu X. (2009) Remediation of chromite ore processing residue by pyrolysis process with sewage sludge, *Bioresource Technology*, 100, 2874-2877.

In: Management of Hazardous Residues Containing Cr(VI) ISBN: 978-1-61470-637-3
Ed: Maria José Balart Murria

Chapter 15

SAFE DISPOSAL AND RE-USE OF CHROMIUM RICH WASTE MATERIALS

Janez Ščančar*[*] *and Radmila Milačič
Department of Environmental Sciences, Jožef Stefan Institute,
Jamova 39, Ljubljana, Slovenia

ABSTRACT

Chromium and chromium chemicals are widely used in the metallurgical and chemical industries, leather tanning, wood treatment etc. Due to its wide use, large amounts of chromium are emitted into the environment. Chromium has been identified both as an essential micronutrient and as a toxic element. The essentiality and toxicity of chromium is a function of its chemical form. In general, Cr(III) compounds are far less toxic than Cr(VI). Under certain environmental conditions e.g. the presence of Mn(IV) oxides, Cr(III) could be oxidised to its toxic hexavalent form. At high alkaline pH's oxidation of Cr(III) with dissolved oxygen may take place. Cr(VI) is quickly reduced to Cr(III) in acidic solutions and in the presence of electron donors. In environmental samples Cr(VI) is reduced with organic matter, sulphides and Fe(II). The process of chromium reduction is 10-times faster than re-oxidation with Mn(IV) oxides. The prevailing oxidizing or reducing conditions in the environment influence the equilibrate concentration of the particular chromium species.

Speciation analysis plays a crucial role in the investigations of transformation and mobility of chromium in the environment. In order to investigate the conditions for safe disposal or re-use of chromium rich waste materials, appropriate speciation techniques regarding specific characteristics of a sample matrix should be applied. Partitioning of chromium between various phases in soils, waste materials and their mixtures with soil, and re-used waste materials gives useful information for the assessment of its environmental hazards. For this purpose, it is necessary to apply long-term experiments and to study possible oxidation-reduction processes of chromium after disposal or re-use of chromium rich waste materials. On the basis of complex investigations estimation can be done on conditions for safe disposal or re-use of chromium rich waste materials.

* E-mail: janez.scancar@ijs.si

1. INTRODUCTION

Chromium is a naturally occurring element found in ores and minerals, soils, volcanic dust, and gases. It is also present in living organisms. The most common chromium forms in the environment are trivalent (Cr(III)) and hexavalent (Cr(VI)). A starting material for production of metallic chromium is a chromite ore (Cr_2O_3 x FeO). Metallic chromium, chromium alloys and different chromium compounds are widely used in various industrial applications like steel manufacturing, production of ferrous and nonferrous alloys, electroplating, the chemical industry and the production of pigments [1]. Chromium is also used in corrosion preventing coatings in the automotive industry [2,3], leather tanning [4-6], wood treatment [7] and is present in cement and cement products [8]. The extensive use of chromium in industrial processes resulted in the release of large amounts of anthropogenic chromium emissions into the air, waters and terrestrial environments. In living organisms, the toxicity and bioavailability of chromium depend primarily on its chemical form. Cr(III) compounds are essential for glucose metabolism and are much less toxic than those of Cr(VI) [9]. Cr(VI) is known to be carcinogenic, mutagenic and an inducer of skin dermatitis [10]. The presence of reducing and/or oxidizing agents and available complexing ligands, influence the concentration of the particular Cr(III) and Cr(VI) compounds in different environmental compartments [11]. In several industrial processes Cr(VI) is formed at high temperatures in the presence of oxygen. In plasma cutting or welding of stainless steel, fumes of Cr(VI) are formed from metallic chromium and emitted into the workplace air [12,13]. In the production of steel and electric arc furnaces, dust is generated that contains up to 10 % of Cr(VI) [1]. In the environment Cr(III) could be oxidised to Cr(VI) under the presence of Mn(IV) oxides. These processes have been demonstrated in different soils containing more than 100 mg kg^{-1} of Mn(IV) oxides [14,15]. Oxidation of Cr(III) may also take place with oxygen at high alkaline pH's. Oxidation is relatively slow, while the reduction of Cr(VI) is a very rapid process. The favourable conditions for reduction are in acidic solutions in the presence of electron donors. In environmental samples like soils, waste waters and different waste materials, Cr(VI) is reduced with organic matter, sulphides and Fe(II) [11]. The hazardous effects of chromium in the environment are related primarily to its hexavalent oxidation state. Most of Cr(VI) compounds are mobile in soils and are readily released into soil solution, and ground or surface waters. On the contrary, the mobility of Cr(III) in soils is limited due to formation of hydroxo precipitates [1]. When estimating chromium toxicity, it is necessary to consider the oxidation-reduction processes that may occur under certain environmental conditions [14-16]. The easily soluble chromium species have the greatest influence on the terrestrial and aquatic environment. Therefore, it is important to use reliable analytical procedures for the determination of exchangeable chromium and the content of exchangeable Cr(VI) in soils [17], waste materials and their mixtures with soils [16]. In studies of chromium environmental impacts it is also very important to determine other moderately and sparingly soluble chromium species. Several fractionation procedures applying a number of successive selective extractions have been developed for that purpose [18,19]. To estimate the environmental hazards, long-term experiments should be applied [20-22] and possible oxidation-reduction processes of chromium after disposal [16,20] or re-use of chromium rich waste materials should be investigated [23,24].

In the present chapter some alternatives for long-term safe disposal of chromium rich waste materials are presented. In addition, the potential of the use of waste materials from steel production in civil engineering is evaluated.

2. Analytical Tools for the Investigation of the Transformation and Mobility of Chromium in the Environment

Investigations of the transformation and mobility of chromium in the environment are of great importance in the assessment of the alternatives for minimisation and waste management of hazardous residues containing Cr(VI). Regarding the specific characteristics of the materials investigated, analytical tools that include speciation analysis, fractionation procedures and/or leachability tests are applied.

2.1. Speciation Analysis

Speciation plays a crucial role in the investigations of transformation and mobility of chromium in the environment. In order to prevent hazardous effects of Cr(VI) to the terrestrial and aquatic environment, and to ensure safe disposal of chromium rich waste materials [16,20-22], it is very important to determine the content of Cr(VI) in soils [17,25-27], waters [28-30], waste materials [4,16,27] and waste amended soils [16]. Reliable determination of Cr(VI) is also essential in investigations of the oxidation-reduction processes of chromium in soils [15] and in evaluation of the efficiencies of remediation of chromium contaminated soils [31-33] and waters [28,34-36]. In general, the procedures for the determination of Cr(VI) in soil samples are based on alkaline extraction for quantification of total Cr(VI) content [37,38] and on leaching procedures for the determination of the easily soluble Cr(VI) concentrations in soils [17,26,37]. Quantitative extraction methods for total Cr(VI) (soluble and insoluble forms) in soils and waste materials are used when the potential hazard of hexavalent chromium to living beings and eco systems is assessed, while soluble and exchangeable fractions of Cr(VI) are useful parameters for estimating soil levels of hexavalent chromium that may be leached to groundwater or be uptaken by plants and micro-organisms [37]. For quantitative extraction of total Cr(VI) from soils, sludges, sediments and solid wastes United States Environmental Protection Agency (EPA) recommended Method 3060A [38]. It consisted of hot alkaline (0.28 mol L^{-1} Na_2CO_3/0.5 mol L^{-1} NaOH) digestion at 90 to 95 °C including also the addition of $MgCl_2$ and phosphate buffer to the alkaline extracting solution in order to prevent the risk of Cr(III) oxidation. For determination of the easily soluble Cr(VI) in soils, phosphate buffer solutions were suggested for extraction of exchangeable Cr(VI) from soil (water-soluble Cr(VI) and Cr(VI) adsorbed on various oxides, clay particles and organic matter). Different concentrations of KH_2PO_4 [14,26] and mixtures of K_2HPO_4/KH_2PO_4 buffer solutions [17,37] were proposed for extraction. The most commonly used method for the quantification of Cr(VI) in soil extracts [14,25,27,37], waste materials [4,23] and environmental water samples [28,32] was spectrophotometry. However, in the last decade much more sensitive methods that are less liable to interferences were applied for chromium speciation. These methods are in general based on ion-exchange chromatography in combination with different detectors e.g. atomic absorption spectrometry (AAS) [8,12,13,17,20,24,40], chemiluminescence [41] or inductively coupled plasma mass spectrometry (ICP-MS) [2,3,8,42]. Whenever speciation of chromium is applied, specific characteristics of the sample matrix should be considered in order to achieve measurements without interferences [2,6,7,8,25]. In addition, the methods should be used that prevent risk of Cr(III) and Cr(VI) species interconversions during the extraction and/or the speciation procedure [2,3,8,43]. The use of stable isotopes is one of the most advanced analytical tools for reliable investigations of species transformations under different environmental conditions and their accurate determinations in various sample matrices [30,44,45]. In these investigations samples are spiked with known amounts of different stable isotope-enriched forms of chromium, like for instance, Cr(III) is added as a ^{53}Cr-enriched standard and Cr(VI) as a ^{50}Cr-enriched standard, or *vice versa*. After separation of both chromium species by ion-exchange chromatography and their determination by ICP-MS, the degree of species conversion between Cr(III) and Cr(VI) is determined. Based on isotope dilution mass spectrometry (IDMS), using equations on different chromium isotope ratios, accurate initial Cr(III) and Cr(VI) concentrations in the sample analysed is calculated [30,44,45]. The choice of the technique employed in the determination of Cr(VI) in various samples depends on its concentration and on the sample matrix.

2.2. Fractionation Procedures

Fractionation procedures based on sequential extractions are commonly used for the assessment of the mobility of metals in different solid matrices and for the evaluation of their impact on the environment [4,16,18,19,21,22]. In sequential extraction procedures various extractants are applied successively to the sediment, soil or waste material for selective leaching of the particular chemical forms of elements from samples analysed. In the first step, extractants with moderate ionic strength are used e.g. 1 mol L^{-1} $MgCl_2$ in Tessier's scheme [18] or 0.11 mol L^{-1} acetic acid in the BCR extraction procedure [19], to leach the water-soluble and exchangeable metal species that have the greatest influence on the aquatic and terrestrial environment and are readily uptaken by plants. Each extractant that follows is different in chemical nature and more vigorous than the previous one. From the results of fractionation studies the partitioning of metals between various easily and sparingly soluble fractions is estimated. Fractionation procedures based on Tessier's sequential extraction scheme were found to be suitable for study of the partitioning of chromium in natural soils and sediments [16,18], waste materials [4] and waste amended soils [16]. This scheme was particularly applicable to highly alkaline samples [21], since it reflects the real solubility of metals in the easily soluble fractions. The widely used BCR scheme was appropriate for acidic samples and samples close to neutral pH [19]. This scheme was not applicable for highly alkaline samples, since 0.11 mol L^{-1} acetic acid, that was used for the extraction of the easily soluble metal fraction, tremendously changed the pH of the sample [19]. Regardless, the use of particular fractionation procedure, aqueous or phosphate buffer extraction should be applied prior to the sequential extraction, so that the easily soluble Cr(VI) may also be determined by the speciation procedures [4,16,20,22].

2.3. Leaching Tests

The re-use of industrial by-products for civil engineering has been widely accepted in developed countries. Materials originating from industrial by-products can only be used when the final products possesses appropriate technical [46-48] and environmental properties [49-51]. Every waste material that can be potentially re-used in civil engineering may contain various toxic substances. Therefore, it is necessary to determine the content of toxic substances in these materials and to estimate the risk to the environment as a result of the release of contaminants from the final products [51,52]. The main environmental impact from re-use of secondary materials is the leaching of potentially toxic substances to soil and water [46]. The amount of substances that can be leached from secondary materials is estimated by applying different leaching tests [53]. Before its actual application, a secondary material should fulfil the requirements of the relevant national regulations. Methods for the preparation of leachates, by leaching of aggregates or granular waste materials and sludges, are prescribed in Standards EN 1744-3:2001 [54] and EN 12457-4:2004 [55]. For the estimation of the environmental burden leaching tests are also performed on final products like asphalt and concrete mixtures [56,57]. In order to estimate the long-term environmental impact, there is the Netherland's NEN 7345 standard prescribed leaching test which is based on diffusion [58]. This test was applied in the leaching protocol developed for concrete [59]. Similar leaching protocol was used also to evaluate the environmental acceptability of the asphalt composites and cement composites with addition of electric arc filter (EAF) dust [23,24].

3. Management of Chromium Rich Waste Materials

Due to the widespread use of chromium in various industrial processes [1-8], huge amounts of chromium rich waste materials are produced. Regarding the specific matrix characteristics of waste materials, the alternatives for their safe disposal or re-use should be investigated. For this purpose studies of the long-term environmental impacts are essential in order to ensure the conditions that would enable reduction of hazardous hexavalent chromium and prevent or minimise chromium oxidation in different environmental compartments. In the following paragraphs some case studies of such investigations are presented.

3.1. The Fate of Chromium in Tannery Waste and Sewage Sludge Amended Soils

Chromium salts are widely used as a tanning agents in leather production. Among them, the most frequently used is basic Cr(III) sulphate ($Cr(OH)SO_4$). Tannery waste, rich in Cr(III) is under environmental scrutiny due to its high chromium content (25-35 g kg^{-1}) [4]. When tannery waste is disposed of on land, the risk of potential oxidation of some Cr(III) species to toxic Cr(VI) exists in the presence of Mn(IV) oxides [14,15]. Sewage sludge from waste water treatment plants may also contain elevated concentrations of chromium that originate from various industrial activities [60]. Due to the high reduction potential of the sewage sludge (high content of organic matter) [21], chromium in this waste material exists predomonantely in its trivalent oxidation state.

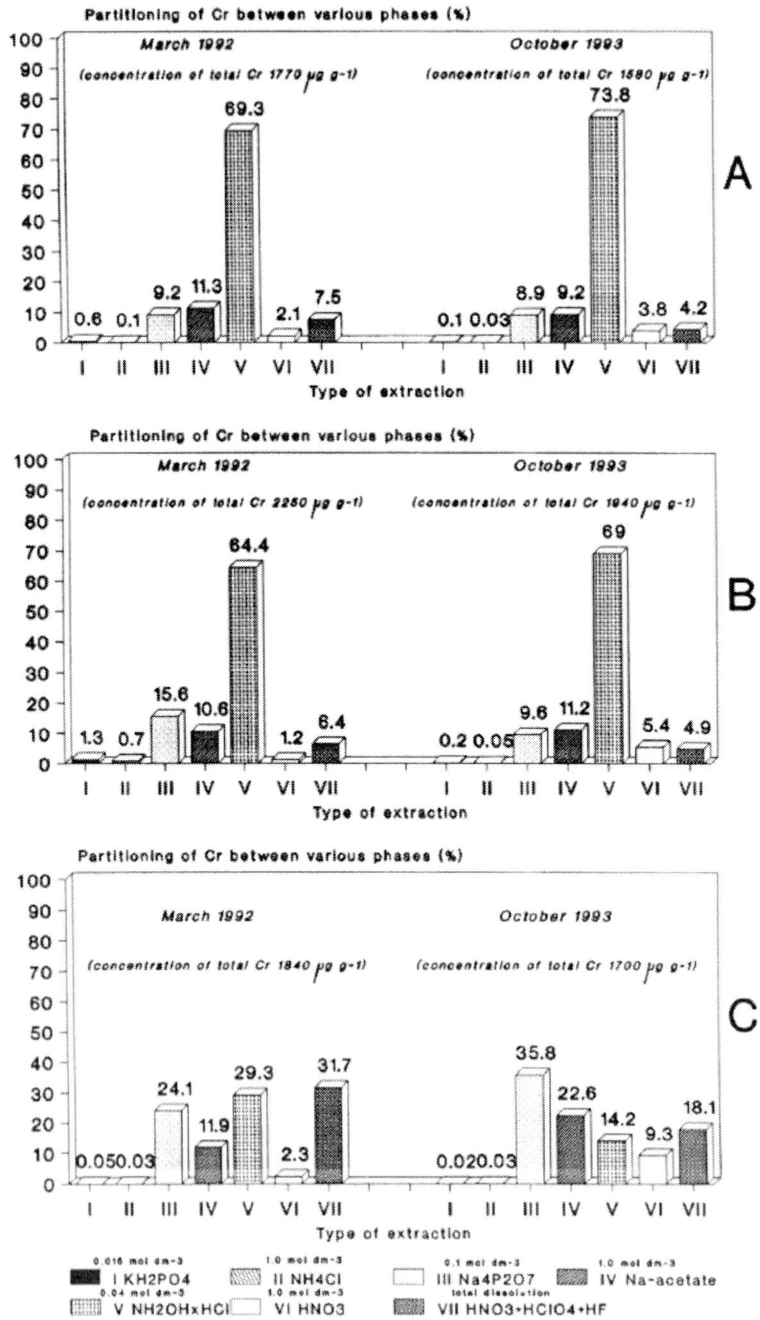

Fig. 1. Partitioning of chromium in tannery waste-amended soils. (A) sand; (B) clay; (C) peat. (Reproduced from reference [16] Milačič, R., & Štupar, J. (1995), *Environ. Sci. Technol., 29, 506-514.*, with permission from American Chemical Society.).

The fractionation and oxidation-reduction processes of chromium were studied in tannery waste and sewage sludge amended soils [16]. Natural clay and sandy soils high in Mn(IV) oxides and low in organic matter, and peat soil high in organic matter and low in Mn(IV) oxides contents and their mixtures with waste materials were investigated. The addition of waste material to a particular soil was in the amount that covered the requirement for nitrogen in one season. Samples were filled into 30 L plastic beakers with perforated bottoms. The experiment was carried out for two years under atmospheric conditions. To study the oxidation-reduction processes of chromium in soils amended with waste materials, Cr(VI) was determined in the water-soluble soil fraction (0.015 mol L^{-1} KH_2PO_4). For estimation of the environmental hazards, partitioning of chromium was also performed by the use of modified Tessier's sequential extraction procedure [16].

The results of fractionation studies in tannery waste amended sand, clay and peat soils [16] are presented in Figure 1.

In tannery waste, amended sand and clay soils two order of magnitude enhancement of chromium in the water-soluble and exchangeable fractions in the first five months after the waste application was observed. In peat soil this increase was moderate. After a five-month period the water-soluble and exchangeable chromium were partially eluted from topsoil with rain waters, while a part of chromium was redistributed into the less soluble soil fractions. In sand and clay tannery waste amended soils the prevailing fraction of chromium was bound to iron and manganese oxides and hydroxides, while in peat soil chromium was distributed predominantly between organic fraction, fraction bound to iron and manganese oxides and hydroxides and the mineral fraction.

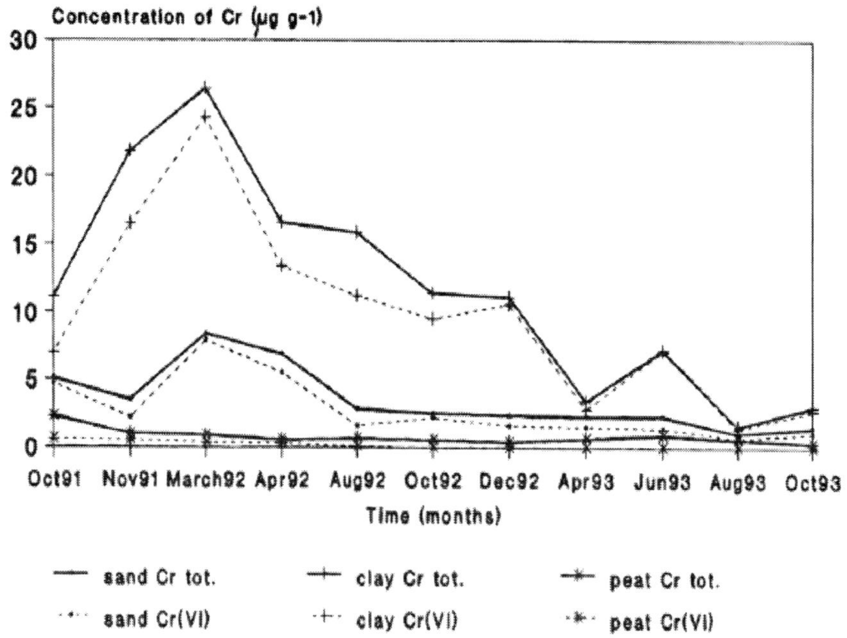

Fig. 2. Variation of total water-soluble (KH_2PO_4, 0.015 mol dm^{-3}) chromium and Cr(VI) with time in various tannery waste-amended soils. (Reproduced from reference [16] Milačič, R., & Štupar, J. (1995), *Environ. Sci. Technol., 29, 506-514.*, with permission from American Chemical Society.).

Based on data of speciation of chromium in the water-soluble fractions (Figure 2), the oxidation of Cr(III) to Cr(VI) was observed in sand and clay soils. Cr(VI) concentrations were gradually increased in the first five months after the tannery waste application and were followed by the decrease in the second period of the experiment. The degree of chromium oxidation depended on the water-soluble content of chromium in soils, concentration of soil Mn(IV) oxides and on the reducing capacity of the soils. In sand and clay soils high in Mn(IV) oxides content (170 and 260 mg kg^{-1}, respectively) and low in organic matter (3.5 and 3.0 %, respectively) [16] water-soluble Cr(III) in tannery waste amended soils was oxidised to concentrations up to 7 and 23 mg kg^{-1} of Cr(VI), respectively.

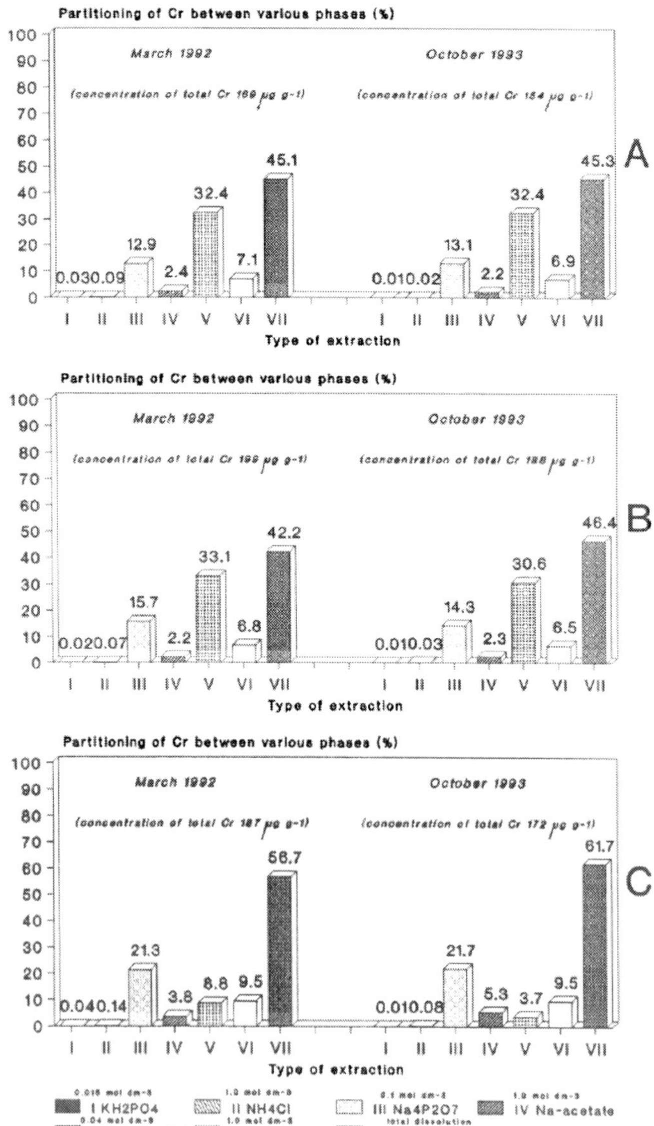

Fig. 3. Partitioning of chromium in sewage sludge-amended soils. (A) sand; (B) clay; (C) peat. (Reproduced from reference [16] Milačič, R., & Štupar, J. (1995), *Environ. Sci. Technol., 29, 506-514.*, with permission from American Chemical Society.).

The oxidation of chromium was found to be negligible in peat soil with high reducing capacity (high content of organic matter 54 %) and low concentration of Mn(IV) oxides (30 mg kg^{-1}). The results of the presented study demonstrated that tannery waste could be safely applied on peat soils or other soils high in organic matter. Care should be taken with soil types low in organic matter and high in Mn(IV) oxides that are able to oxidise chromium. Rates of application on such soils should be small and concentration of soluble Cr(VI) should be monitored in the top soil a few months after the application in order to prevent the potential contamination of the surface and ground waters.

Data of fractionation of chromium in sewage sludge amended soils [16] (Figure 3) demonstrated that chromium in the water-soluble and exchangeable fractions was only slightly increased in the first five months after the sludge application. In the continuation of the experiment, a decrease of chromium concentrations in these fractions to almost the natural level in soils was observed. Due to the reducing characteristics of sewage sludge (about 50 % of organic matter [21]) and low concentrations of total water-soluble chromium, the processes of oxidation of Cr(III) was not observed in sewage sludge amended soils. Regarding chromium content, sewage sludge could be disposed on land without hazardous effects on the terrestrial environment as far as the rate of sludge application covers the requirement for nitrogen in one season.

3.2. Environmental Impacts of the Use of Sawdust and Quicklime for Stabilisation of Sewage Sludge: Oxidation-Reduction Processes of Chromium

Environmental consequences of the use of sawdust and quicklime for sewage sludge stabilisation were investigated [20,21]. Liming has been traditionally used for abiotic remediation to increase the alkalinity and hence in general reduce the mobility of metals. Sewage sludge containing pathogenic microorganisms was amended with quicklime (CaO) for disinfection and with sawdust for post-stabilisation prior to disposal or reuse in the environment. The amount of quicklime added to the sludge or sludge mixed with sawdust was selected according to the criterion of raising the pH above 12. The mobility of chromium and other metals in the dewatered sewage sludge and its mixtures with quicklime, sawdust and lime-treated sludge mixed with sawdust by applying a modified Tessier's sequential extraction procedure was investigated [21]. To the original Tessier partitioning scheme [18], extraction with water as a first step was added in order to perform speciation analysis of Cr(VI). It was experimentally proven that water and extracting solutions, e.g., 1 mol L^{-1} $MgCl_2$ (pH 7), do not change the original pH of sludge and were therefore the most convenient for use in the estimation of the easily soluble portions of metals in the lime-treated sludge. The possible oxidation-reduction processes of chromium in highly alkaline lime-treated sludge and lime-treated sludge mixed with sawdust samples were also investigated in the time span of six months by determination of Cr(VI) in sludge extracts [20]. On the basis of the experimental data, estimation was made of whether lime-treatment is environmentally acceptable for safe sludge disposal.

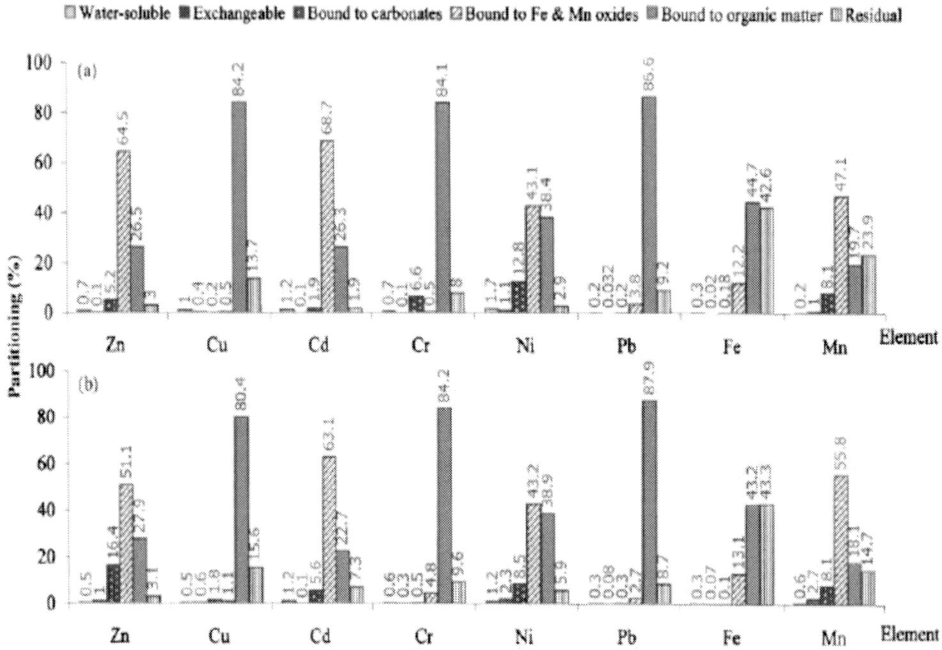

Fig. 4. Partitioning of heavy metals in the dewatered sludge 48 h (a) and 3 months after the start of the experiment (b). (Reproduced from reference [21] Ščančar, J., Milačič, R., Stražar, M., Burica, O., & Bukovec, P. (2001), *J. Environ. Monit., 3, 226-231.*, with permission from The Royal Society of Chemistry.).

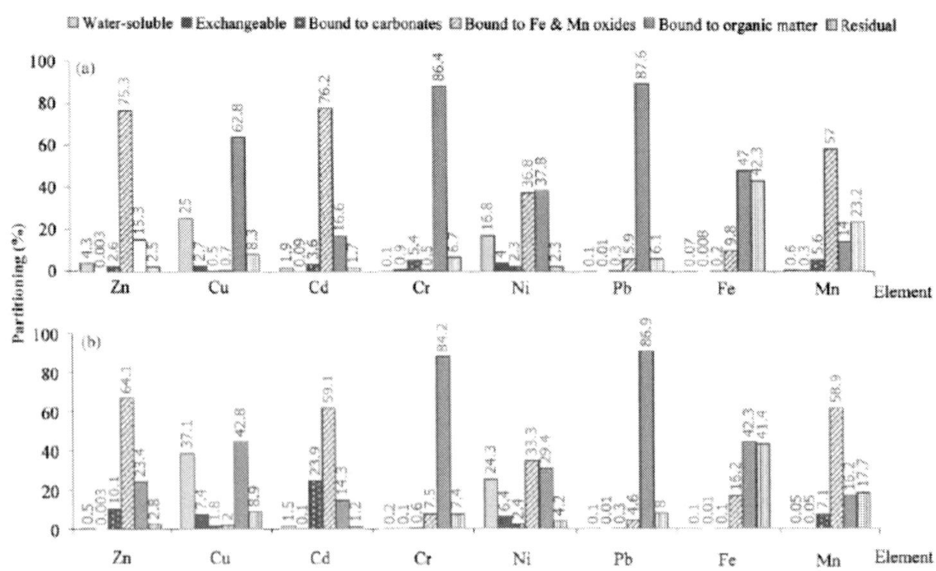

Fig. 5. Partitioning of heavy metals in the lime-treated sludge 48 h (a) and 3 months after liming (b). (Reproduced from reference [21] Ščančar, J., Milačič, R., Stražar, M., Burica, O., & Bukovec, P. (2001), *J. Environ. Monit., 3, 226-231.*, with permission from The Royal Society of Chemistry.).

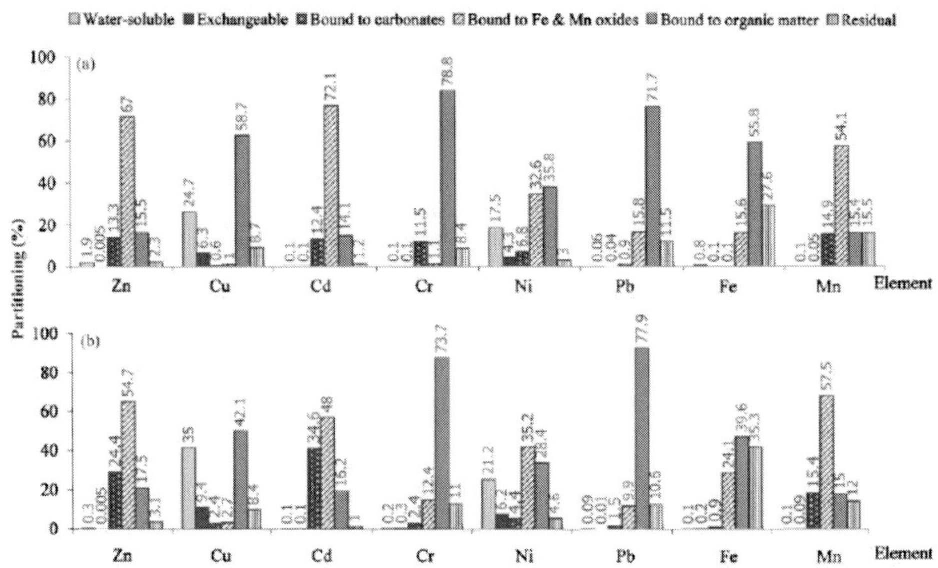

Fig. 6. Partitioning of heavy metals in the lime-treated sludge mixed with sawdust 48 h (a) and 3 months after liming (b). (Reproduced from reference [21] Ščančar, J., Milačič, R., Stražar, M., Burica, O., & Bukovec, P. (2001), *J. Environ. Monit., 3, 226-231.*, with permission from The Royal Society of Chemistry.).

Data of fractionation study indicated that liming slightly decreased the total water-soluble chromium in the dewatered sludge (compare data from Figures 4 and 5) and in the lime-treated sludge mixed with sawdust (compare data from Figures 4 and 6) 48 h and 3 months after the start of the experiment, but has no influence on the redistribution of chromium to other sludge phases. On the contrary, application of sawdust slightly increased the total water-soluble chromium in dewatered sludge 48 h after the start of the experiment (compare data from Figures 4 and 7). Regarding other elements, liming significantly increased solubility of nickel and copper, so it cannot be recommended for sludge stabilisation if these two metals are present at elevated concentrations in the sludge [21].

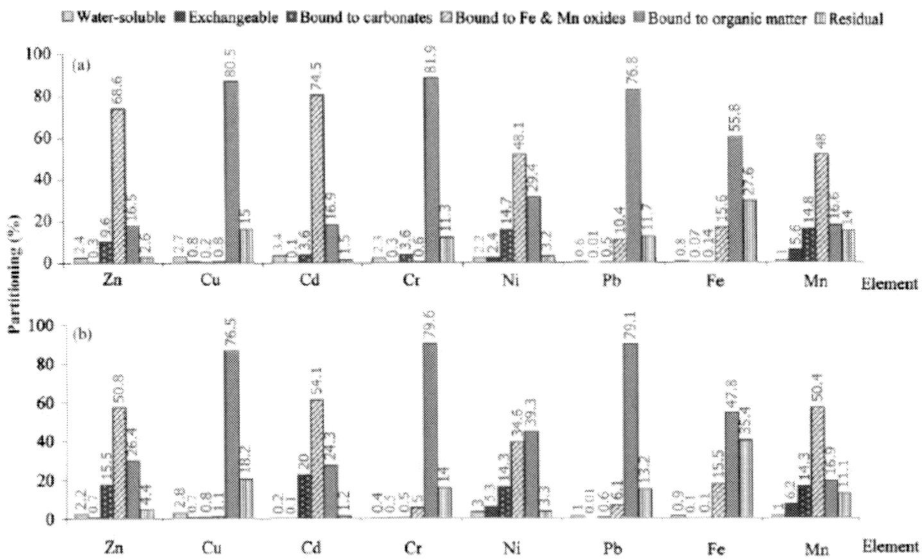

Fig. 7. Partitioning of heavy metals in the sludge mixed with sawdust 48 h (a) and 3 months after liming (b). (Reproduced from reference [21] Ščančar, J., Milačič, R., Stražar, M., Burica, O., & Bukovec, P. (2001), *J. Environ. Monit., 3, 226-231.*, with permission from The Royal Society of Chemistry.).

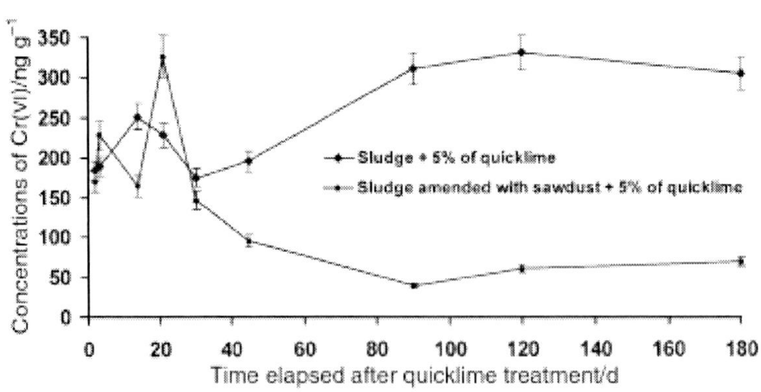

Fig. 8. Variation of Cr(VI) concentrations (dry mass basis) with time elapsed after quicklime treatment in extracts of sludge samples treated with 5% of quicklime, and sludge samples amended with sawdust and treated with 5% of quicklime. Cr(VI) was determined by the FPLC-ETAAS technique, $n = 2$. (Reproduced from reference [20] Milačič, R., & Ščančar, J. (2000), *Analyst, 125, 1938-1942.*, with permission from The Royal Society of Chemistry.).

Analysis of Cr(VI) in untreated, sawdust-treated, lime-treated and sawdust amended lime-treated sludge indicated that in aqueous extracts of dewatered and sawdust-treated sewage sludge Cr(VI) was not detected [20]. In aqueous extracts of lime-treated and sawdust amended lime-treated sewage sludge Cr(VI) was determined 48 h after liming (Figure 8). Despite the highly reducing characteristics of the sewage sludge, liming provoked oxidation of Cr(III) to Cr(VI) at pH 12, most probably through the action of $MnOOH_{(s)}$ and dissolved oxygen. The amount of hexavalent chromium in lime-treated sludge slightly increased and after three months reached an almost constant value of about 300 ng g^{-1} Cr(VI). During the course of the experiment no significant changes in pH of these sludge samples were observed. The concentration of hexavalent chromium in the lime-treated sludge amended with sawdust increased in the first three weeks after the lime treatment to about 300 ng g^{-1} Cr(VI), and appreciably decreased after one month to a concentration of about 50 ng g^{-1} Cr(VI). In the latter sludge samples a decrease of pH from 12.0 at the start of the experiment to 9.6 after one month to the end of experiment was observed. The decrease in pH and reduction of Cr(VI) to Cr(III) was presumably caused by the release of organic acids from decaying sawdust.

Taking into consideration possible environmental hazards and, in particular, formation of Cr(VI) at high pH, liming cannot be recommended for sludge disinfection prior its disposal on land. On the other hand, beneficial effect was observed on sludge stabilisation when sludge was amended with sawdust.

3.3. Re-Use of Waste Materials From Steel Production: Environmental Concerns

Waste materials from steelworks such as black steel slag, ladle slag and electric arc filter dust (EAF dust) can successfully substitute natural raw materials and be used as alternative materials in civil engineering or for road construction. The main advantages of re-use of these waste materials are preservation of natural resources, extensive decrease in landfill load, protection of the environment and the positive economic effects. Whenever waste materials from steel production are disposed or re-used, the potential leachnig of hazardous Cr(VI) should be considered. As an example, some alternatives for re-use of EAF dust that represents an appropriate material for incorporation into asphalt or cement are demonstrated. To evaluate the long-term environmental impacts NEN 7345 standard based on diffusion [58] and the leaching protocol that was developed for concrete [59] were adopted. Leaching experiments were performed for six months in water and salt water (3.8 % NaCl), following the concentrations of total chromium and Cr(VI) in leachates. Compact and ground asphalt composites with addition of 2 % by mass of EAF dust [23] and compact and ground cement composites and cement composites with addition of 1.5 % of EAF dust by mass were studied [24]. Ground composites simulated decomposition in the environment with time, while salt water as a leaching solution simulated the salting of roads during the winter time [23] and the sea water environment [24]. The pH of leachates of the samples investigated were about 7. The results of the leaching tests for four different asphalt composites with addition of EAF dust [23] indicated that Cr(VI) was not leached with water and salt water from compact asphalt composites with addition of EAF dust.

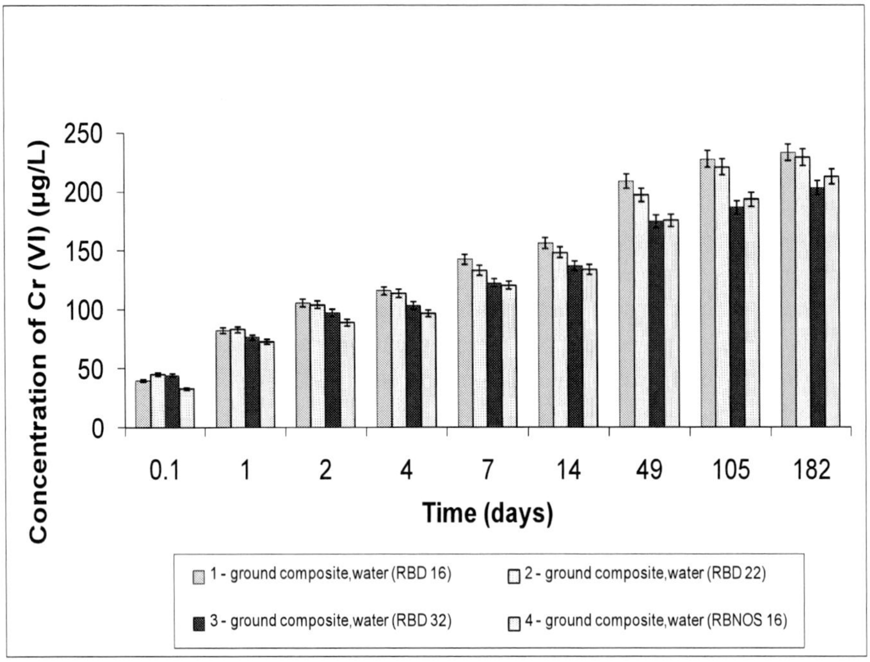

Fig. 9. Leaching of Cr(VI) in water with time from ground composites with addition of filter dust. Results represent the average of two parallel samples. In each bar the two concentrations that characterize the mean value are indicated. (Reproduced from reference [23] Vahčič, M., Milačič, R., Mladenovič, A., Murko, S., Zuliani, T., Zupančič, M., & Ščančar, J. (2008). *Waste Manage., 28, 2667-2674.*, with permission from Elsevier.).

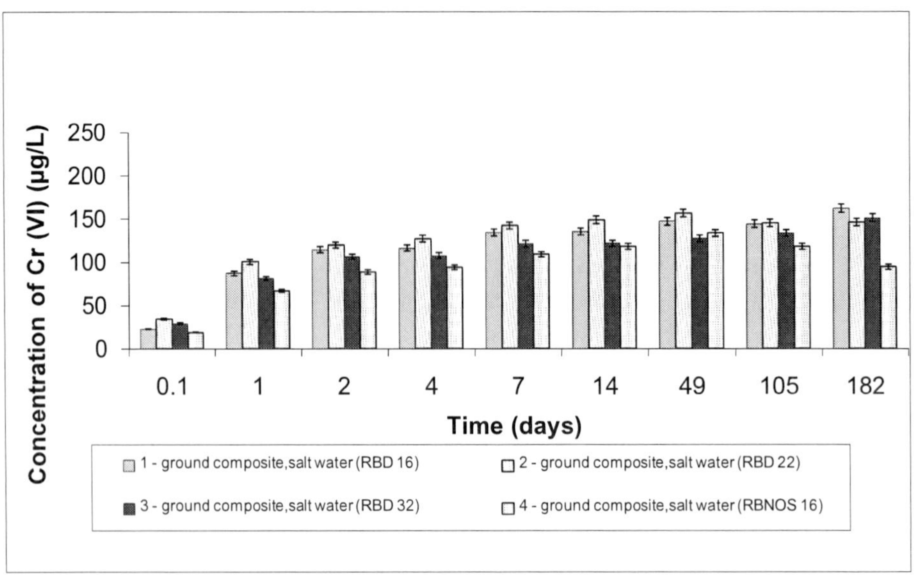

Fig. 10. Leaching of Cr(VI) in salt water with time from ground composites with addition of filter dust. Results represent the average of two parallel samples. In each bar the two concentrations that characterize the mean value are indicated. (Reproduced from reference [23] Vahčič, M., Milačič, R., Mladenovič, A., Murko, S., Zuliani, T., Zupančič, M., & Ščančar, J. (2008). *Waste Manage., 28, 2667-2674.*, with permission from Elsevier.).

However, in ground asphalt composites with addition of EAF dust Cr(VI) was leached with water and salt water (Figure 9 and Figure 10, respectively). Leaching of Cr(VI) from water and salt water was similar for four different samples of asphalt composites with addition of EAF dust. During the course of the experiment gradual increase in Cr(VI) concentration in water from ground asphalt composites with addition of EAF dust was observed. The constant Cr(VI) concentrations (about 220 μg L^{-1}) were reached after approximately 100 days. In salt water, lower concentrations from ground composites with addition of EAF dust were leached in comparison to leaching in water. The constant Cr(VI) concentrations in leachates of salt water (about 150 μg L^{-1}) were reached faster, after 7 days. The phenomenon of lower leaching levels of chromium in salt water than in water was also observed in other studies [59], where the leaching of metals from concrete in water and sea water was investigated. Even in ground asphalt composites with addition of EAF dust, that simulated the worst case scenario when asphalt decomposes with time, relatively low Cr(VI) concentrations were leached. These Cr(VI) concentrations may possibly represent the environmental hazard only if the leachates reach the ground and/or drinking water reservoirs. Taking into consideration the latter environmental consequences, EAF dust can be used as an additive to asphalts for road construction.

In the investigation of the potential use of EAF dust as additive to cement [24] the results of the leaching tests for compact and ground cement composites and cement composites with addition of 1.5 % of EAF dust indicated that the pH of aqueous and salt water leachates of cement composites and cement composites with addition of EAF dust was, due to dissolved $Ca(OH)_2$ and $CaCO_3$, highly alkaline and ranged between 11 and 13. In ground composites the pH was for one unit higher than in compact composites due to higher specific surface and consequently higher leachability of $Ca(OH)_2$ and $CaCO_3$. In comparison to aqueous leachates, the pH was also one unit higher in salt water composites. This indicates that due to the higher ionic strength of the leaching solution, the leachability of $Ca(OH)_2$ and $CaCO_3$ from cement composites is more efficient in salt water. Data of the leaching test further indicated that during the course of the experiment the total chromium and Cr(VI) were not leached with water in compact cement composites and compact cement composites with addition of EAF dust. In ground cement composites and ground cement composites with addition of EAF dust, chromium was leached with water. Although it was, at present, exclusively in its hexavalent form, the concentrations that were leached were very low, below 5.5 μg L^{-1} and did not represent environmental hazard.

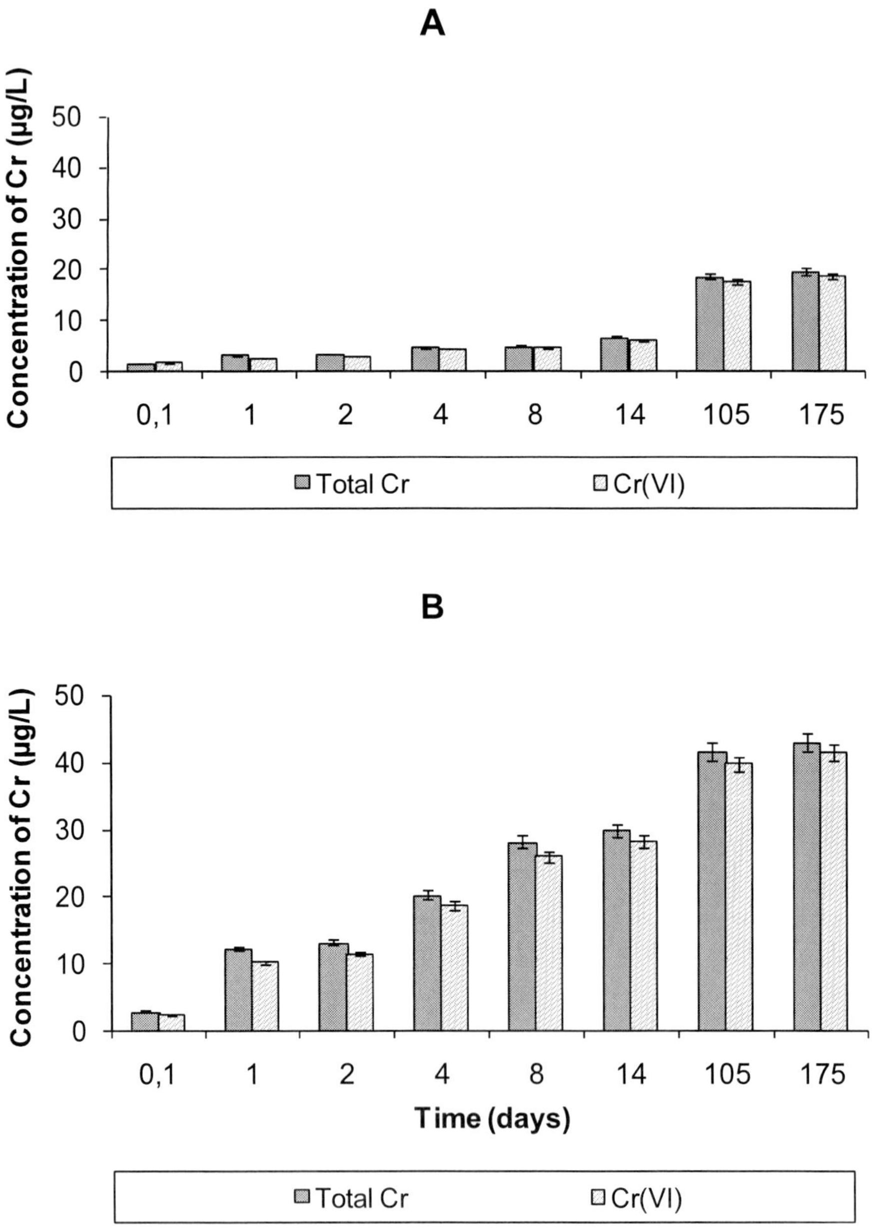

Fig. 11. Leaching of total Cr and Cr(VI) in salt water with time from compact cement composite (A) and compact cement composite with addition of filter dust (B). Results represent the average of two parallel samples. In each bar the two concentrations that characterize the mean value are indicated. (Reproduced from reference [24] Šturm, T., Milačič, R., Murko, S., Vahčič, M., Mladenovič, A., Strupi-Šuput, J., & Ščančar, J. (2009). *J. Hazard. Mater.*, 2009, 166, 277-283., with permission from Elsevier.).

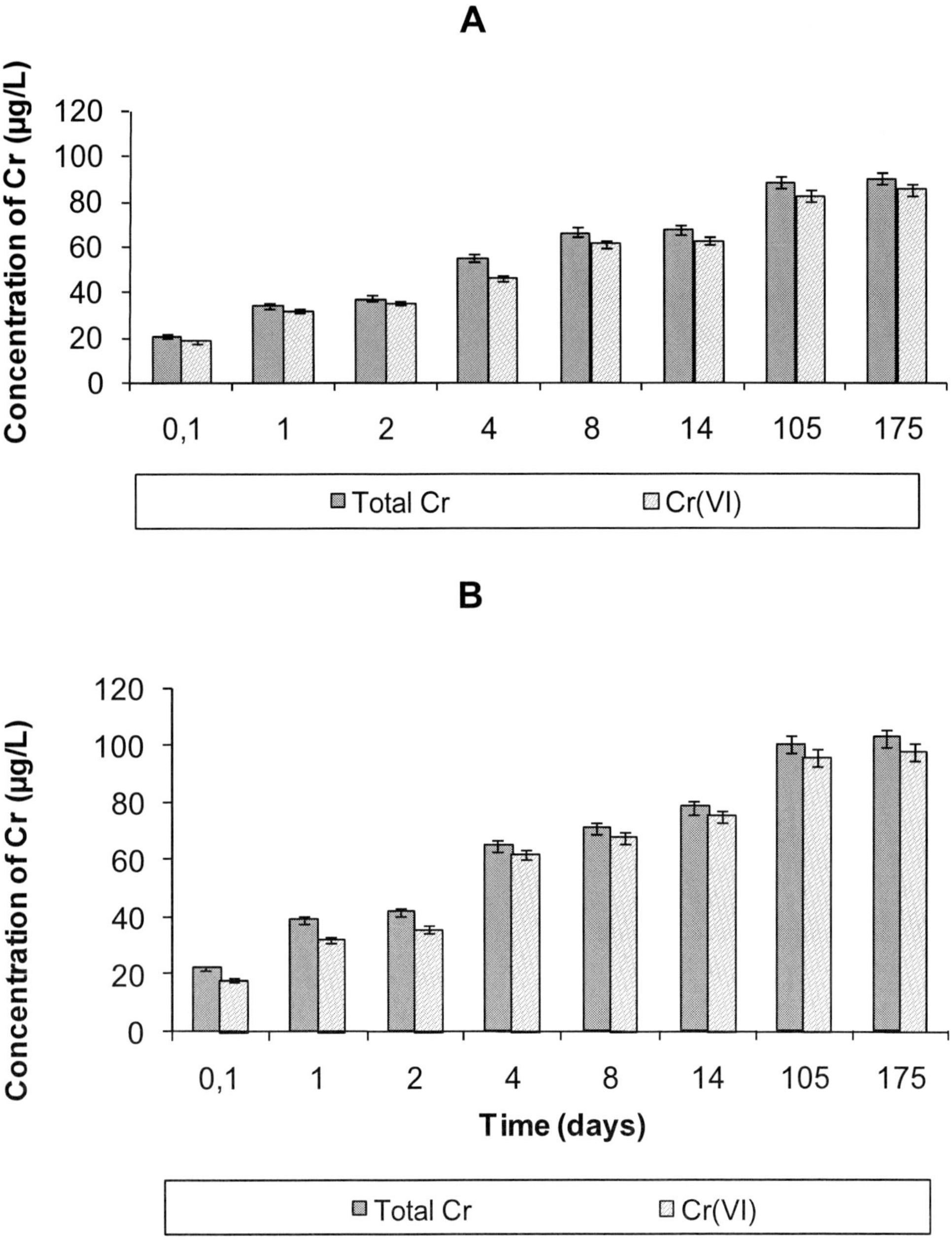

Fig. 12. Leaching of total Cr and Cr(VI) in salt water with time from ground cement composite (A) and ground cement composite with addition of filter dust (B). Results represent the average of two parallel samples. In each bar the two concentrations that characterize the mean value are indicated. (Reproduced from reference [24] Šturm, T., Milačič, R., Murko, S., Vahčič, M., Mladenovič, A., Strupi-Šuput, J., & Ščančar, J. (2009). *J. Hazard. Mater.*, 2009, 166, 277-283., with permission from Elsevier.).

In salt water, the extent of leaching of Cr(VI) was higher than in water (Figure 11 and Figure 12) since the higher pH of leachate (pH 13) favoured the existence of Cr(VI). From the data of Figure 11 it is evident that during the course of the experiment the concentrations of Cr(VI) in leachates from compact cement composites in salt water did not exceed 20 μg L^{-1} and in leachates from compact cement composites with addition of EAF dust 40 μg L^{-1}. The difference (20 μg L^{-1} of Cr(VI)) corresponded to the leachability of Cr(VI) from EAF dust added to cement composite. More intensive leaching in salt water was observed from ground cement composites (up to 80 μg L^{-1} of Cr(VI)) and ground cement composites with addition of EAF dust (up to 100 μg L^{-1} of Cr(VI)) (Figure 12). Again, the difference (20 μg L^{-1} of Cr(VI)) corresponded to the leachability of Cr(VI) from EAF dust added to cement composite. Therefore, the main contribution to the leachability represented Cr(VI) that originates from cement. The highest Cr(VI) concentrations that were observed in salt water leachates from ground cement composite with addition of EAF dust did not exceed 100 μg L^{-1}, that is according to the national Slovenian legislation the maximal allowed Cr(VI) concentration in leachates from disposals of inert waste [24]. Based on these experimental data it may be concluded that considering Cr(VI) concentration, cement composites with the addition of EAF dust do not represent an environmental hazard neither in the worst case scenario when the cement composite with addition of EAF dust is decomposed with time in the marine environment nor when it is disposed of in a landfill. Regarding the physico-mechanical properties, as well as from the environmental point of view, cement composites to which a maximum of 1.5 % by mass of EAF dust has been added can be used in civil engineering for all purposes, including applications in the external environment. Cement composites with the addition of EAF dust can also be used as balances in washing machines which, at the end of their service life, are disposed of on landfills.

CONCLUSION

The most common chromium oxidation states are trivalent and hexavalent. The equilibrium concentration between Cr(III) and Cr(VI) species is a result of the prevailing environmental conditions. Under the acidic pH's and in the presence of reducing agents like Fe(II), sulphides and organic matter, chromium trivalent oxidation state is favored, since Cr(VI) is readily reduced to Cr(III). On the contrary, at high alkaline pH's Cr(III) may be oxidised to Cr(VI) by atmospheric oxygen. Oxidation of Cr(III) may be also provoked by the presence of Mn(IV) oxides. When the alternatives of safe disposal and re-use of chromium rich waste materials are investigated the specific behaviour of chromium under different environmental conditions should be considered. Complex experiments need to be performed by the use of appropriate analytical tools. Regarding the particular characteristics of waste materials and chromium behaviour in different matrices, fractionation procedures that enable estimation of the distribution of chromium between easily and sparingly soluble fractions, speciation procedures for the determination of Cr(VI), and/or leachability tests for the evaluation of the long-term environmental impacts should be applied. The design of the experiments for the assessment of chromium environmental impacts depends on the specific characteristics of the materials investigated.

Experimental data of described case studies demonstrated that sewage sludge could be disposed of on land without hazardous effects on the terrestrial environment concerning chromium concentration if the rate of sludge application covers the requirement for nitrogen in one season. Disposal of tannery waste to soils high in Mn(IV) oxides and low in organic matter may provoke oxidation of Cr(III) to its hazardous hexavalent form. Rates of application should be small and Cr(VI) monitored in the top soil to prevent pollution of groundwater. Reasonable amounts of tannery waste could be safely applied on soils low in Mn(IV) oxides and high in organic matter. Liming cannot be recommended for sewage sludge disinfection prior to its disposal on land due to possible environmental hazards caused by formation of Cr(VI) at high pH. Beneficial effects for sludge stabilisation were observed when sludge was amended with sawdust.

Waste materials from steelworks may be used as alternative materials in civil engineering or for road construction. Based on the experimental data of leaching tests EAF dust can be used as an additive to asphalts for road construction when the concentration of EAF dust does not exceed 2 % by mass. EAF dust may also be used as an additive to cement. Cement composites that contain up to 1.5 % by mass of EAF dust can be used in civil engineering for all purposes and as balances in washing machines. Black steel slag and ladle slag also represent potential materials for their application in composites for road construction and in civil engineering. The re-use of waste materials from steel production represents opportunity to save natural resources. However, before any chromium rich waste materials are disposed of or re-used, complex investigations for the estimation of long-term environmental consequences have to be carried out.

REFERENCES

[1] Metze, D., Jakubowski, N., & Klockow, D. (2005). Speciation of Chromium: Speciation of Chromium in Environment and Food. In R. Cornelis, J. Caruso, H. Crews, & K. Heumann (Eds.), *Handbook of Elemental Speciation II – Species in the Environment, Food, Medicine and Occupational Health* (p.p. 120-135). The Atrium, Southern Gate, Chichester, West Sussex PO19 8SQ, England, John Wiley & Sons, Ltd.

[2] Séby, F., Gagean, M., Garraud, H., Castetbon, A., & Donard, O.F.X. (2003). Development of analytical procedure for determination of total chromium by quadrupole ICP-MS and high-resolution ICP-MS, and hexavalent chromium by HPLC-ICP-MS, in different materials in the automotive industry. *Anal. Bioanal. Chem., 377, 685-694.*

[3] Séby, F., Castetbon, A., Ortega, R., Guimon, C., Niveau, F., Barrois-Oudin, N., Garraud, H., & Donard, O.F.X. (2008). Development of analytical procedures for the determination of hexavalent chromium in corrosion prevention coatings used in the automotive industry. *Anal. Bioanal. Chem., 391, 587-597.*

[4] Milačič, R., Štupar, J., Kožuh, N., Korošin, J., & Glazer I. (1992). Fractionation of Cr and determination of Cr(VI) in blue shavings. *J. Am. Leather Chem. Assoc., 87, 221-234.*

[5] Milačič, R., Ščančar, J., & Urbanc, M. (1998). Is spectrophotometry a reliable technique for determination of Cr(VI) in leather samples? *J. Soc. Technol. Chem., 82, 91-94.*

[6] Ščančar, J., Osterman, T., Bukovec, N., & Milačič, R. (2007). Critical appraisal of analytical procedures for the determination of Cr(VI) in dyed leathers by 1.5

diphenylcarbazide spectrophotometry after sample dilution or color removal. *J. Am. Leather Chem. Assoc., 102, 85-92.*

[7] Cruz, F.G., Katz, S.A., & Milačič, R. (1995). Determination of hexavalent chromium in CCA-treated building timbers. *J. Environ. Sci. Health A. 30, 299-306.*

[8] Ščančar, J., Milačič, R., Séby, F., Donard, O.F.X. (2005). Determination of hexavalent chromium in cement by the use of HPLC-ICP-MS, FPLC-ETAAS, spectrophotometry and selective extraction techniques. *J. Anal. At. Spectrom., 20, 871-875.*

[9] Katz, S.A., & Salem, H. (1994). *The Biological and Environmental Chemistry of Chromium,* (pp. 1-10). 220 East 23rd Street, New York: VCH Publishers, Inc.

[10] Katz, S.A., & Salem, H. (1994). *The Biological and Environmental Chemistry of Chromium,* (pp. 195-210). 220 East 23rd Street, New York: VCH Publishers, Inc.

[11] Rai, D., Eary, L.E., & Zachara, J.M. (1989). Environmental chemistry of chromium. *Sci. Tot. Environ., 86, 15-23.*

[12] Ščančar, J., & Milačič, R. (2002). A novel approach for speciation of airborne chromium by convective-interaction media fast-monolithic chromatography with electrothermal atomic-absorption spectrometric detection. *Analyst, 127, 629-633.*

[13] Milačič, R., Ščančar, J., & Tušek, J. (2002). Determination of Cr(VI) in welding fumes by anion-exchange fast protein liquid chromatography with electrothermal atomic absorption spectrometric detection. *Anal. Bioanal. Chem., 372, 549-553.*

[14] Bartlett, R.J., & James, B. (1979) Behaviour of chromium in soils, 3. Oxidation. *J. Environ. Qual. 8, 31-35.*

[15] Kožuh, N., Štupar, J., & Gorenc, B. (2000). Reduction and oxidation processes of chromium in soils. *Environ. Sci. Technol., 34, 112-119.*

[16] Milačič, R., & Štupar, J. (1995). Fractionation and oxidation of chromium in tannery waste and sewage sludge amended soils. *Environ. Sci. Technol., 29, 506-514.*

[17] Ščančar, J., Zupančič, M., & Milačič, R. (2007). Development of analytical procedure for the determination of exchangeable Cr(VI) in soils by anion-exchange fast protein liquid chromatography with electrothermal atomic absorption spectrometry detection. *Water Air Soil Pollut., 185, 121-129.*

[18] Tessier, A., Campbell P.G.C., & Bisson, M. (1979). Sequential extraction procedure for the speciation of particulate trace metals. *Anal. Chem., 51, 844-851.*

[19] Rauret, G., Lopez-Sanches, J.F., Sahuquillo, A., Rubio, R., davidson, C., Ure, A., & Quevauiller, Ph. (1999). Improvement of the BCR three step sequential extraction procedure prior to the certification of new sediment and soil reference materials. *J. Environ. Monit. 1, 57-61.*

[20] Milačič, R., & Ščančar, J. (2000). Determination of hexavalent chromium in lime-treated sewage sludge by anion-exchange fast protein liquid chromatography with electrothermal atomic absorption spectrometry detection. *Analyst, 125, 1938-1942.*

[21] Ščančar, J., Milačič, R., Stražar, M., Burica, O., & Bukovec, P. (2001). Environmentally safe sewage sludge disposal: the impact of liming on the behaviour of Cd, Cr, Cu, Fe, Mn, Ni, Pb, and Zn. *J. Environ. Monit., 3, 226-231.*

[22] Milačič, R., & Ščančar, J. (2003). Safe disposal of chromium-rich waste materials. *Crit. Rev. Anal. Chem., 33, 69-75.*

[23] Vahčič, M., Milačič, R., Mladenovič, A., Murko, S., Zuliani, T., Zupančič, M., & Ščančar, J. (2008). Leachability of Cr(VI) and other metals from asphalt composites with addition of filter dust. *Waste Manage., 28, 2667-2674.*

[24] Šturm, T., Milačič, R., Murko, S., Vahčič, M., Mladenovič, A., Strupi-Šuput, J., & Ščančar, J. (2009). The use of EAF dust in cement composites: assessment of environmental impact. *J. Hazard. Mater.,* 2009, 166, 277-283.

[25] Milačič R., Štupar, J., Kožuh, N., Korošin, J., & Glazer, I. (1992). Critical evaluation of three analytical techniques for the determination of chromium (VI) in soil extracts. *Analyst, 117, 125-130.*

[26] Kožuh, N., Štupar, J., Milačič R., & Gorenc, B. (1994). Optimization of extraction procedure for determination of total water-soluble chromium and chromium(VI) in various soils. *Int. J. Environ. Anal. Chem., 56, 207-217.*

[27] Marqués, M.J., Salvador, A., Morales-Rubio, A. & de la Guardia, M. (1998). Analytical methodologies for chromium speciation in solid matrices: a survey of the literature. *Fresenius J. Anal. Chem. 362, 239-248.*

[28] Lu, A., Zhong, S., Chen, J., Shi, J., Tang, J., & Lu, X. (2006). Removal of Cr(VI) and Cr(III) from aqueous solutions and industrial wastewaters by natural clino-pyrrohotite. *Environ. Sci. Technol., 40, 3064-3069.*

[29] Marqués, M.J., Salvador, A., Morales-Rubio, A. & de la Guardia, M. (2000). Chromium speciation in liquid matrices: a survey of the literature. *Fresenius J. Anal. Chem. 367, 601-613.*

[30] Ma, H.-L., & Tanner, P.A. (2008). Speciated isotope dilution analysis of Cr(III) and Cr(VI) in water by ICP-DRC-MS. *Talanta, 77, 189-194.*

[31] James, B.R. (1996). The challenge of remediating chromium-contaminated soil. *Environ. Sci. Technol., 30, 248A-251A.*

[32] Jeyasingh, J., & Philip, L. (2005). Bioremediation of chromium contaminated soil: optimization of operating parameters under laboratory conditions, *J. Hazard. Mater., 118, 113-120.*

[33] James, B.R. (2001). Remediation-by-reduction strategies for chromate-contaminated soils, *Environ. Geochem. Health, 23, 175-179.*

[34] Smith, W-L. (2000). Hexavalent chromium reduction and precipitation by sulphate-reducing bacterial biofilms. *Environ. Geochem. Health, 23, 297-300.*

[35] Quintelas, C., Sousa, E., Silva, F., Neto, S., & Tavares, T. (2006). Competitive biosorption of ortho-cresol, phenol, chlorophenol and chromium(VI) from aqueous solution by bacterial biofilm supported on granular activated carbon. *Process Biochem., 41, 2087-2091.*

[36] Quin, G., McGuire, M.J., Blute, N.K., Seidel, C., & Fong, L. (2005). Hexavalent chromium removal by reduction with ferrous sulfate, coagulation, and filtration: A pilot-scale study. *Environ. Sci. Technol., 39, 6321-6327.*

[37] James, B.R., Petura, J.C., Vitale, R.J. & Mussoline, G.R. (1995). Hexavalent chromium extraction from soil: A comparison of five methods. *Environ. Sci. Technol., 29, 2377-2381.*

[38] United States Environmental Protection Agency, Method 3060A (1996). In: Test Methods for Evaluating Solid Wastes, Physical/Chemical Methods, SW846, Update, (pp. 1-15.) Office of Solid Waste and Emergency Response Washington DC.)

[39] Bartlett, R.J., & James, B.R. (1996). (In: Sparks, D.L., Page, A.L., Helmke, P.A., Loeppert, R.H., Soltanpour, P.N., Tabatabi, M.A., Johnston, C.T. & Sumner, M.E. (Eds.), Chromium. Methods of Soil Analysis. Part 3 – Chemical methods. (pp. 692-694). Madison, Wisconsin, USA: Soil Science Society of America, Inc.)

[40] Milačič, R., & Štupar, J. (1994). Simultaneous determination of chromium(III) complexes and chromium(VI) by fast protein anion-exchange liquid chromatography - atomic absorption spectrometry. *Analyst, 119, 627-632.*

[41] Gammelgaard, B., Liao, Y, & Jons, O. (1997). Improvement on simultaneous determination of chromium speceis in aqueous solution by ion chromatography and chemiluminescence detection. *Anal. Chim. Acta, 354, 107-113.*

[42] Sakai, T., McCurdy, E., & Wilbur, S. (2005). Ion chromatography (IC) ICP-MS for chromium speciation in natural waters. *Agilent Technologies, Inc., 5989-2482EN, 1-7.*

[43] Pettine, M., & Capri, S. (2005). Digestion treatments and risks of Cr(III)-Cr(VI) interconversions during Cr(VI) determination in soils and sediments-a review. *Anal. Chim. Acta, 540,231-238.*

[44] Kingston, H.M., Huo, D., Lu, Y., & Chalk, S. (1998). Accuracy in species analysis: speciated isotope dilution mass spectrometry (SIDMS) exemplified by the evaluation of chromium species. *Spectrochim. Acta B, 53, 299-309.*

[45] Huo, D., & Kingston, H.M. (2000). Correction of species transformation in the analysis of Cr(VI) in solid environmental samples using speciated isotope dilution mass spectrometry. *Anal. Chem., 72, 5047-5054.*

[46] Hartlen, J. (1996). Environmental consequences of using residues. *Waste Manage. 16, 1-6.*

[47] Motz, H. & Geiseler, J. (2001). Products of steel slags an opportunity to save natural resources. *Waste Manage. 21, 285-293.*

[48] Pioro, L.-S., & Pioro, I. L. (2004). Reprocessing of metallurgical slag into materials for the building industry. *Waste Manage. 24, 371-379.*

[49] Lind, B. B., Fallman A.-M., & Larsson, L.B. (2001). Environmental impact of ferrochrome slag in road construction. *Waste Manage. 21, 255-264.*

[50] Valls, S., & Vazquez, E. (2002). Leaching properties of stabilised / solidified cement-admixtures-sewage sludges systems. *Waste Manage. 22, 37-45.*

[51] Marion, A.-M., De Laneve, M., De Grauw, A. (2005). Study of leaching behaviour of paving concretes: quantification of heavy metal content in leachates issued from tank test using demineralized water. *Cement Concrete Res. 35, 951-957.*

[52] Flyhammar, P., & Bendz, D. (2006). Leaching of different elements from subbase layers of alternative aggregates in pavement constructions. *J. Hazard. Mater., B137, 603-611.*

[53] U. S. Environmental Protection Agency, Guide for Industrial Waste Management. http://www.epa.gov/epawaste/nonhaz/industrial/guide/index.htm

[54] EN 1744-3. 2002. Tests for chemical properties of aggregates – part 3: Preparation of eluates by leaching of aggregates.

[55] EN 12457-4. 2004. Characterisation of waste – Leaching – Compliance test for leaching of granular waste materials and sludges – Part 4: One stage batch test at a liquid to solid ratio of 10 L/kg for materials with particle size below 10 mm (without or with size reduction).

[56] Rossetti, V.A., Di Palma, L., & Medici, F. (2006). Production of aggregate from non-metallic automotive shredder residue. *J. Hazard. Mater., B137, 1089-1095.*

[57] Sukontasukkul, P., & Chaikaew, C. (2006). Properties of concrete pedestrian block mixed with crumb rubber. *Construct. Build. Mat. 20, 450-457.*

[58] NEN 7345 (1995). Leaching characteristics of solid earthy and stony building and waste materials. Leaching tests. Determination of the leaching inorganic components from buildings and monolithic waste materials with diffusion test.
[59] Hohberg. I., de Groot, G.J., van der Veen, A.M.H., & Wassing, W. (2000). Development of a leaching protocol for concrete, *Waste Manage., 20, 177-184.*
[60] Ščančar, J., Milačič, R., Stražar, M., & Burica, O. (2000). Total metal concentrations and partitioning of Cd, Cr, Cu, Fe, Ni and Zn in sewage sludge. *Sci. Tot. Environ. 250, 9-19.*

In: Management of Hazardous Residues Containing Cr(VI) ISBN: 978-1-61470-637-3
Ed: Maria José Balart Murria

Chapter 16

CONTROL OF CR(VI) CONTENT IN WASTE COAL FLY ASH

Eleonora Sočo* and Jan Kalembkiewicz
Department of Inorganic and Analytical Chemistry, Faculty of Chemistry,
Rzeszow University of Technology, Powstańców Warszawy Av. 6,
PL-959 Rzeszów, Poland

ABSTRACT

Waste coal fly ash is a source of environmental contamination by heavy metals. Evaluation of metals' mobility from fly ash to environmental solution and soil is performed by sequential extraction and total concern of metals also Cr. A difficult task is evaluation of content and mobility of chemical forms of metals, in that Cr(VI) in the system ash-solution for the sake of the course of the redox reaction. Recognizing Cr(VI) stability in conditions takes studying its content in ashes and the migration degree of Cr(VI) from ash to solution. This has been determined as an important research problem. The work is devoted to this issue. Main elements of work are: (i) One-step digestion of fly ash and total content of Cr. (ii) Sequential extraction of ash and content fraction of Cr. (iii) Stability of Cr(III) and Cr(VI). (iv) Extraction of Cr(VI) from ash eluates. (v) Content of Cr(VI) in ash and its mobility in the ash-solution system. (vi) Mobility of Cr and Cr(VI).

1. INTRODUCTION

The production of fly ash (50 million Gg/year in Poland [1],) resulting from coal combustion is of major interest due to its environmental impact [2]. The largest ash quantity is stored in slag heap (Poland - approximately 12 million Gg/year [1]). Ash contamination poses a serious threat to the environment. Water extracts from fly ash are commonly alkaline and contain excessive amounts of dissolved solids [2]. Leachability of heavy metals presents a hazard for natural waters [3]. Since many trace elements existing in fly ash can leach out

* Corresponding authors. Tel.: +48 17 8651508, Fax.: +48 17 8543655,
E-mail addresses: eleonora@prz.edu.pl

into groundwater resources, the study of them has been regarded in recent years as important in connection with protection of the environment. Trace metals, though, present as a relatively small fraction in fly ash, and are of special interest, due to their cumulative build up, long life and high toxicity to man, plants and animals through air, water, and soil intake [2]. Several studies [2-16], show the efficiency of fly ash in the removal of these heavy metals. Nowadays, speciation of metals in the waste [2-13], especially of Cr [14-16] are acquiring essential importance. The waste fly ash is particularly important for the sake of production and storing present objects with diverse chemical composition.

Chromium exists in the environment and in soil around stores of fly ash in two species, as Cr(III) and Cr(VI). Above, are two oxidation species of Cr which have various properties and biological activity on living organisms [17-22]. Trivalent chromium is considered a trace element essential for the proper functioning of living organisms, because it is needed for proper glucose metabolism. Hexavalent chromium has toxic and mutagenic properties related to its oxidizing activity [20-22]. Presence of excess chromium(VI) damages circulatory systems and causes carcinogenic changes, and is harmful due to its bioactivity and biotoxicology, its ability to accumulate in plants especially, and its mobility in the environment. These were the main reasons for research of Cr chemical fractions in the coal fly ash, along with its ability to migrate from the ash to water solutions in environmental conditions (pH ~ 5-7).

The confirmed presence of Cr in fly ashes (max. content about 100 mg kg^{-1}) [14] is why it is a very important problem to evaluate the (i) chemical transformation of different Cr speciation, mainly Cr(VI) in the ash-solution system and (ii) the possibility analytical control these process. The control of Cr(VI) content in fly ash and all its chemical fraction require assurance of stability of the Cr(VI) and Cr(III) forms. Problems of concern are (i) Cr(VI) to Cr(III) reduction or (ii) Cr(III) to Cr(VI) oxidation in conditions of Cr leaching from fly ash to solution.

Monitoring Cr(VI) in fly ash requires precise methods of separating the element from the fly ash into a solution and its determination. Conditions of correct analytical progress is the lack of a reduction reaction of Cr(VI) or oxidation of Cr(III) to Cr(VI) in these systems. The methods used to date did not specify the conditions of Cr(VI) in fly ash or its fractions. Others include: (i) conditions of separation of Cr(VI) or Cr(III) from fly ash or its eluates, (ii) information about coexistence and stability of Cr(VI) and Cr(III) forms in conditions of sequential extraction of fly ash. The problems were described in this work.

The researches of Cr(VI) content in ash or its fractions is frequently made difficult by the complex inorganic matrix of obtained eluates found after mineralization or sequential extraction. Elimination of this problem can be done by the extraction of Cr(VI) from solution and its determination by independence methods. The chromium element determination in fly ash eluate obtained by sequential extraction is possible on the basis of direct spectrophotometric determination of Cr(VI) with the use of 1,5-diphenylcarbazide [23,24], and indirect determination of Cr(III) by total chromium determination. There are some methods of separation of Cr(VI) and Cr(III) and direct determination of those metal ions in final solutions [25]. Preliminary separation of Cr(VI) and Cr(III) enables independent determination of those metal ions in solutions, whereas total chromium analysis allows one to control the mass balance in ash eluates and find the correctness of obtained determinations for metal ions.

For one-step digestion of fly ash and multi-step leaching (sequential extraction) [2-16] procedures were applied to solutions with different chemical compositions and conditions of extraction (time, temperature). Water eluates of ash have rich inorganic matrices with oxidizing conditions (conc. $H_2SO_4/HClO_4$, $HNO_3/HClO_4$) or reducibles (e. g. $NH_2OH{\cdot}HCl$). During extraction of Cr(VI) from ash to an aqueous solution and in ash eluate Cr(III) can be oxidizing to Cr(VI) however Cr(VI) – is reducible.

Previous examinations have not resulted in determination of explicit conditions for carrying out Cr(VI) analysis in fly ash for many reasons, among others because of the influence of several parameters on the preparation process of an ash sample, the necessity of creating determined conditions of Cr(VI) or Cr(III) separation from ash or ash eluates and selecting a defined method of determination of an unrestricted chemical form of chromium in the prepared sample. Determination of Cr forms in ash eluates, obtained after the sequential extraction of ash, is possible by the direct determination of Cr(VI) and by the indirect determination of Cr(III) by determination of the total Cr. The initial separation of Cr(VI) and Cr(III) enables independent determination of these forms in the obtained solutions, however determination of the total Cr can serve a purpose for checking the Cr mass balance in ash eluates and correctness of acquired results.

The monitoring of Cr(VI) content in fly ash and ash eluate was performed: (i) to evaluate chemical changes of Cr(III) and Cr(VI) in the ash-solution system, (ii) for separation of Cr(VI) from ash eluate before its final determination, (iii) and for application and obtained results in the practice (for control of Cr(VI) content in fly ash and interpretation of results).

2. Ash Preparation and Conditions of Researches

The coal fly ash obtained from coal combustion in the Heat and Power Plant (Rzeszow, Poland) was used in the experiment. The coal fly ash was obtained by fluidized combustion of coal and collected on electrostatic precipitators. The coal fly ash was collected from freshly raised ash slag heap. Five samples were collected by multiple sampling according to the BN-81/0623-01 norm (by Polish Standards). All of them were collected at a depth of about 20 cm. The ash samples were put into polyethylene bags and transported to the laboratory. The wet weight of each sample was about 2.5 kg. Air-dry ash subsamples weighing 0.5 kg were prepared from the 2.5 kg sample. In the laboratory, the ashes were air-dried at ambient-temperature for two weeks. The dried material was sieved through a 0.3 mm sieve. The obtained material was homogenized by shaking and then placed in the dark. Then, the ash samples were kept in polyethylene containers until extraction. The containers were rinsed with acetonitrile. Composition and characteristic of coal fly ash are shown in Table 1.

Prepared ash for research was applied:

One-step coal fly ash digestion
Sequential extraction of ash
Stability of Cr(III) and Cr(VI)

Table 1. Composition and characteristic of coal fly ash

No.	Determination	Value
1.	pH	9.2
2.	Bulk density, kg m^{-3}	614.2
3.	Moisture content, % weight	1.04
4.	Loss on ignition (LOI), %	7.62
5.	Size distribution, % weight:	
	0.315 – 0.2 mm	2.7
	< 0.2 mm	97.3
6.	Content of combustible part, % weight	10.1
7.	Content of SiO_2 and metals oxide, % weight:	
	SiO_2	47
	Al_2O_3	24
	Fe_2O_3	14
	CaO	2.5
	MgO	3
	K_2O	0.2
	Na_2O	0.3
8.	Content of heavy metals, mg kg^{-1}:	
	lead	137
	cadmium	5
	copper	73
	nickel	39
	zinc	93
	manganese	939
	cobalt	34

Cr(VI) extraction from ash eluates
Determination of Cr/Cr(VI) and verification of results

Analysis of the total Cr was carried out in reference fly ash (certified reference material CTA-FFA-1). Settled conditions of carrying out Cr analysis for the CTA-FFA-1 fly ash were used to determine total Cr in the research ash.

3. One-Step Coal Fly Ash Digestion

The total content of Cr has been studied in the research fly ash and reference fly ash by means of digestion in the mixture of conc. acids $H_2SO_4/HClO_4/HF$. The research was carried out on four parallel samples (n = 4). The reagent blank was simultaneously run. The determined Cr concentration in ash eluates was recalculated to Cr concentration per ash dry weight. Details of carried out digestion of ash was described in works [12,13]. The results are shown in Table 2.

Table 2. The results of Cr analyzed in coal fly ash (certified value for certified reference material CTA-FFA-1: $[Cr]_{total}$ - 156 mg kg^{-1}). Average chromium concentration for n = 4 and p = 95%

On-step digestion	**$H_2SO_4/HClO_4/HF$**	**79.4 ± 3.4**
Sequential extraction of ash and kind of fraction	exchangeable fraction F(1),	2.5 ± 0.6
	adsorption fraction F(2),	4.0 ± 1.2
	oxide fraction F(3)	16.0 ± 2.5
	sulfide fraction F(4)	8.5 ± 2.2
	residual F(5)	41.6 ± 6.8
	$\sum_{i=1}^{5}[Cr]_i$	72.6 ± 4.4

Reprinted from Journal of Fuel, Vol. 88, E. Sočo, J. Kalembkiewicz, Investigations of Cr mobility from coal fly ash, pp. 1513-1519, Copyright (2009), with permission from Elsevier.

3.1. Total Content of Cr

Analysis of the total Cr was carried out in reference fly ash (certified reference material CTA-FFA-1) by the FAAS method, obtaining recovery $\eta > 98$ %. In established conditions ($H_2SO_4/HClO_4/HF$) for CTA-FFA-1 was carried out analysis of fly ash and obtained $[Cr]_{total} = 79.4 \pm 3.4$ (Table 2).

4. Sequential Extraction of Ash

The experiment was carried out on four parallel samples of research fly ash and reference fly ash. The compositions of extracting solutions and research conditions are shown in Figure 1.

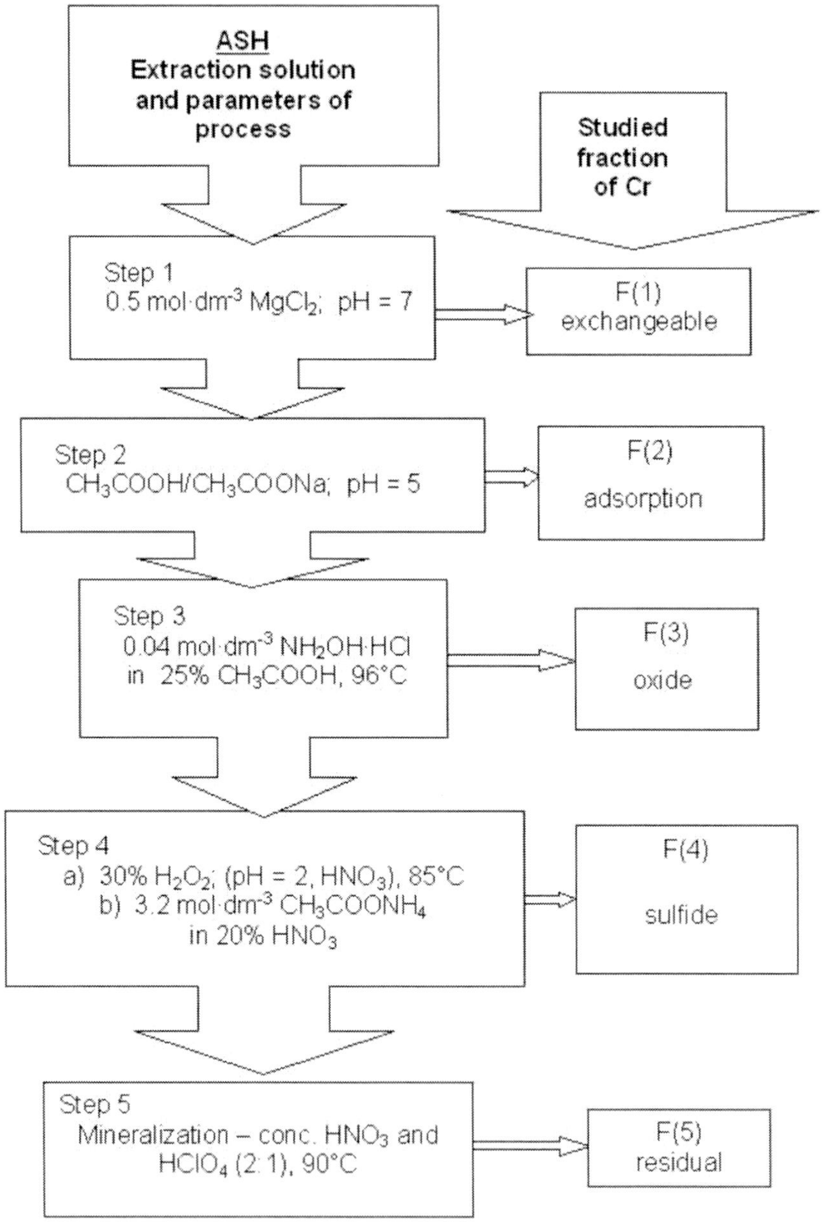

Figure 1. Conditions of sequential extraction of Cr from fly ash.

Step One

The ash sample was transferred to an Erlenmeyer flask and 50 ml 0.5 mol l^{-1} $MgCl_2$ solution (pH = 7) was added. The obtained mixture was shaken for 30 min at ambient temperature in a horizontal rotary mechanical shaker at a speed of 200 rpm. The extract was separated from the solid residual by centrifugation at 3000 rpm for 5 min and decantation of the supernatant liquid into a polyethylene container. The metals were measured in the supernatant. The solid residual was retained for the next step.

Step Two

In the next step the residual was treated with 50 ml acetate buffer: CH_3COOH/CH_3COONa at pH = 5, shaking for 1 hour and centrifuging. The ash extract and the ash residual were obtained. The metals were measured in the supernatant. The solid residual was retained for the next step.

Step Three

In the next step, the residual was treated with 20 ml of 0.04 mol l^{-1} $NH_2OH \cdot HCl$ in 25 % CH_3COOH and digestion was performed for an hour on a heating plate at 96°C. A subsequent portion of the solution of 30 ml hydroxylamine hydrochloride of concentration 0.04 mol l^{-1} in 25 % CH_3COOH was added, digestion was continued for the next hour and centrifuged for 5 min. The metals were measured in the supernatant. The solid residual was retained for step four.

Step Four

To the obtained residual small portions of 20 ml of 30% H_2O_2 (adjusted to pH 2 with HNO_3) were added and heated at 85°C. The flask was covered with watch glass and the content was digested for an hour. The glass was taken off and the flask content was evaporated to a small volume (1 – 2 ml). Later, a successive portion of 30% H_2O_2 was added, heated for an hour in a temperature of 85°C and the flask content was evaporated again. To the cooled residual after vaporization of the solution, 20 ml of 3.2 mol l^{-1} CH_3COONH_4 in 20% HNO_3 was added and agitated continuously for 30 min. After centrifugation, the obtained supernatant was analyzed for Cr.

Step Five

The residual was treated with a mixture of concentrated HNO_3 and $HClO_4$ acids (2:1). The flask was covered with a watch glass, put on a heating plate (temp. 90°C) and heated for an hour. After a lapse of one hour the flask content was evaporated to dryness. To the dry, cooled residual 15 ml of HCl solution (1 : 1) was added, it was put again on a heating plate and heated as before. The flask content was hot-filtered by a medium quantitative filter paper, decanting sediment traces with the help of the hot 1% HCl. Residual of Cr were determined in the obtained solution.

Blank extractions were carried out for each set of analysis using the same reagent as described above. The results are shown in Figure 2 [15].

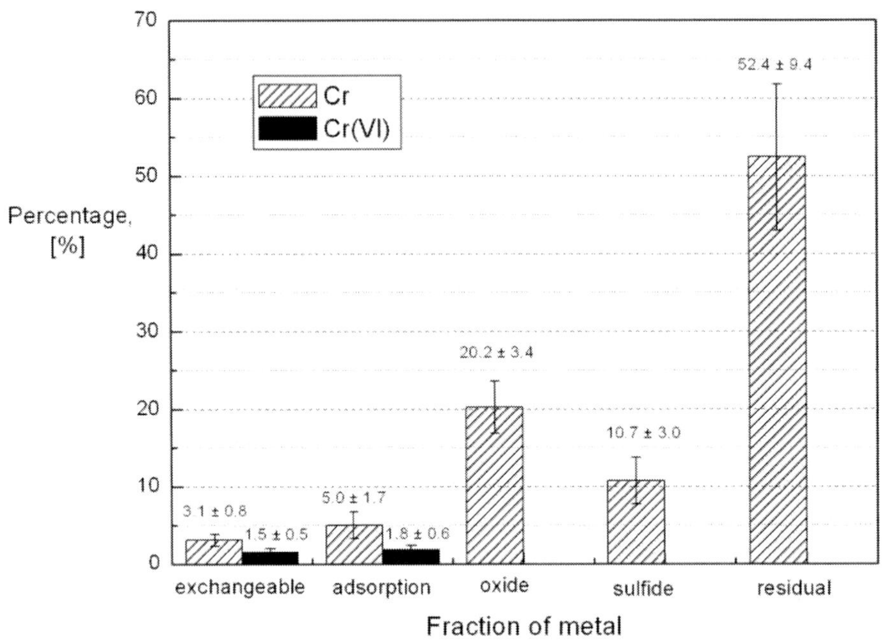

Reprinted from Journal of Fuel, Vol. 88, E. Sočo, J. Kalembkiewicz, Investigations of Cr mobility from coal fly ash, pp. 1513-1519, Copyright (2009), with permission from Elsevier.

Figure 2. Distribution of chemical fraction of Cr and percentage share of mobility fractions of Cr and Cr(VI) in the coal fly ash: F(1) – exchangeable (pH = 7), F(2) – adsorption (pH = 5) and hard mobility: F(3) – oxide, F(4) – sulfide, F(5) – residual.

4.1. Content Fraction of Cr

Research of metal migration from environment samples and waste (ashes) to solutions were carried out by sequential extraction method [29,30] which is a leaching process of ash by means of water solution with different chemical composition. Among various methods of the sequential extraction, the Tessier's procedure [26,27] is most often used both in cases of environmental samples as well as sediments. Ash eluates (1) – (5) are obtained by treating ash with solutions in turn: F(1) - $MgCl_2$ (exchangeable fraction), F(2) - CH_3COOH/CH_3COONa (adsorption fraction), F(3) - $NH_2OH{\cdot}HCl$, (oxide fraction), F(4) - $H_2O_2/HNO_3/CH_3COONH_4$ (sulfide fraction) and F(5) – $HNO_3/HClO_4$ (residual). A fraction of Cr in ash not extracted in the 5th step of sequential extraction was called "indigestive fraction." Obtained results were presented per ash dry weight. The percentage content of each fraction of Cr in the ash are shown in Figure 2. It was found that Cr occurs first of all in the sulfide fraction (10.7%), oxide (20.2%) and the ash residual (52.4%) as well as in lower concentrations in the exchangeable (3.1%) and adsorption fraction (5.0%). The indigestive fraction of Cr ($[Cr]_{total} - \sum_{i=1}^{5}[Cr]_i$) equal 8.6%.

5. Analysis of Cr(III) and Cr(VI)

Oxidation of Cr(III) to Cr(VI) and coexistence of Cr(III) and Cr(VI) were researched (i) in solutions applied to sequential extraction of ash and (ii) in ash eluates after sequential extraction. The researches were carried out in solution and eluates F(1)/F(2) obtained according to the Tessier method [26,27], i.e. in the systems with $MgCl_2$ (pH=7) and CH_3COOH/CH_3COONa (pH = 5).

The influence (for extractants solution) of pH (1–12), mixing time (3 min–5 h), temperature (20-50°C), concentration of Cr(III) - 10–500 mg·kg^{-1} on oxidation of Cr(III) and coexistence of Cr(III) and Cr(VI) was studied.

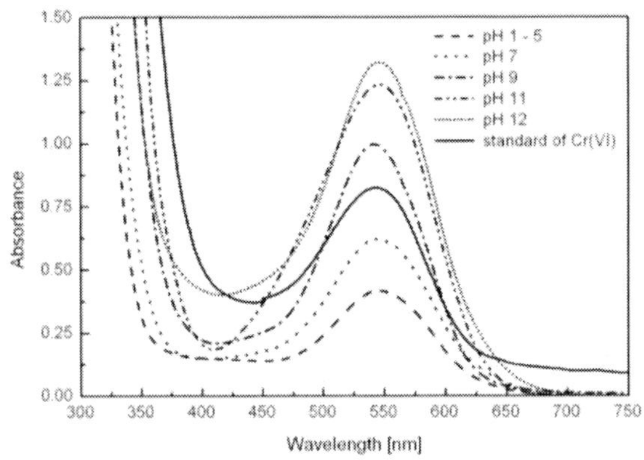

(A)

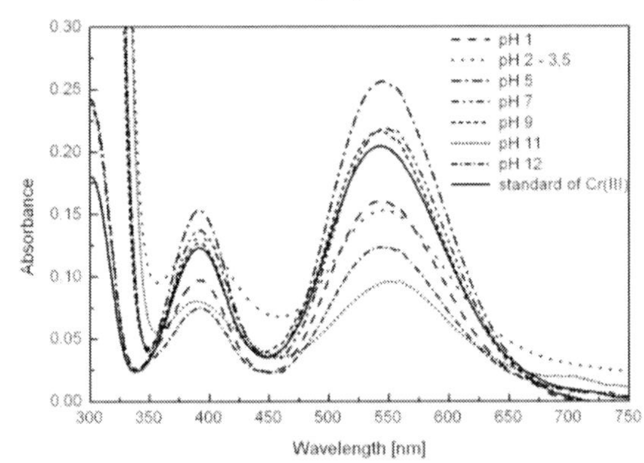

(B)

Figure 3. Influence of pH on oxidation of Cr(III) to Cr(VI) in solution with CH_3COOH/CH_3COONa, $t_{mix.}$ = 15 min; $[Cr(III)]_{initial.}$ = 100 mg·kg^{-1}, t = 20°C; a) the spectra of Cr(VI) with DPC, b) the spectra of Cr(III) with Na_2EDTA.

Spectrophotometric analysis of solution and ash eluates with $MgCl_2$ and CH_3COOH/CH_3COONa after the addition of the DPC and Na_2EDTA was performed, in order to characterization of Cr(VI) and Cr(III) ions. The spectra were performed for (a) standard solutions ($K_2Cr_2O_7$ - DPC – H_2O; $Cr(NO_3)_3$ – Na_2EDTA – H_2O); (b) model solutions ($K_2Cr_2O_7$ – $MgCl_2$ or CH_3COOH/CH_3COONa – DPC – H_2O; $Cr(NO_3)_3$ – $MgCl_2$ or CH_3COOH/CH_3COONa – Na_2EDTA – H_2O) and (c) ash eluates with $MgCl_2$ or

CH_3COOH/CH_3COONa – Cr(VI) – DPC/ Na_2EDTA. As an example the results are shown in Figs 3-6 for acetate solutions.

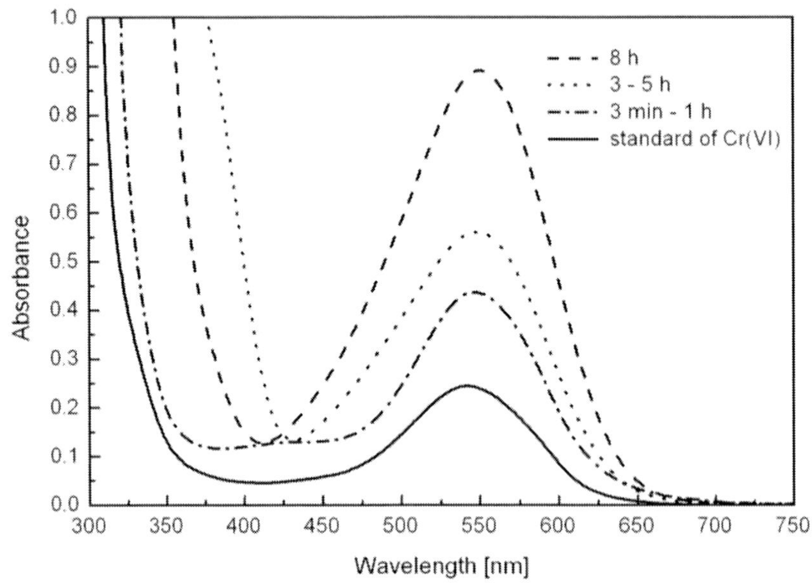

(A)

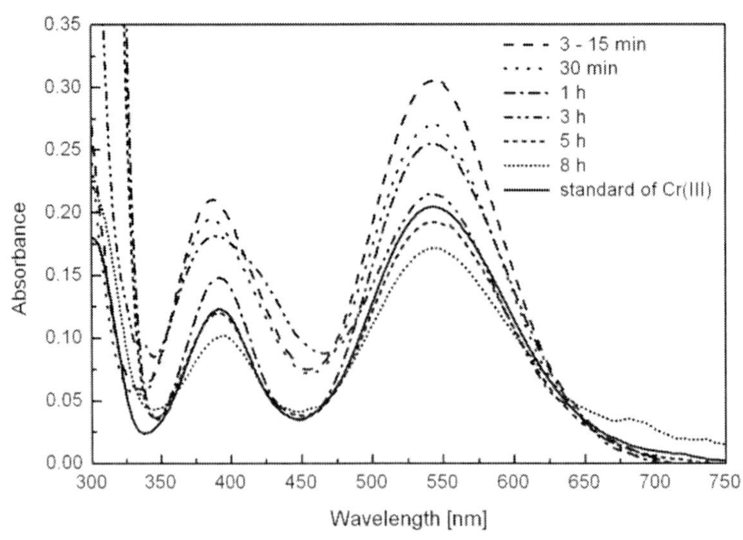

(B)

Figure 4. Influence of mixing time on oxidation of Cr(III) to Cr(VI) in solution with CH_3COOH/CH_3COONa; pH = 5; $[Cr(III)]_{initial}$. = 100 $mg{\cdot}kg^{-1}$, t = 20°C, a) the spectra of Cr(VI) with DPC, b) the spectra of Cr(III) with Na_2EDTA.

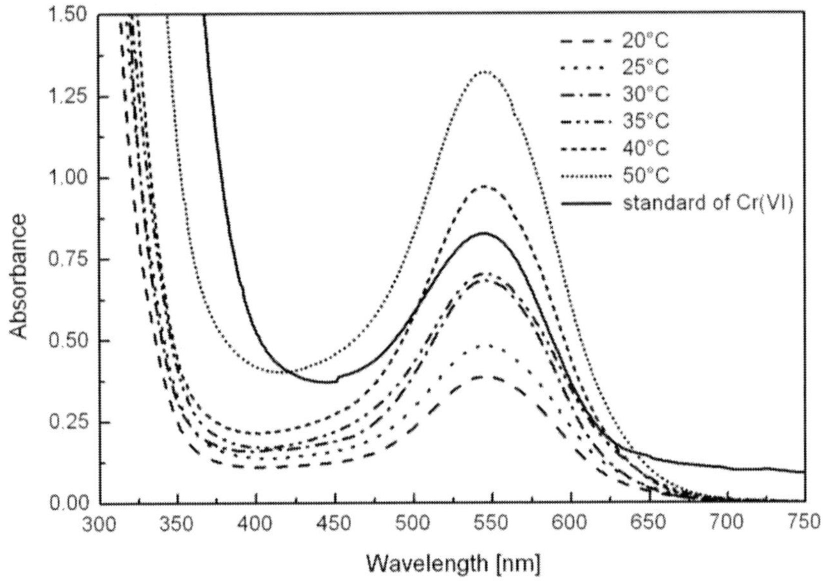

(A)

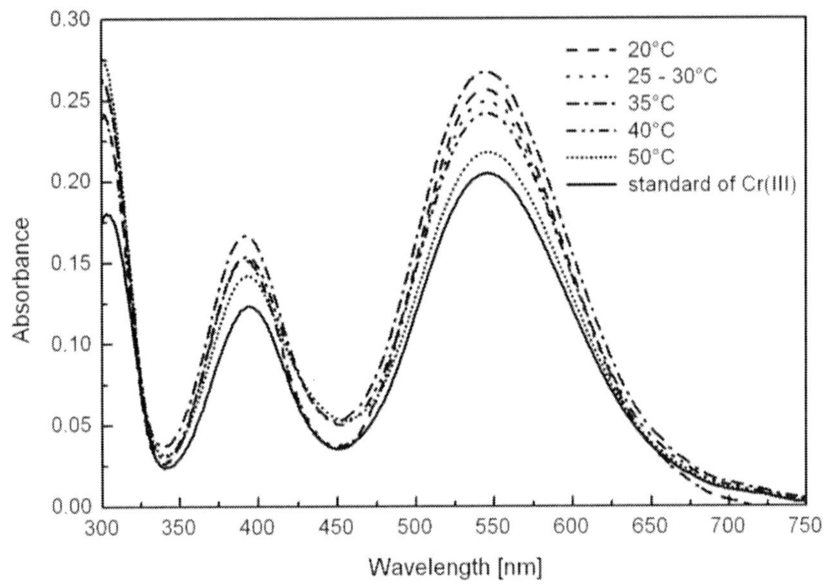

(B)

Figure 5. Influence of temperature on oxidation of Cr(III) to Cr(VI) in solution with CH_3COOH/CH_3COONa; pH = 5; $[Cr(III)]_{initial.}$ = 100 mg·kg^{-1}, $t_{mix.}$ = 15 min, a) the spectra of Cr(VI) with DPC, b) the spectra of Cr(III) with Na_2EDTA.

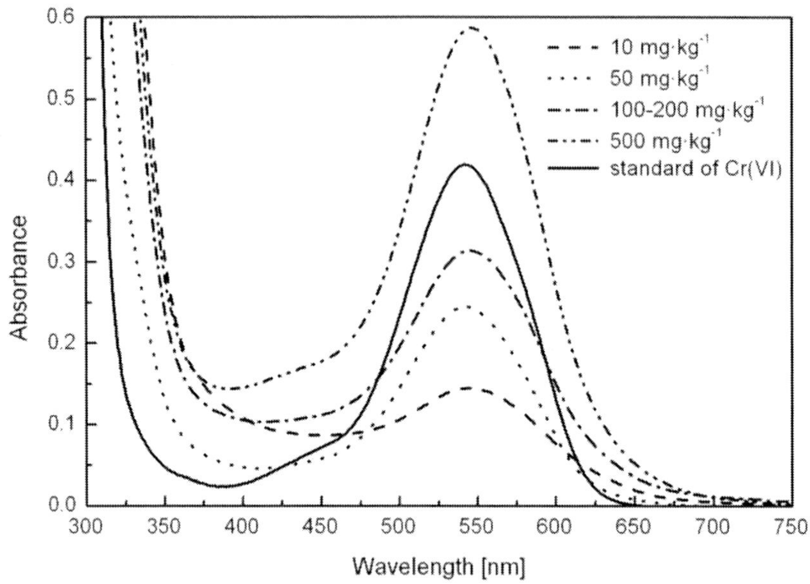

(A)

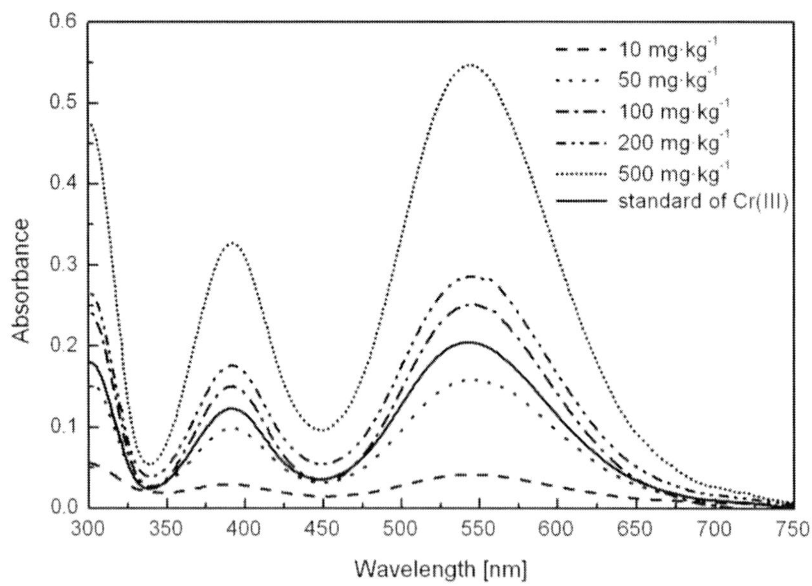

(B)

Figure 6. Influence of Cr(III) concentration on oxidation of Cr(III) to Cr(VI) in solution with CH_3COOH/CH_3COONa; pH = 5; t = 20°C, $t_{mix.}$ = 15 min, a) the spectra of Cr(VI) with DPC, b) the spectra of Cr(III) with Na_2EDTA.

5.1. Stability of Cr(III) and Cr(VI)

Investigations of Cr(III) to Cr(VI) oxidation in solution of $MgCl_2$ and CH_3COOH/CH_3COONa were carried out. The influence of: pH 1-12, initial concentration of

Cr(III) 10 – 500 mg kg^{-1}, temperature 20 - 50°C and mixing time 5 – 300 min on the course of reaction was studied. Spectrophotometric analysis of solution and ash eluates with $MgCl_2$ and CH_3COOH/CH_3COONa after the addition of the DPC and Na_2EDTA was performed, in order to characterize Cr(VI) and Cr(III) ions. The spectra were performed for (a) standard solutions ($K_2Cr_2O_7$ – DPC – H_2O; $Cr(NO_3)_3$ – Na_2EDTA – H_2O); (b) model solutions ($K_2Cr_2O_7$ – $MgCl_2$ or CH_3COOH/CH_3COONa – DPC – H_2O; $Cr(NO_3)_3$ – $MgCl_2$ or CH_3COOH/CH_3COONa – Na_2EDTA – H_2O) and (c) ash eluates with $MgCl_2$ or CH_3COOH/CH_3COONa – Cr(VI) – DPC/Na_2EDTA. As an example, the results are given in Figs 3-6 for acetate solutions.

Obtained results indicate that: (i) in the studied solutions proceed weak oxidation of Cr(III) to Cr(VI) (to 5%), this fact point to spectral band for Cr(VI) with DPC at λ = 545 nm. (ii) in the solutions coexistence of Cr(III) and Cr(VI), influence on their concentrations have pH, temperature and mixing time, (iii) Cr(VI) shows stability in studied solutions – spectral band for Cr(VI) with DPC at λ = 545 nm is presence at: pH 2-5, time reaction to 5 h, temperature 20-50°C.

Comparison of spectra of Cr(III) and Cr(VI), indicate that in ash eluates with $MgCl_2$ (exchangeable fraction) and CH_3COOH/CH_3COONa (adsorption fraction) appear characteristic band of Cr(III) and Cr(VI). Obtained results indicate, that ash eluates with $MgCl_2$ and CH_3COOH/CH_3COONa have weak oxidizing properties in conditions sequential extraction of Tessier's and chromium occurs as Cr(III) and Cr(VI). In these conditions, there is a rise to 3.5% of Cr(VI) in relation to total concentration of Cr in solution, and this fact cannot neglect control and evaluate of Cr(VI) content in ash. Presence of Cr(VI) at levels of 0.7 mg·kg^{-1} in ash eluate ($MgCl_2$, CH_3COOH/CH_3COONa), can effect Cr(III) oxidation in conditions of sequential extraction of ash.

Obtained results indicate also, that additives of Cr(VI) to ash eluates F1 ($MgCl_2$) and F2 (CH_3COOH/CH_3COONa) do not cause change of concentration of Cr(VI) additive. It was found, that ash eluates with $MgCl_2$ and CH_3COOH/CH_3COONa do not have reducible properties.

Analysis of conditions for running several steps of the sequential extraction of soil i. e. F(3) – $NH_2OH{\cdot}HCl$, F(4) – $H_2O_2/HNO_3/CH_3COONH_4$ and F(5) – $HNO_3/HClO_4$ indicates that reduction of Cr(VI) to Cr(III) can take place among others in case of oxides and sulfides fraction as well as oxidation of Cr(III) to Cr(VI) in case of residual.

Performed studies indicate, that research and evaluate of Cr(VI) concentration in ash and ash eluates can be in conditions which oxidation reaction of Cr(III) to Cr(VI) or reduction of Cr(VI) do not proceed or are limited (< 3.5%). This conditions assure sequential extraction of 1-est (exchangeable) and 2-end (adsorption) fraction of Cr from ash.

6. Cr(VI) Extraction From Ash Eluates

The conditions of sequential extraction and separation of Cr(VI) from Cr(III) in chloride and acetate solutions (extractants of 1st and 2nd step of sequential extraction of ash) were studied. The extraction of Cr(VI) was carried out in the systems: tri-n-octylamine (TOA) – organic solvent - Cr(VI) and Cr(III) - aqua in a temperature of 20 ± 0.1 °C, V_o/V_{aq} = 1:1 and $[Cr(VI)]_{initial}$ = 100 mg kg^{-1}, $[Cr(III)]_{initial}$ = 100 mg kg^{-1}. The influence of pH (0.5 - 13), a kind of organic solvent (toluene, carbon tetrachloride, chloroform, pentan-1-ol, MIBK), with concentration of TOA (0.005 – 0.1 mol l^{-1}), extraction time of ($t \leq 6$ h) and number of extraction steps (n = 3) on extraction of Cr(VI) was studied.

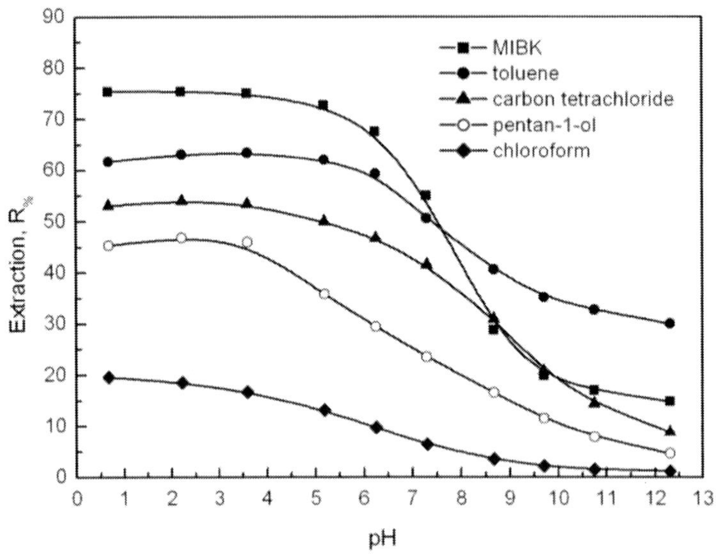

Figure 7. The influence of pH and organic solvent on Cr(VI) extraction in the system: TOA – organic solvent (MIBK, toluene, carbon tetrachloride, pentan-1-ol, chloroform) – Cr(VI) – aqueous solution CH_3COOH/CH_3COONa; [Cr(VI)] = 100 mg kg^{-1}; V_o:V_{aq} = 1:1; [TOA] = 0.1 mol l^{-1}; t = 20 °C; I = 0.5; t_{extn} = 15 min; reextractant 4 mol l^{-1} HNO_3.

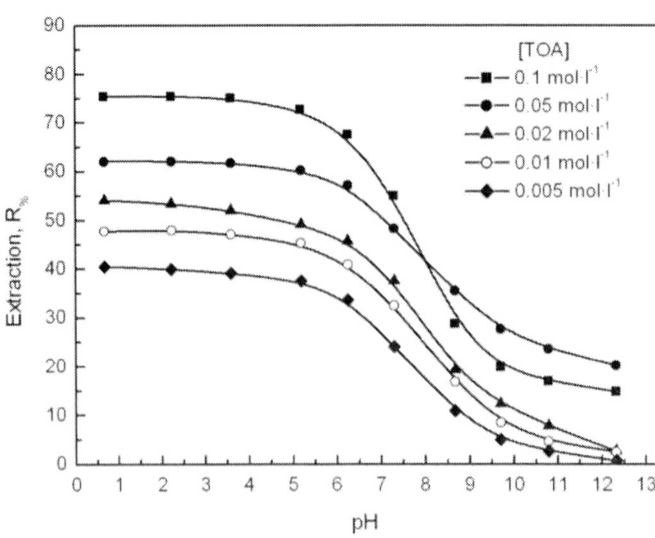

Figure 8. The influence of TOA concentration on Cr(VI) extraction in the system: TOA – MIBK – Cr(VI) – aqueous solution CH_3COOH/CH_3COONa; [Cr(VI)] = 100 mg kg^{-1}; V_o:V_{aq} = 1:1; t = 20°C; I = 0.5; t_{extn} = 15 min; reextractant 4 mol l^{-1} HNO_3; concentration of [TOA] = 0.005; 0.01; 0.02; 0.05; 0.1 mol l^{-1}.

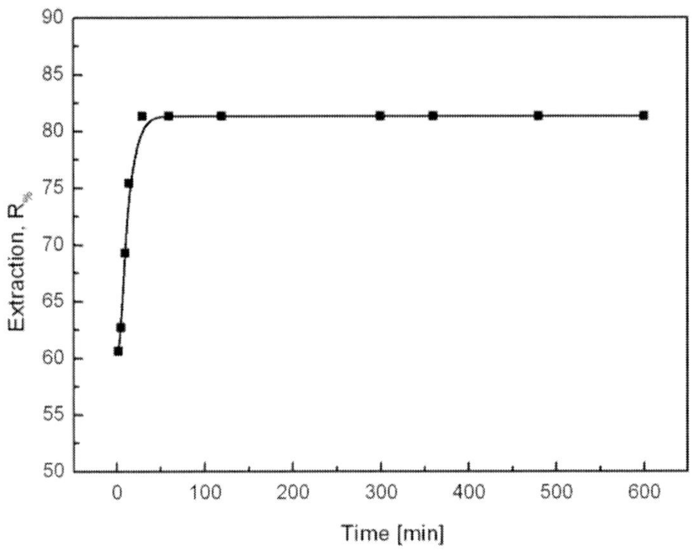

Figure 9. The influence of time on Cr(VI) extraction in the system: TOA – MIBK – Cr(VI) – aqueous solution CH_3COOH/CH_3COONa; [Cr(VI)] = 100 mg kg^{-1};V_o:V_{aq} = 1:1; [TOA] = 0.1 mol l^{-1}; I = 0.5; pH = 2; t = 20°C; reextractant 4 mol l^{-1} HNO_3; extraction time (t_{extn} = 5, 10, 15, 30, 45, 60, 120, 180, 240, 300, 360 min.).

The results are shown in Figs 7-9. Cr(VI) was extracted from ash eluates F(1) – $MgCl_2$ and F(2) - CH_3COOH/CH_3COONa into the organic solvent with TOA using optimal conditions for Cr(VI) separation in model two-phase systems (Figure 10). To the water F(1) or F(2) eluates 30 ml (4 × 7.5 ml) extracting phase (0.1 mol l^{-1} TOA in MIBK) was added and solution was acidified to pH = 1. The system was shaken for 30 minutes and Cr(VI) was reextracted from organic phase into aqueous4 mol l^{-1} HNO_3. The Cr(VI) was determined in those final water solutions.

6.1. Content of Cr(VI) in Ash Eluates After Solvent Extraction

Here we present the optimum extraction of Cr(VI) from ash extracts. The aim of this research was to establish the conditions of speciation analysis of Cr(III) and Cr(VI). Then two-phase extracting systems for Cr(VI) and Cr(III) separation from eluate were studied. We have found that Cr(VI) separation and speciation of Cr(VI) and Cr(III) has given reliable results when the F(1) – $MgCl_2$; F(2) – CH_3COOH/CH_3COONa extracting solution has been used. For the remaining ash eluates, i.e. F(3) – $NH_2OH{\cdot}HCl$, F(4) – H_2O_2 / HNO_3 / CH_3COONH_4 and F(5) – HNO_3 / $HClO_4$, because of reducing or oxidizing properties of solutions and course of redox reaction with the Cr(VI)/Cr(III) participation researches extraction of Cr(VI) was not performed. Then two-phase extracting systems for Cr(VI) separation from F(1) and F(2) eluates were studied. The influence of: pH (Figure 7), a kind of organic solvent (Figure 7), concentration of TOA (Figure 8), and extraction time (Figure 9) on recovery factor of Cr(VI) from aqueous solution to organic phase was analyzed. Investigation was carried out in a system: tri-n-octylamine (TOA) – organic solvent – aqua ($MgCl_2$; CH_3COOH/CH_3COONa) using as an organic solvent (methyl isobutyl ketone

(MIBK), penthan-1-ol, toluene, chloroform, carbon tetrachloride). We have focused on tri-n-octylamine (TOA) as the extracting reagent based on reported data [31]. This solvent was proven to be poor extractor for Cr(III). Established optimum conditions of chromium(VI) extraction in liquid two-phase systems with $MgCl_2$ or CH_3COOH/CH_3COONa (pH of system 2-4, type of organic solvent – MIBK, TOA concentration – 0.1 mol l^{-1}, extraction time – 15 min and number of extraction steps – 3) were used for chromium(VI) separation from ash solution (Figure 10). In studied conditions, the separation factor of Cr(VI) from Cr(III) was 99%.

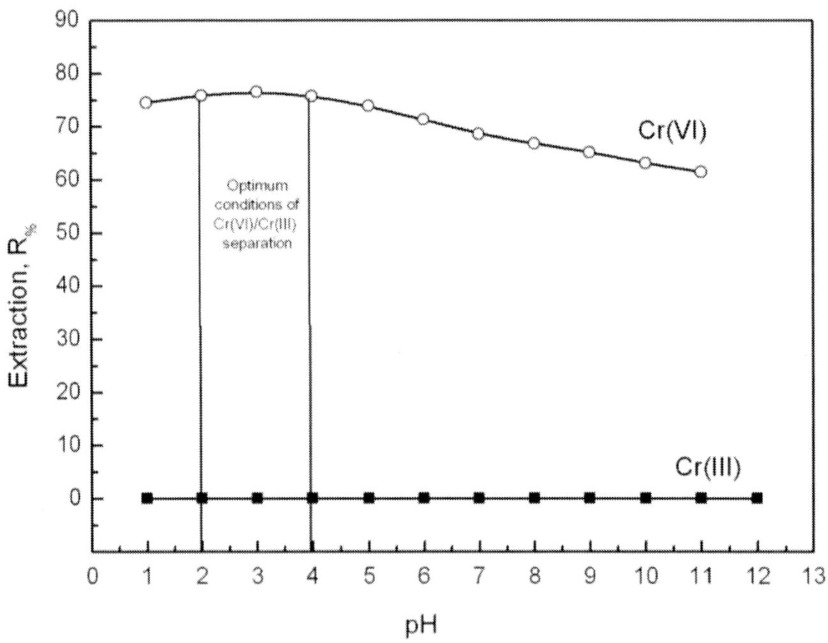

Figure 10. Conditions of Cr(VI)/Cr(III) extraction in the system: TOA – MIBK – Cr(VI) – aqueous solution; $[Cr(VI)]_{initial}$ = 100 μg g^{-1}, $[Cr(III)]_{initial}$ = 100 mg kg^{-1}, $V_o:V_{aq}$ = 1:1; [TOA] = 0.1 mol l^{-1}; t_{extn} = 15 min; reextractant 4 mol l^{-1} HNO_3.

7. Determination of Cr/Cr(VI) and Verification of Results

Concentrations of Cr in ash eluates (after sequential extraction and total ash mineralization) were determined by using the flame atomic absorption spectrometry method (FAAS). The calibrant solutions containing 0.0; 3.5; 5,0; 7.0 mg kg^{-1} Cr were prepared. The interferences of other metals were eliminated by adding 1 ml: 1% sodium azide and 1% ammonium chloride [28].

After ash mineralization, sequential extraction, extraction into organic phase and reextraction, the Cr was determined by FAAS and, independently,the Cr(VI) by the diphenylcarbazide method (in ash eluates F(1) and F(2)).

The results of Cr concentrations for reagent blank were corrected. All analytical determinations were carried out in triplicate. The sequential extraction of ash and digestion of the studied samples and reference ash was carried out on four parallel samples. Obtained results were presented per ash dry weight. The statistical analysis of the results was performed for the studied chemical fraction as well as for the total metal content in the ash. The arithmetic mean of the obtained results, after each extraction step are shown in Tables 2 and 3. The obtained results have the following value of standard deviation s_R = 2.5%. The standard deviation (s_R) value was estimated according to the literature. The relative standard deviation (RSD) % and Student correlation coefficient were calculated, significance was set at p = 95% and n = 4.

Table 3. Cr, Cr(VI) and Cr(III) determination in ash eluates with $MgCl_2$ or CH_3COOH/CH_3COONa after sequential extraction of ash (n = 4, p = 95%)

Method of determination	$MgCl_2$		CH_3COOH/CH_3COONa	
	eluate	ash	eluate	ash
FAAS	$[Cr]_{total}$, mg kg^{-1}			
	1.2 ± 0.3	2.4 ± 0.6	2.0 ± 0.6	4.0 ± 1.2
Cr(VI) extraction by means of TOA in MIBK FAAS	[Cr(VI)], mg kg^{-1}			
	0.6 ± 0.2	1.2 ± 0.4	0.7 ± 0.4	1.4 ± 0.8
Diphenylcarbazide method	[Cr(VI)], mg kg^{-1}			
	0.7 ± 0.4	1.4 ± 0.8	0.6 ± 0.3	1.2 ± 0.5
after extraction of Cr(VI) FAAS	[Cr(III)], mg kg^{-1}			
	0.6 ± 0.3	1.2 ± 0.6	1.4 ± 0.3	2.8 ± 0.6

7.1. Content of Cr(VI) in Ash and its Mobility in the Ash-Solution System

Cr(VI) was extracted from soil eluates obtained by treating soil with solutions of $MgCl_2$ or CH_3COOH/CH_3COONa. Extraction of Cr(VI) from soil to aqueous soil fractions and extraction of Cr(VI) by means of TOA was found. Amount of Cr(VI) separated from the aqueous soil fraction to the exchangeable phases (determined by the FAAS method) equals 1.2 mg kg^{-1} ($MgCl_2$) and 1.4 mg kg^{-1} (CH_3COOH/CH_3COONa) respectively. Obtained results are consistent with results of Cr(VI) determination in the aqueous soil fraction F(1) and F(2) by the diphenylcarbazide method ([Cr(VI)] = 1.4 mg kg^{-1} for $MgCl_2$ solution and [Cr(VI)] = 1.2 mg kg^{-1} for CH_3COOH/CH_3COONa solution) - Table 3. The studies of Cr(VI) ions extraction by means of TOA from real soil solutions indicate that in applied conditions Cr(VI) ions can be extracted and be subjected to determination in solutions by FAAS and diphenylcarbazide methods. Obtained results indicated that Cr(VI) content in mobility fraction of ash i.e. F(1) and F(2) (after calculate per ash dry weight) amount 2.4 and 4.0 mg kg^{-1} adequately.

7.2. Mobility of Cr and Cr(VI)

Essential, from a practical point of view, is the quantitative evaluation of leachable Cr from industrial ash to surface waters in environmental conditions and metal contamination of the environment. Chromium, mobile in environmental conditions (pH ~ 5 – 7) are bound to: (i) exchangeable fraction (pH ~ 7) and (ii) adsorption. The other metal fractions present in ashes (oxide, sulfide) are bound to hardly mobile fractions and can migrate to the environment in specific conditions i.e. aggressive environments (F(3) and F(4) fraction).

In order to characterize Cr and Cr(VI) mobility in fly ash, the mobility parameter was calculated as:

$$\alpha\,[\%] = \frac{\sum_{i=1}^{2}[M]_i}{[M]_{total}}\,100\% \qquad (1)$$

α parameter presents the percentage share of metal mobile forms (F(1) and F(2)) (Eq. 1) in relation to content close total in the ash ($[M]_{total}$). The calculations of the percentage contents of each fraction of Cr and Cr(VI) in the ash are shown in Figure 11. The results of the performed research indicate that the mobility of Cr and Cr(VI) fractions in environmental conditions is adequately: 8.1% and 3.3%.

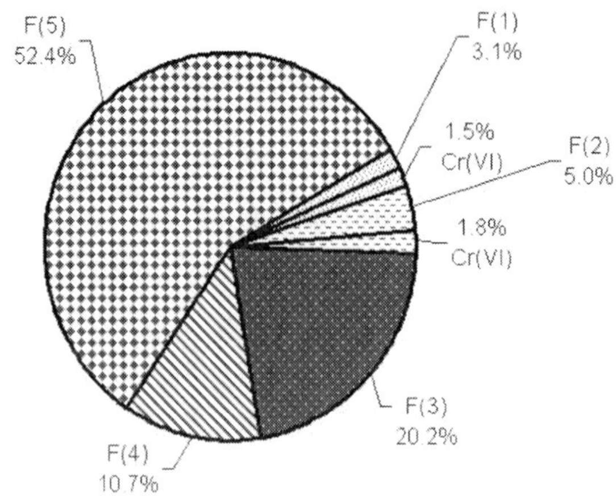

Figure 11. Percentage share of mobility fractions of Cr and Cr(VI) in the coal fly ash: F(1) – exchangeable, F(2) – adsorption, F(3) –oxides, F(4) – sulfide, F(5) – residual.

8. Reagents and Apparatus

8.1. Reagents

Magnesium chloride, acetate buffer, hydroxylamine hydrochloride, 30% solution of H_2O_2, tri-n-octylamine, chloroform, carbon tetrachloride, toluene, pentan-1-ol, methyl isobutyl ketone, 1,5-diphenylcarbazide (DPC), **ethylenediaminetetraacetic acid disodium salt** (Na_2EDTA), sulfuric acid, perchloric acid, hydrofluoric acid, acetic acid, ammonium acetate, ammonium chloride, sodium azide, and standard chromium solution (1000 mg l^{-1}) were used, as well as the certified reference material CTA-FFA-1. All reagents were Fluka or Aldrich analytical grade or suprapur grade. More diluted working solutions were prepared daily with the corresponding final medium for each of the steps of the sequential extraction procedure.

All calibrants and reagent solutions were prepared using double deionized water. If the analysis could not be performed immediately, the extracts were stored in polyethylene flasks. All glassware and plastic ware were soaked in 5% (v/v) nitric acid for at least 24 h and rinsed with deionized water before use.

8.2. Apparatus

Total chromium determination in the extracts was carried out by means of a PERKIN-ELMER Model 3100 atomic absorption spectrometer. Hollow cathode lamp at 10 mA for Cr was used. The analytical lines were selected using a slit width of 0.7 nm. Measurements were performed at 359,4 nm using a yellow fuel-rich air-acetylene flame (o/f ratio 1.5) and burner height of 3 mm. The Cr(VI) determination in the extracts was carried out by means of spectrophotometer of the BECKMAN company, type DU-640 with $\lambda = 545$ nm. The Elmetron Company (Poland), CPI-551 model pH-meter was used for pH adjustment of the extracts. A universal shaker Model 358 A of the ELPIN-PLUS company (Poland) operating at 200 rpm in horizontal position and electric hotplate with temperature control and digital display of 1°C Model SD 300 of the STUART company (England) were used for the extraction. A centrifuge tube test Model WE-6 (Precision Engineering, Poland) was used for the centrifugation of the ash extracts at the speed of 2500 rpm for 5 min. The deionizator of the Cobrabid-AQUA Company (Poland) was used for the deionization of water.

Conclusion

1) The total contents of Cr in the investigated coal fly ash was 79.4 mg kg^{-1}.
2) After the following sequential extraction, the Cr content in the studied fractions was found to be: exchangeable (2.5) < adsorption (4.0) < sulfide (11.7) <oxide (22.0) < residual (57.3 mg kg^{-1}). Indigestive fraction with $HNO_3/HClO_4$ – 8.2 mg kg^{-1}.
3) The researches of Cr(III) and Cr(VI) oxidizing, its coexistence and stability of Cr(VI) in solutions and ash eluates were performed. Several parameters were studied: pH of solution (1-12), mixing time (3 min – 8 h), temperature (20-50°C), concentration of

Cr(III) in the system (10-500 mg kg^{-1}). It was found, that conditions of obtaining ash eluates F(1) - $MgCl_2$ and F(2) - CH_3COOH/CH_3COONa as well eluates with no reducible conditions, however have weakly oxidizing conditions. The time and temperature range should be specified and does proceed (3%) oxidizing of Cr(III) to Cr(VI), coexistence of Cr(III) and Cr(VI) and the stability of Cr(VI) form. In these conditions, one can control the Cr(VI) content in fly ash.

4) The conditions of studying several Cr fractions in ash: F(3) - $NH_2OH{\cdot}HCl$ (96°C), F(4) - H_2O_2/HNO_3 (pH 2)/CH_3COONH_4 (85°C) and F(5) – $HNO_3/HClO_4$ (90°C) have reducible or oxidizing properties and analysis of Cr(VI) in this fractions of Cr is impossible.
5) It was found that in the established conditions Cr(VI) and Cr(III) ions can be separated and then they can be quantitatively determined in a direct way. The established conditions should be specified.
6) In conditions of the optimum Cr(VI) extraction from $MgCl_2$ or CH_3COOH/CH_3COONa solutions (pH ~ 2 - 4, using 0.1 mol l^{-1} of TOA in MIBK as extraction reactant and extraction time $t_{extrn} \geq 15$ minutes), Cr(III) does not extract to the organic phase.
7) The mobility fraction of Cr and Cr(VI) that can be leached from the ash in environmental conditions was 8.1% and 3.3%.

REFERENCES

[1] Statistical yearbook (2008), Condition and environmental protection, Warsaw, Poland.

[2] Van Herck, P; Van der Bruggen, B; Vogels, G; Vandecasteele, C. Application of computer modeling to predict the leaching behaviour of heavy metals from MSWI fly ash and comparison with a sequential extraction method, *Waste Manage.*, 2000, 20, 203-210.

[3] Hansen, HK; Pedersen, AJ; Ottosen, LM; Villumsen, A. Speciation and mobility of cadmium and wood combustion fly ash, *Chemosphere*, 2001, 45, 123-128.

[4] Landsberger, S; Cerbus, JF; Larson, S. Elemental characterization of coal ash its leachates using sequential extraction techniques, *J. Radioanal. Nucl. Ch.*, 1995, 192, 265-274.

[5] Fernández-Turiel, JL; deCalvalho, W; Cabanas, M; Querol, X; López-Soler, A. Mobility of heavy metals from coal fly ash, *Environ. Geol.*, 1994, 23, 264-270.

[6] Praharaj, T; Powell, MA; Hart, BR; Tripathy, S. Leachability of elements from sub-bituminous coal fly ash from India, *Environ. Int.*, 2002, 27, 609-615.

[7] Kida, A; Noma, Y; Imada, T. Chemical speciation and leaching properties of elements in municipal incinerator ashes, *Waste Manage.*, 1996, 16, 527-536.

[8] Kirk, DW; Chan, ChCY; Marsh, H. Chromium behavior during thermal treatment of MSW fly ash, *J. Hazard. Mater.*, 2002, B90, 39-49.

[9] Hong KJ, Tokunaga S, Ishigami Y, Kajiuchi T. Extraction oh heavy metals from MSW incinerator fly ash using saponins, *Chemospere,* 2000; 41: 345-352.

[10] Shoham-Frider, E; Shelef, G; Kress, N. Chemical changes in different types of coal ash during prolonged large scale, contact with seawater, *Waste Manage.*, 2003, 23, 125-134.

[11] Fytianos, K; Tsaniklidi, B. Leachability of heavy metals in Greek fly ash from coal combustion, *Environ Int.,* 1998, 24, 477-486.

[12] Sočo, E; Kalembkiewicz, J. Investigations of chemical fraction of Co and Ni in industrial fly ash and mobility of metals in environmental conditions, *Chemosphere*, 2007, 67, 359-364.

[13] Sočo, E; Kalembkiewicz, J. Investigations of sequential leaching behaviour of Cu and Zn from coal fly ash and their mobility in environmental conditions, *J. Hazard. Mater.*, 2007, 145(3), 482-487.

[14] Hwang, J; Wang, W-J. Determination of hexavalent chromium in environmental fly ash samples by an inductively coupled plasma-atomic emission spectrometer with ammonium ion complexation, *Appl. Spectrosc.*, 1994, 48, 1111-1117.

[15] Sočo, E; Kalembkiewicz, J. Investigations of Cr mobility from coal fly ash, *Fuel*, 2009, 88, 1513-1519.

[16] Kalembkiewicz, J; Sočo, E. Determination of Chromium(VI) in the Soil Organic Fraction, *J. Anal. Chem.*, 2009, 64(5), 475-478.

[17] Chmielnicka, J; Świetlicka, E; Nasiadek, M. Essential elements as early indicators of hexavalent chromium nephrotoxicity, *Extoxicol. Environ.*, 2002, 53(1), 20-26.

[18] Huvinen, M; Makitie, A; Jarventaus, H; Wolff, H. Nasal cell micronuclei, cytology and clinical symptoms in stainless steel production workers exposed to chromium, *Mutagenesis*, 2002, 17(5), 425-429.

[19] Dayan, A; Paine, A. Mechanisms of chromium toxicity, carcinogenicity and allergenicity, *Human Experimental Toxicol.*, 2001, 20(9), 439-451.

[20] Cieślak-Golonka, M. Toxic and mutagenic effects of chromium(VI), *Polyhedron*, 1995, 15, 3667-3689.

[21] Błasik, J; Kowalik, J. A comparison of the in vitro genotoxicity of tri- and hexavalent chromium. *Mutat. Res.*, 2000, 469, 135-145.

[22] Dayan, A; Paine, A. Mechanisms of chromium toxicity, carcinogenicity and allergenicity, *Hum. Toxicol.*, 2001, 20, 439-451.

[23] Standard Test Method for Chromium in Water (1995). In 1995 Annual Book of ASTM Standards, *Water and Environmental Technology. American Society for Testing and Materials*, Philadelphia.

[24] *3500-Cr D. Colorimetric Method in Standard Methods for the Examination of Water and Wastewater* (1995), 19th edition., American Public Health Association, Washington D.C.

[25] Barałkiewicz, D; Siepak, J. Speciation Analysis of Chromium(III) and (VI) in Natural Water Samples by Graphite Furnace Atomic Absorption Spectrometry, *Chem. Anal. (Warsaw)*, 1999, 44, 879-884.

[26] Tessier, A; Campbell, PGC; Bisson, M. Particulate trace metal speciation in stream sediments and relationships with grain size: implication for geochemical exploration, *J. Geochem. Explor.* 1982, 16, 77-104.

[27] Tessier, A; Campbell, PGC; Bisson, M. Sequential extraction procedure for the speciation of particulate trace metals, *Anal. Chem.*, 1979, 51, 844-851.

[28] The Perkin-Elmer Corporation, Analytical Methods for atomic absorption spectrophotometry (1982), Norwalk Connecticut, USA.

[29] Asikainen Jay, M.; Nikolaidis Nikolaos, P. Sequential extraction of chromium from contaminated aquifer sediments, *Ground Water Monit. Rem.*, 1994, 14(2), 185-91.

[30] Kalembkiewicz, J; Sočo, E. Sequential extraction of metals from environmental samples. *Wiad. Chem.* (in Polish) 2005, 59, 697-721.

[31] Huei, HY; Yung, ChCh; Feng, KJ. Extraction equilibria of chromium(VI) from hydrochloric acid solutions with triisooctylamine in o-xylene, *Hydrometallurgy*, 1992, 28(30), 297-307.

In: Management of Hazardous Residues Containing Cr(VI) ISBN: 978-1-61470-637-3
Ed: María José Balart Murria

Chapter 17

EVIDENCE AND STUDY OF THE OXIDATION OF CR(III) TO CR(VI) AND ITS MOBILISATION FROM SOLID CR-CONTAINING WASTE MATERIALS

Jochen Petersen*

Department of Chemical Engineering, University of Cape Town, Private Bag X6, Rondebosch 7701, South Africa

ABSTRACT

While the environmental and health hazards of Cr(VI) are well documented, it is generally accepted that Cr(III) is much less hazardous and hence waste materials containing only Cr(III) are considered safe for disposal or re-use as construction material. Evidence gathered during long-term leaching studies on certain solid waste materials arising from the stainless steel industry shows that Cr(VI) continues to emerge in the leachate long after any initial concentrations should have been removed, and it may even increase in concentration as time progresses. This points to a potential oxidation of Cr(III) phases in these materials to Cr(VI). A subsequent study of similar materials confirmed slow natural oxidation of Cr(III) to Cr(VI) in the presence of certain CaO phases and oxygen to form $CaCrO_4$. In a leach environment this can become mobilised and washed from its original source. The implications of these effects are potentially significant for the long-term stability of any waste materials that contain both Cr(III) oxide and CaO phases, and their disposal or re-use must be evaluated accordingly.

1. INTRODUCTION

Trivalent chromium (Cr(III)) and hexavalent chromium (Cr(VI)) are the two predominant oxidation states of chromium in the natural environment. Cr(VI) is generally considered far more toxic than the trivalent form, with negative health effects ranging from skin irritations to

---

* Tel: +27 21 650 5766; Fax: +27 21 650 5501; Email: jochen.petersen@uct.ac.za

lung cancer (Yassi and Nieboer, 1988; Sheehan et. al., 1991). Hence permissible Cr(VI) levels in any form of industrial effluent are strictly regulated in environmental codes.

Cr(III) has traditionally been of much less concern, as it is itself not considered as toxic as Cr(VI) and because it tends to be insoluble in natural aqueous environments and hence unlikely to become mobilised from its source. Disposal of Cr(III) containing solid wastes by landfill, or its use as road fill and building material, are considered to be of acceptable environmental risk, as it is generally assumed that this oxidation state will be maintained indefinitely. However, the potential and extent of Cr(III) to become oxidised to Cr(VI) in such waste materials is not well appreciated to this day and merits a more comprehensive investigation.

Rai et al. (1987) describe the thermodynamics of Cr(III) and C(VI) in aqueous solution. Cr(III) shows amphoteric behaviour with soluble species forming in acidic (Cr^{3+}, $Cr(OH)^{2+}$, $Cr(OH)_2^+$) and alkaline environment ($Cr(OH)_4^-$). $Cr(OH)_3$ has a solubility of around 7.5 μg/L Cr in the pH range 6-11 (Richard and Bourg, 1991).

Thermodynamically, aqueous Cr(III) should be oxidised by atmospheric oxygen to Cr(VI) at pH values exceeding pH 4 (Nyholm, 1947; reaction (1)), and, consequently, by dissolved oxygen at a pH exceeding 8, given the low solubility of oxygen at atmospheric pressure. At very high pH values (>12), where Cr(III) becomes more soluble again, it should prevail in the Cr(VI) state in the presence of oxygen (reaction (2)). A proposed industrial process involving the alkali pressure leaching of Cr(III) oxide utilises this phenomenon (Farrow and Burkin, 1973).

$$2\,Cr^{3+} + 1.5\,O_2 + 5\,H_2O = 2\,HCrO_4^- + 8\,H^+ \tag{1}$$

$$2\,Cr(OH)_4^- + 1.5\,O_2 = 2\,CrO_4^{2-} + 3\,H_2O + 2\,H^+ \tag{2}$$

In natural environments such high pH values are unlikely to be encountered, however, and the oxidation of Cr(III) to Cr(VI) is kinetically inhibited by the low solubility of $Cr(OH)_3$. Nonetheless, the reaction has been observed by a number of researchers albeit at rates with reaction half lives in the order years (Bartlett and Kimble, 1976; van der Weijden and Reith, 1982; Saleh et al., 1989; Nakayama et al., 1981; Schroeder and Lee, 1974). Much more research has focussed on the role of MnO_2 as a catalyst in the oxidation reaction in natural aqueous environments (Eary and Rai, 1988; Johnson and Xyla, 1991; Fendorf and Zasoski, 1992), and it is also used as an oxidation catalyst in the pressure leaching process mentioned above:

$$Cr(OH)^{2+} + 1.5\,MnO_2 \rightarrow HCrO_4^- + 1.5\,Mn^{2+} \tag{3}$$

Very limited information appears in the literature on the aerial oxidation of Cr(III) to Cr(VI) in the solid state. Kilau and Shah (1984) report the presence of Cr(VI) in the leachate from chromium containing slags. They propose the following reaction equation, which is well known from the glazing of ceramics:

$$Cr_2O_3 + 2\,CaO + 1.5\,O_2 \rightarrow 2\,CaCrO_4 \tag{4}$$

Reaction (1) is thermodynamically feasible even at ambient temperatures with ΔG_R^0 = -299.6 kJ/mol. The authors showed furthermore that in slags under certain conditions a stable calcium chromite ($CaO \cdot Cr_2O_3$) phase can form during solidifcation, which effectively represents an intimate mixture of Cr_2O_3 and CaO. Hattori et al. (1978) have observed the reaction occur at ambient temperatures in a sludge produced from the neutralisation of an acidic Cr(III) containing tannery waste water.

The studies presented here arose from a series of tests done to assess the environmental stability of a variety of waste materials from a stainless steel smelter. During a long-term leach test on an electric arc furnace (EAF) slag continued generation of Cr(VI) was observed, well beyond what was initially found in the material. A follow-on study then focussed on exploring the potential for this to be attributable to reaction (4) and its kinetic quantification. The discussion and conclusion draw together the insights gained from these studies and their implication for the waste management of this and similar materials.

2. Study 1: Long-term Cr(VI) Oxidation in an EAF Slag Material

In the course of a comprehensive study of the environmental stability of a range of waste materials produced at a stainless steel smelter located in South Africa, the long-term leachate generation behaviour of an EAF slag was investigated in a column test. A 225 mm diameter column was packed with the slag material and irrigated periodically with a simulated acid rain solution over a two-year period.

The material was a blend of electric arc furnace (EAF) and argon-oxygen decarburiser (AOD) slags, representing the two steps of the stainless steel making process. Each of these came in two forms, relating to whether austenitic or ferritic type steel was being produced in a given campaign. As all the materials are co-disposed in the industrial operation, the test work was conducted on a production related blend of these. The key elemental composition of the blended slag is shown in Table 1 with the balance consisting of O, which was not analysed separately. The high concentration of Ca in this material relates to the use of CaO as fluxing agent in the process, which entirely reports to the slag phase.

Table 1. Elemental composition (wt%) of blended slag material tested in column test

Ca	Mg	Si*	Fe	Cr	Mn	Ti
22.0	8.99	18.0	1.36	2.20	0.75	1.28

Na	K	Zn	Ni	Sn	Pb	Cd
**	0.021	0.062	0.013	<0.01	<0.01	<0.0001

* estimated ** not measured

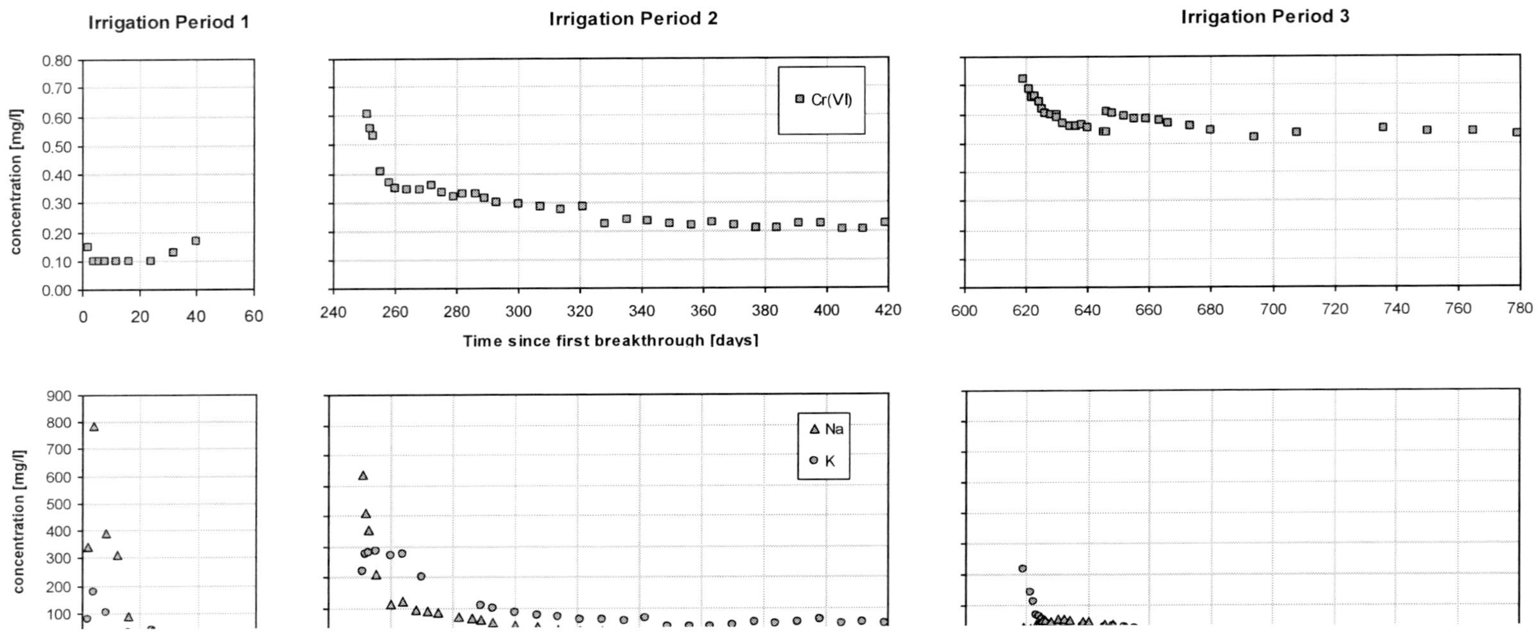

Figure 1. Leachate concentrations of Cr(VI) (top row) and Na and K (bottom row) measured over time in the slag column experiment. The periods between day 40 and 240, and day 420 and day 600 were idle periods in which there was no flow through the column (data from Petersen, 1998).

Initially, the blended material was leached in a 24 hour water leach at 1:10 solid-liquid ratio and also through a standard Toxicity Characteristic Leaching Procedure (TCLP; US EPA, 1992). The results of the water leach, expressed as mg leached per 100 g slag are shown in the first column in Table 2. The TCLP essentially yielded the same results, with its acid buffer capacity rapidly neutralised by the CaO in the material. The suitability of the TCLP for analysing this type of material has been questioned previously (Cohen et al., 1999) as its organic acid lixiviant is of no relevance in the disposal context of this type of waste material.

Table 2. Amounts of of Cr(VI), K and Na removed in water wash and column leach

	mg washable per 100 g slag	mg washable in column	mg leached after 780 days	% of estimated washable
TDS	859			
Cr(VI)	0.019	15.6	26.1	167%
Na	3.27	2707	2263	84%
K	6.73	5565	4691	84%
final pH	12.1			

The water wash indicates that a very small concentration of Cr(VI) does become solubilised, although this is well below the 50 ppb level conventionally considered acceptable in such tests. The wash resulted in a fairly alkaline pH value, relating primarily to the strong buffering capacity of the CaO phase. Moderate amounts of Na and K are the only other species solubilised during the water wash.

The column test involved filling approximately 83 kg of the dry slag material into a column, 225 mm in diameter, to a height of approximately 1 m. The column was irrigated from the top using a dripper on a moving arm, which ensured even distribution of the solution over the column cross section. The irrigation solution was an artificial acid rain mixture (distilled water acidified to pH 4.00 ± 0.05 with a 3:2 mixture of sulphuric and nitric acids and well aerated). The initial irrigation rate was 500 mL/day supplied to the column in 4 periods of 5 hour duration each day (with a one hour rest period in between). Later, this was reduced to 200 mL/day by supplying solution to the column during 4 periods of 2 hours duration each day (with a 4 hour rest period in between). The column was housed in a temperature controlled enclosure at 22±2 °C. Due to evaporation and absorption of water into the solid matrix, the amount of leachate collected at the bottom of the column remained consistently at around 80% of the amount fed to the top.

The initial irrigation period extended over 6 weeks from the first drop of leachate collected at the bottom. The column was then left to rest for 6 months, before irrigation resumed at the reduced flow rate for a further 6 months. The column was then allowed to rest again for 6 months with a final 6 months of leaching. Solution break-through was achieved after 19 days of irrigation in each period.

The concentrations of Cr(VI), Na and K in the leachate during the three leach intervals are plotted in Figure 1 as a function of time since the first leachate was collected during the first interval. All three sets of data show the same behaviour in that concentrations peak at the beginning of each interval and then exponentially decline, as would be expected when washing out a readily soluble constituent. However, the Cr(VI) data shows a decidedly different trend from both Na and K in terms of concentration levels in each interval. For Na and K, concentrations at the peak and at the tail decrease for each interval, indicating a progressively higher degree of removal of a fixed quantity of these components. In contrast, these levels *increase* for Cr(VI) in each interval – peak levels rise from 15 to 60 to 80 ppb whereas the tails level at 10, 20 and finally 50 ppb.

The total amounts of Na, K and Cr(VI) removed from the column over the entire 2 year leach period were calculated and compared to the expected soluble amounts of these species indicated by the water wash test (Table 2). This suggests that 84% removal was achieved for both Na and K, while the amount for Cr(VI) was 167%. This, together with the ever increasing levels of Cr(VI) detected in the leachate, would suggest that Cr(VI) is actively being generated in the column over time.

3. Study 2: Oxidation of Cr(III) to Cr(VI) in EAF Slag and Cr_2O_3

The study by Pillay (2001) (see also Pillay et al. (2003)) followed from the results of the initial column test. A new sample of EAF slag was obtained from the same operation as in the previous test work, however, the original steel making process had been redesigned in the interim and the composition of the slag material changed substantially (Table 3). The material was crushed and separated into 3 size fractions, < 75 μm, 75-300 μm and 300-1000 μm. These were separately analysed for the key composition, as shown in Table 3. This data indicates that the Cr and Si content had been substantially reduced and the Ca content substantially increased from the previous sample.

The slag material was tested for its tendency to become oxidised in ambient air in two different types of test:

powder experiments: 25 g portions of the material were placed in plastic weighing-boats and left to stand open in a laboratory, kept at 20 °C, for up to 320 days.

ball experiments: 25 g portions of material were mixed with distilled water to form a stiff paste, which were subsequently rolled into spherical balls. These were also left to stand open in a laboratory kept at 20 °C for up to 320 days

For comparison, a simulated slag material was created by preparing pellets from a mixture of 10 g each of reagent grade CaO and Cr_2O_3 pressed in a 10 bar pellet press. Again these pellets were left to stand in the open in a laboratory kept at 20 °C for up to 90 days.

Table 3. Key chemical composition of EAF slag sample used in oxidation study

Particle Size Class	Chemical Composition (wt.%)			
	Cr	Ca	Mg	Si
< 75 μm	0.67	38	5.4	
75-300 μm	0.89	37	ND	8.7
300-1000 μm	2.04	32	4.4	

Data reprinted from Pillay et al., 2003, with permission from Elsevier.

From time to time powder, ball or pellet samples were collected in duplicate. Each sample was weighed, dried where necessary, and crushed with pestle and mortar, then leached in exactly 1 L of 1 M NaOH for 1 hour by rotating a 2 L bottle ends-over at 30 rpm. A 50 mL sample of the leach suspension was filtered through a 0.45 μm filter, acidified and stored for Cr(VI) analysis. For all tests a set of reference samples was stored under nitrogen in a closed container.

The results of these oxidation tests are shown in Figure 2, expressed as mg Cr(VI) formed per g Cr present in the sample. These clearly indicate that significant and rapid oxidation of Cr(III) to Cr(VI) is taking place in the slag samples, but only marginal oxidation in the Cr_2O_3-CaO pellet. The results were consistent and repeat samples showed variations of less than 10%. The reference samples kept under nitrogen showed no Cr(VI) formation after as much as 100 days.

All samples showed a limited extent of reaction, with the degree of Cr(VI) formation stabilising after 60 days. The effect of particle size on the rate of oxidation is very clear, with the finest size reacting the most rapidly and to the furthest extent. The ball experiments performed more slowly, but to a similar extent, as the powder experiments in the same size class (only the 300-1000 μm class shown here, more data can be found in Pillay et al. (2003)).

The results clearly illustrate that Cr oxidation does proceed in ambient air and that the exposure to air plays a critical role. Increasing particle size and agglomeration of particles into balls all contribute to a lower rate of oxygen diffusion to sites where the oxidation reaction can take place. The much lower degree of Cr oxidation (by 2 orders of magnitude) in the Cr_2O_3-CaO pellets indicates that such sites are less prevalent here due to limited contact between the two phases. Pillay et al (2003) report further tests conducted with Cr_2O_3-CaO mixtures in powder and ball experiments where rate and extent of reaction were even less.

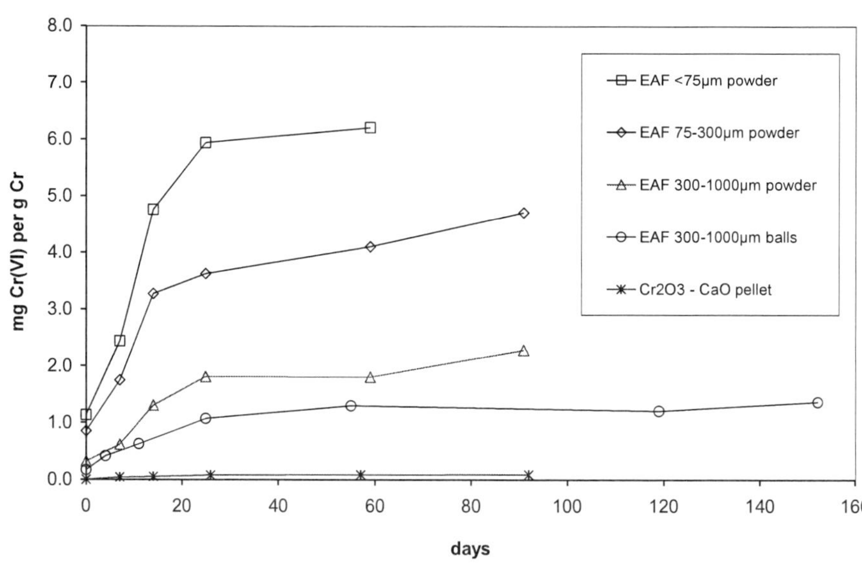

Redrawn from data from Pillay et al., 2003, with permission from Elsevier.

Figure 2. Cr(VI) formation with time in the ball, powder and pellet samples prepared from EAF slag and Cr_2O_3-CaO mixture.

4. Discussion

Pillay et al. (2003) proposed a model of the oxidation reaction proceeding at the grain or phase boundaries between Cr_2O_3 and CaO in materials where these exist as separate phases. This model suggests that close proximity of the two phases is critical, and that as the layer of $CaCrO_4$ forms in the course of the oxidation reaction, it contributes to the physical separation of its generating phases and the reaction ceases. This was used as an explanation as to why the formation of Cr(VI) slowed down and eventually ceased altogether after some time in the ball and powder studies (see Figure 2). However, there exists the possibility that in the presence of water the $CaCrO_4$ is dissolved, enabling the continued oxidation reaction at the phase boundary.

Table 4. maximum specific rate of Cr oxidation observed in the various tests

Test	max rate [mg/g Cr/day]
EAF < 75 μm powder	0.33
EAF 75-300 μm powder	0.22
EAF 300-1000 μm powder	0.098
EAF 300-1000 μm balls	0.042
Cr_2O_3 - CaO pellet	0.006
column tail rate	$4.7 \cdot 10^{-5}$

The oxidation reaction is dependent on the availability of contact sites and to some degree also on the rate of oxygen transfer to the site. It appears that more such contact sites occur for the fine powder (< 75 mm) fraction than in the coarse fraction or that ingress into larger particles is more limited by oxygen diffusion than in the fine fraction. Rolling the slag mixture into a ball only marginally reduces the rate of oxidation due to the increased diffusion path.

According to the analysis presented in Table 4, the relatively most rapid oxidation occurred when exposing fine material (<75 μm) to ambient air with initial conversion rates of as much as 0.33 mg/g Cr/day. By contrast, in the column study described here only $3.15 \cdot 10^{-5}$ mg/g Cr/day were produced, if one assumes that the rate of Cr(VI) leaching at the tail of the third irrigation period (days 680 onwards) corresponded to the rate of oxidation of Cr(III) to Cr(VI) within the column. This rate is almost four orders of magnitude less than that determined for fine particles. However, direct comparison is limited due to the following reasons:

- the column, although open to the atmosphere, allowed limited ingress of air into the depth of the bed
- the fine material in the column became cemented over time, forming a porous solid block, rather than a finely divided open surface for the oxidation reaction
- the material tested in the column contained significantly less Ca than the material used for the oxidation tests. Kilau and Shah (1984) point out that a CaO: SiO_2 ratio of 2 or

larger is required to form the critical $CaO{\cdot}Cr_2O_3$ phase – therefore it is possible that substantially less of this phase was present in the column material.

Nonetheless, the fact that a relatively short column containing an apparently inert material tended to release Cr(VI) consistently at concentration levels at or exceeding prescribed limits for long periods of time, must give rise to some concern over the long-term stability of deposits containing such materials.

CONCLUSION

Thermodynamically, Cr(VI) is a stable phase in solution exposed to oxygen, at all pH > 3.0. Solution phase oxidation of Cr(III) to Cr(VI) is however kinetically extremely slow and proceeds at appreciable rates only at very high pH values or in the presence of a catalyst. This is linked to the extremely limited solubility of Cr(III) in the 4-10 pH interval, and even at higher pH values the solubility of Cr(III) phases remains limited.

Also, the solid state oxidation of Cr(III) to Cr(VI) in the presence of oxygen and CaO is thermodynamically favourable, leading to the formation of $CaCrO_4$, which is moderately soluble in water. The EAF slag materials tested in the context of this work present ideal conditions for the formation of $CaCrO_4$ if exposed to air, which was clearly demonstrated in the experimental studies. The rate and extent of the oxidation reaction seem to depend very strongly to the incidence of CaO-Cr_2O_3 contact (grain contact, grain boundary or mixed phase) and the ingress of oxygen to the contact site (fully exposed, within agglomerate or within a packed bed). Some further evidence suggests that the reaction has a limited extent if the formed $CaCrO_4$ is not removed by dissolution – possibly by creating a physical barrier between the CaO and Cr_2O_3 phases.

Cr oxidation is not limited to the type of stainless steel making slags studied here, but is possible for other materials for which above criteria may hold. Such materials include slag produced in the production of ferro-chromium (Petersen, 1998), coal fly ash (Glasser, 2004), Cr containing sludges produced from neutralising acidic waste waters from the metal finishing and tannery industries with an excess of lime (Hattori, 1978), and possibly also Cr(III) containing solid waste materials stabilised by their mixing with limed cement (Cohen, 1997; Jacobs, 1992).

The implication for safe waste management of these materials are immediately obvious: although oxygen ingress deep into a deposit is likely to be limited, some penetration near surfaces exposed to air is likely to occur. Unfortunately, these surfaces also tend to be exposed to the ingress and run-off of rain water, which may solubilise Cr(VI) that may have formed and subsequently mobilise it away from its source. Although the concentrations of Cr(VI) in such run-off waters is likely to remain small and below acceptable environmental limits, its production is nonetheless undesirable if it continues in perpetuity. Capping of deposits or subterranean disposal should be considered, as well as appropriate leachate and run-off collection systems. Use of such materials as road-fill or building material should be done only with caution.

Overall, any Cr(III) bearing material should be carefully evaluated for its potential to become oxidised to the thermodynamically favoured hexavalent form in the long-term, and its waste management or reuse must be planned accordingly.

ACKNOWLEDGMENTS

The author would like to thank Columbus Joint Venture and the National Research Foundation of South Africa for their support of this work, and acknowledge the contributions made to this body of work by Kriveshini Pillay, Harro von Blottnitz, Brett Cohen and Jim Petrie.

REFERENCES

Bartlett, R.J.; James, B.R. In *Chromium in the Human and Natural Environment*; Nriagu, J.; Ed.; John Wiley and Sons: New York, 1988, pp 267-304.

Bartlett, R.J.; Kimble, J.M. *J. Environ. Qual.* 1976, 5, 383-386.

Burke, T.; Fagliano, J.; Goldoft, M.; Hazen, R.E.; Iglewize, R.; McKee, T. *Environ. Health Persp.* 1991, 92, 131-137.

Calder, L.M. In *Chromium in the Human and Natural Environment*; Nriagu, J.; Ed.; John Wiley and Sons: New York, 1988, pp 215-229.

Cohen, B.; Lewis, A.E.; Petersen, J.; von Blottnitz, H.B.; Drews S.C.; Mahote, S.I. *Adv. Environ. Res.* 1999, 3, pp 152-165.

Cohen B. *Cement-based Solidification of Ferro-alloy Solid Wastes*; PhD Thesis, University of Cape Town, 1997.

Eary, L.E. ; Rai, D. *Environ. Sci. Technol.* 1988, 22, 1187-1193.

Farrow, C.J.; Burkin, A.R. *Leaching Reduct. Hydrometall. [Hydrometall. Meet.]* 1973, pp 20-27.

Fendorf, S.E.; Zasoski, R.J. *Environ. Sci. Technol.* 1992, 26, 79-85.

Glasser, F.P. In *Energy, Waste and the Environment*; Gieré, R. ; Stille, P.; Eds.; The Geological Society: London, 2004, p 200.

Hattori, M.; Yaku, K.; Nakaya , K. *Environ. Sci. Technol.* 1978, 12, 1431-1434.

Jacobs, J.H. *Environ. Prog.* 1992, 11, 123-126.

James, B.R.; Bartlett, R.J.; *J. Environ. Qual.* 1983, 12, 169-172.

Johnson, C.A.; Xyla, A.G.; *Geochim. Cosmochim. Acta* 1991, 55, 2861-2866.

Kilau, H.W., Shah, I.D. In: *Third Symposium, Hazardous Industrial Waste Management and Testing;* Jackson, L.P.; Rohlik, A.R.; Eds.; ASTM STP 851: Philadelphia, 1984, pp 61-81

Nakayama, E. *Anal. Chim.* 1981, 130, 401-404.

Nyholm, R.S. *J.S.C.I* 1947, 66, 449-451.

Petersen, J. *Assessment and Modelling of Chromium Release in Minerals Processing Waste Deposits*; PhD Thesis, University of Cape Town, 1998.

Pillay, K. *A Kinetic and Mechanistic Study on the Oxidation of Chromium Oxide in Pure Chemicals and in Ferrometallurgical Slags;* MSc Thesis, University of Cape Town, 2001.

Pillay, K.; von Blottnitz, H.B.; Petersen, J. *Chemosphere* 2003, 52, pp 1771-1779.

Rai, D.; Sass, B.M.; Moore, D.A. *Inorg. Chem.* 1987, 26, 345-349.

Richard, F.C.; Bourg, A.C.M. *Wat. Res.* 1991, 25, 807-816.

Saleh, F.; Parkerton, T.; Lewis, R.; Huang, J.; Dickson, K.L. *Sci. Total Environ.* 1989, 86, 25-41.

Schroeder, D.C.; Lee, G.F. *Water Air Soil Poll.* 1975, 4, 355-365.

Sheehan, P.J.; Meyer, D.M.; Sauer, M.M.; Paustenbach, D.J. *J. Toxicol. Env. Health* 1991, 12, 161-201.

US EPA *The Toxicity Characteristics Leaching Procedure*. USA Code of Federal Regulations 40th Edn. Washington, D.C , 1992, Part 261, app. II.

van der Weijden, C.H.; Reith, M. *Mar. Chem.* 1982, 11, 565-572.

Yassi, A.; Nieboer, E. In *Chromium in the Human and Natural Environment*; Nriagu, J.; Ed.; John Wiley and Sons: New York, 1988, pp 443–479.

In: Management of Hazardous Residues Containing Cr(VI) ISBN: 978-1-61470-637-3
Ed: Maria José Balart Murria

Chapter 18

The Diffusion of Cr and Cu in Peat and the Effect of Adsorption on Diffusion

Minna P. Koivula[a,*] ***Kauko Kujala***[b] ***and Hannu Rönkkömäki***[c]
[a] Ramboll Finland Oy, Water and Environment, Terveystie 2, 15870 Hollola, Finland
[b]Water Resources and Environmental Engineering Laboratory, Department of Process and Environmental Engineering, Faculty of Technology, P.O.Box 4300, 90014 University of Oulu, Finland
[c] Finnish Institute of Occupational Health, Oulu Regional Office, Aapistie 1, 90220 Oulu, Finland.

Abstract

In this study we evaluated the rate of diffusion of Cr (VI) and Cu (II) through a compacted peat sample. The research was conducted to both discover if diffusion significantly increases migration through hydraulic barrier layers, increasing the risk of pollution of the sub-soil; and to find out how adsorption affects the diffusion.

In this study, similarly to previous research, peat was found to adsorb large amounts of heavy metals. Peat adsorbed approximately 2700 mg/kg of chromium and 3400 mg/kg of copper. The concentrations that leached through the sample were small during the test; at the highest the copper concentration was 0.8 mg/l and the chromium concentration only 0.01 mg/l. Almost all of the Cr (VI) was reduced to Cr (III) during the test, which was expected. Based on the results, the strong adsorption of the heavy metals on peat diminishes the influence of the diffusion on migration greatly compared to a non-adsorptive material. Adsorptive materials would therefore increase the safety of landfill structures.

* Corresponding author. Tel.: +358-20-7556 702; Fax: +358-20-7557 801,
E-mail address: minna.koivula@ramboll.fi

1. INTRODUCTION

1.1. The Sources , Treatment and Properties of Chromium and Copper Waste

Industrial activity in general and mining industry in particular is the most comprehensive waste producer in industrial countries. For example, in the EU countries the solid waste of the mining industry has been estimated to reach 400 million tons per year, which is ca. 29% of the total waste amount. The EU directive (2006/21/EC) setting guidelines for waste management in the mining industry came into effect in 2006. Particular impulsion for the development of this legislation was given by the serious mine tailing accidents in Spain in 1998 (Aznalcóllar) and in Romania in 2000 (Baia Mare and Baia Borsa).

To prevent environmental impacts landfills and tailing impoundments contain barriers which prevent the migration of contaminants into ground water. In tailings and industrial landfills the solute can contain only few contaminants whose concentration can be very high and vary considerably whereas in domestic landfills the variety of contaminants can be very wide. In this paper our research is focused on chromium and copper. Chromium is widely used in various industries such as refractories, and metallurgical and chemical industries. Chromium and copper appear together in, for example, wood preservatives and mining wastes. There are also lots of soils that have been contaminated with chromium and copper (Kotas & Stasicka, 1999).

Cr (III) and Cr (VI) are drastically different in charge, physicochemical properties and chemical and biochemical reactivity. Cr (III) is considered to be a trace element essential for living organisms in small amounts, whereas Cr(VI) exerts toxic effects on biological systems, including human health. Cr (VI) can cause a variety of health problems from asthma and bronchitis to cancers and skin allergies. (Kotas & Stasicka, 1999).

Like Cr (III), Copper is an essential trace element in humans with health effects described by a "U-shaped" dose-response curve. However, large doses of copper may cause a variety of health problems from gastrointestinal disturbances to liver toxicity and may even be fatal. (Sadhra et al., 2007).

1.2. The Use of Peat In Landfills

There are several studies concerning the adsorptive properties of peat (Wei & Tobin, 2003; Dean & Tobin, 1999; Gabaldon et al., 2006 and Ho & McKey, 1998). Due to its adsorptive and physical properties peat has been found to be a suitable material for hydraulic barrier layers in landfills (Koivula et al., 2009). In structures with low hydraulic conductivity, however, diffusion becomes the ruling migration mechanism. Diffusion itself has been studied a lot, for example in clays (Fityus et al., 2006; Malusis et al., 2003; Malusis & Shackelford, 2002 and Myrand et al., 1992) and bentonite clays (Leroy et al., 2005). However the diffusion in peat and the effect of adsorption on diffusion especially in landfills has not been much researched.

The aim of this research was to evaluate the rate of diffusion of Cr (VI) and Cu (II) through compacted peat samples. The research was done to find out if diffusion significantly increases the migration through hydraulic barrier layers and that way increases the risk of pollution of the sub-soil and ground water and to find out how the adsorption affects the rate of diffusion.

2. Materials and Methods

2.1. Peat

Table 1. Chemical composition and the ash content of the peat

Chemical element	Concentration in dry matter (mg/kg)
Al	3480.0
As	6.0
B	<4.0
Ba	42.0
Be	<0.5
Ca	4550.0
Cd	<0.3
Co	2.0
Cr	6.0
Cu	<10.0
Fe	6460.0
K	350.0
Mg	1140.0
Mn	180.0
Mo	1.0
Na	<30.0
Ni	4.0
P	1000.0
Pb	6.0
S	1950.0
Sb	<4.0
Se	<4.0
Sn	<2.0
Ti	140.0
V	12.0
Zn	11.0
total N	2.4 % of dry matter
total C	52.9 % of dry matter
Ash content	8.1 %

Reprinted from Journal of Hazardous Materials, 164 (2009), Koivula, M.P., Kujala, K., Rönkkömäki, H. and Mäkelä, M. , Sorption of Pb(II), Cr(III), Cu(II), As(III) to peat, and utilization of the sorption properties in industrial waste landfill hydraulic barrier layers, 345-352, Copyright (2009), with permission from Elsevier.

Peat was chosen as the study material, because it is a cheap material that is readily available almost everywhere in Finland. Peat has physical properties that make it suitable for hydraulic barrier layers and there have been many promising results with its heavy metal adsorption (Wei & Tobin, 2003; Dean & Tobin, 1999; Gabaldon et al., 2006 and Ho &

McKey, 1998). The same peat that was used in the diffusion tests was previously used in the adsorption test by Koivula et al. (2009).

The peat studied, supplied by the Finnish peat producer Vapo Corp., was dark, medium rough and highly mineralised. Its degree of decomposition was H7 on the von Post scale and its moisture content was 74.8 % of the mass of the wet sample. The plant species determination showed 30 % of the peat to consist of an unidentified mass together with secondary particles, 25 % was root tissue of *Carex* plants (*Carex chorrdorrhiza*) and approximately 25 % were leaf parts of *Sphagnum* and *Bryales* mosses. The peat also included some rudiments of cotton grass (*Eriophorum sp*), wood, and slightly massed twigs.

The pH of the peat, determined according to the ISO 10390 standard, was 4.1 on average. The cation exchange capacity, determined with 1M ammonium acetate diffusion at pH 7 for a sample that was air dried according to the ASTM D2974-00 standard method B, was 112.3 meq/100g. The chemical elements and the ash content in the peat are presented in table 1. The ash content was determined by burning the peat sample at a temperature of 550 °C, according to the CEN/TS 14775: fi –standard. (Koivula et al., 2009).

2.2. Chemicals

The chemicals used were CrO_3 and CuO, both at analytical grades. The chemicals were dissolved into distilled water without any pH adjustment. The heavy metal concentrations used were similar to those that can typically leach from contaminated soils and by-products; 15 mg/l of Cr (VI) and 10 mg/l of Cu (II).

The conditions in the solution (pH 1.9 in aqueous solution) were such that almost all of the Cr (VI) was reduced to Cr (III) (fig 1). This was confirmed by analysing both the total chromium and Cr (VI) from the solution and the soil sample. The analyses showed that less than 0.5 % of the chromium was in the form of Cr (VI).

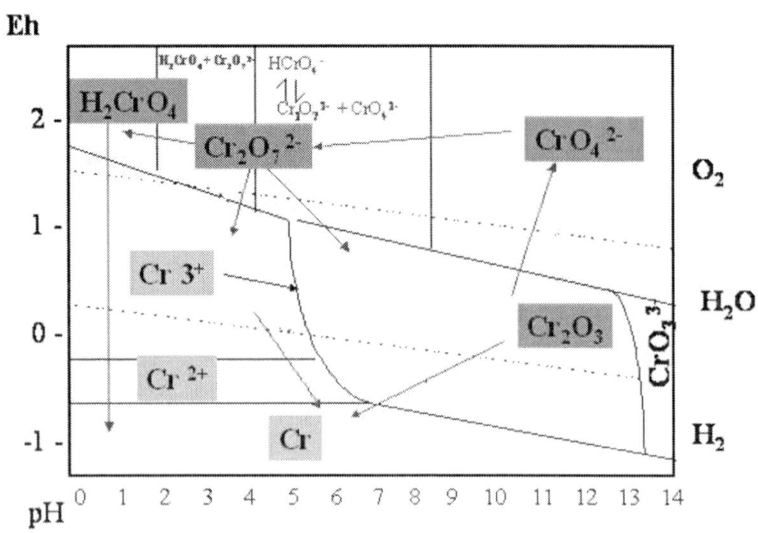

Pourbaix 1974.

Figure 1. Stability Diagram Showing the Stability Domains of Various Chromium Species – based on diagrams by Pourbaix.

2.3. The Diffusion Test

The principle of the diffusion test was the same as in tests done by T. Itakura (2009). However, the diffusion cells were specially designed to suit the conditions and demands of this research. The equipment consists of two water chambers between which the compressed soil sample is situated. The principle drawing of the diffusion equipment is in Figure 1 and the principle drawing of the cell containing the soil sample is in Figure 2.

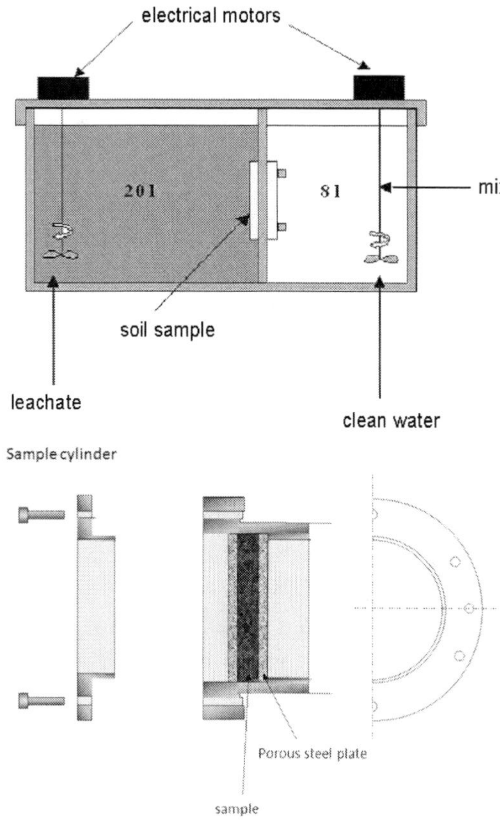

Figure 2. The principle drawings of the diffusion equipment and the sample cell.

All the parts of the equipment were made from acid proof steel to minimize the sorption and desorption on the surfaces. The lid was sealed to prevent vaporization of water. The cylinder containing the sample was also sealed so that the contaminants could only migrate from one chamber to the other through the sample.

There were slow rotating mixers (25 rpm) in both chambers to keep the concentrations in the water even. The chamber containing the contaminants was built to be larger than the fresh water chamber to mimic the situation in a landfill, where there usually is a large reservoir of contaminants and the concentration decreases slowly. The volume of the larger chamber was 20 l and it contained 18 l of water. For the smaller fresh water chamber the numbers were 8 l and 7 l respectively. The hydraulic gradient between the chambers was set to 0, which prevented any advection.

The thickness of the soil sample in the sample cylinder was 15 mm and the diameter of the cylinder 100 mm. Porous, sintered steel plates were put on both sides of the sample. The sample was compressed to match the volume weight of the peat under 50 kPa and saturated with distilled water for 7 days before starting the test to prevent any suction inside the sample. Distilled water was also used in the clean water tank. To determine the diffusion coefficients the data was processed with the POLLUTEv7 © GAEA Technologies Ltd. program (Rowe & Booker, 1987). Both the diffusion coefficient and the distribution coefficient were calculated from the data.

2.4. Analytical Methods

The soil analyses were carried out at the laboratory of the Finnish Institute of Occupational Health in Oulu and at Suomen ympäristöpalvelu Oy. The water analyses were made in the Geochemical Laboratory at the University of Oulu.

Before the metal analyses, 1g of peat was weighed out into a decanter. Subsequently, 5 ml of concentrated HNO_3 and 5 ml of concentrated HCl was added. The sample was allowed to bubble under a glass cover and was carefully warmed until the foaming stopped. The liquid was then heated in a hot sand bath (120 ^oC) until the sample was nearly dry, after which 2.5 ml of concentrated HNO_3 was added, and the sample was washed into a 25 ml measuring flask. The analyses were performed using a Perkin-Elmer AAnalyst 800 ™ AAS by using the flame technique. The metal analyses conformed to the standards OSHA ID-121, NIOSH 7300 and NIOSH 7301.The metal concentrations from the water were analysed with the Perkin-Elmer™ atom absorption spectrophotometer (AAS) by using the flame technique. The analyses were done according to the standards SFS 3044 and SFS 3018 from filtered samples at the Geochemical Laboratory of The University of Oulu.

The electron microscope images were taken at the Institute of Electron Microscopy of the University of Oulu with a Jeol JSM-6400™ Scanning Electron Microscope.

3. Results and Discussion

3.1. The Diffusion of Chromium and Copper in Peat

The analysed copper and chromium concentrations and the fitted curves calculated with the POLLUTE modelling program are presented in figures 3 and 4. The used parameters are presented in table 2 and the analysed concentrations are presented in table 3. The regression factor R^2 for chromium was 0.80 and for copper $R^2 = 0.89$ (for the end of the graph, because the fitting was done excluding the fluctuation at the start of the graph which is probably due to errors when taking the sample or because the copper concentration was not even in the water).

The concentration of chromium decreases quickly in the contaminated water, but increases very little in the fresh water. This indicates a high adsorption rate on the peat. With copper the concentration decreases much slower and some of the copper also migrates through the sample. The highest measured copper concentration with pure water was 0.77

mg/l and the chromium concentration 0.008 mg/l. Both values are smaller than the limits set for drinking water in Finland (2 mg/l and 0.05 mg/l, respectively).

Table 2. Parameters used in the POLLUTE program

Deposit data:	**Values**
Layers	1
Laplace parameters	$\tau = 1$, $n = 20$ RNU = 2
Darcy velocity	0 m/s
Layer data:	
# sub layers	5
Thickness	1,5 cm
dry density	0.33 g/cm^3
Porosity	0.8
D^*	Cr: $6 \cdot 10^{-10}$ m^2/s Cu: $7 \cdot 10^{-10}$ m^2/s
K_d	Cr: 3000 cm^3/g Cu: 600 cm^3/g

Table 3. Concentrations of chromium and copper during the diffusion test in the contaminated water (Cr in and Cu in) and in the fresh water (Cr out, Cu out)

Day	Cr in (mg/l)	Cr out (mg/l)	Day	Cu in (mg/l)	Cu out (mg/l)
0	14.09	0.000	0	10.00	0.000
1	10.79	0.005	1	5.98	0.231
4	10.31	0.004	4	7.04	0.249
8	9.96	0.010	8	7.11	0.373
11	10.18	0.007	14	8.74	0.430
14	10.13	0.006	18	6.78	0.383
18	10.40	0.005	22	6.52	0.642
20	9.92	0.006	25	6.87	0.584
22	9.83	0.008	27	7.59	0.634
25	9.84	0.006	29	7.40	0.674
27	9.16	0.003	32	7.29	0.597
29	9.29	0.003	35	7.59	0.699
32	9.12	0.004	39	7.49	0.706
35	8.99	0.004	43	7.59	0.720
39	10.47	0.003	49	7.64	0.688
43	9.59	0.003	60	7.16	0.718
49	9.30	0.003	67	7.12	0.717
60	6.65	0.005	75	6.93	0.728
67	6.24	0.006	83	7.02	0.760
75	8.56	0.006	90	6.84	0.743
83	5.44	0.008	98	6.80	0.752
90	5.14	0.007	106	6.60	0.767
98	4.79	0.007			
106	4.03	0.008			

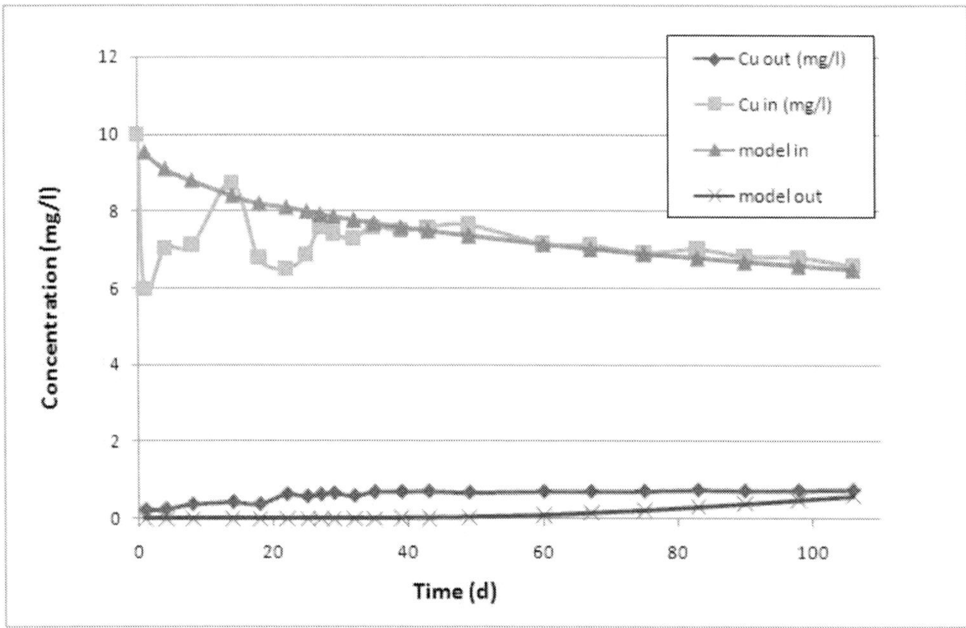

Figure 3. Diffusion of copper through peat and the graphs fitted with the Pollute modelling program.

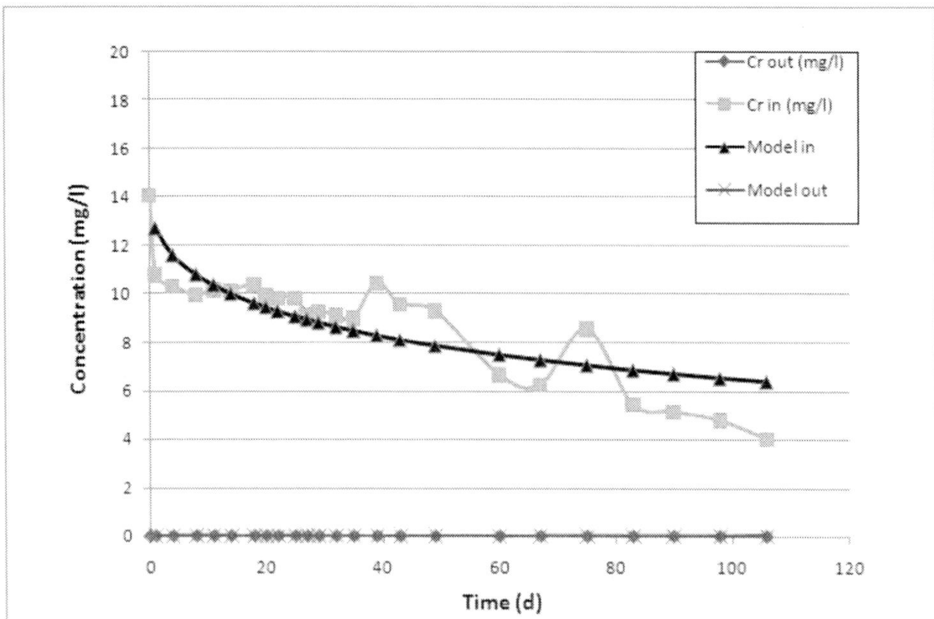

Figure 4. Diffusion of chromium through peat and the graphs fitted with the Pollute modelling program.

3.2. The Effect of the Adsorption on Diffusion

The effect of the adsorption on diffusion was great: if the adsorption is calculated separately, the diffusion coefficient of chromium in peat would be $6{\cdot}10^{-10}$ m^2/s as the distribution coefficient for chromium is 3000 cm^3/kg. When the adsorption is taken into account within the effective diffusion coefficient, the coefficient is $9.2{\cdot}10^{-13}$ m^2/s (table 4.).

Similarly for copper without the adsorption, the diffusion coefficient is $7{\cdot}10^{-10}$ m^2/s and the distribution coefficient is 600 cm^3/kg. When the effect of the adsorption is taken into account in the diffusion coefficient, the value of the coefficient becomes almost a hundred times smaller; $3.4{\cdot}10^{-12}$ m^2/s (table 4). The diffusion coefficients without the adsorption are very similar to those measured in clays and bentonite materials (Leroy et al., 2006; Lorenzetti et al., 2005) The effect of sorption on the rate of diffusion has been noted by others as well (Shackelford, 1991).

Table 4. The pH and diffusion and distribution coefficients of chromium and copper in the diffusion test

Chromium	**D^* (m^2/s)**	**K_d (cm^3/kg)**	**pH**
Without adsorption	$6{\cdot}10^{-10}$	3000	1,95
With adsorption	$9.2{\cdot}10^{-13}$	-	
Copper			
Without adsorption	$7{\cdot}10^{-10}$	600	1,95
With adsorption	$3.4{\cdot}10^{-12}$	-	

Peat adsorbed 3400 mg/kg of copper during the diffusion test (fig 5.) and 2700 mg/kg of chromium. The amount of the chromium analysed from the peat sample is smaller than what the water analyses and the distribution coefficient would suggest (which would be approximately 5000 mg/kg). This is because the chromium adsorbs very strongly right on the surface of the sample and as the sample was compressed between the porous plates, some of the peat got into the plate inside the pores and was caught there. The missing amount of chromium was adsorbed already on the peat inside the porous plate, which was not included in analyses. There was no solid precipitate to be seen in the chambers, so the chromium did not precipitate on the chamber walls. For future tests, it would be important to also treat the porous plates similarly as the peat samples are treated to get the total amount of the adsorbed metals.

The metals were adsorbed mostly on the first 0.5 cm layer of the peat sample, and copper clearly migrated deeper than chromium. This can also be seen from the leachate concentrations which are greater with copper. Very similar results were gotten from the adsorption column tests in the previous study. (Koivula et. al, 2009).

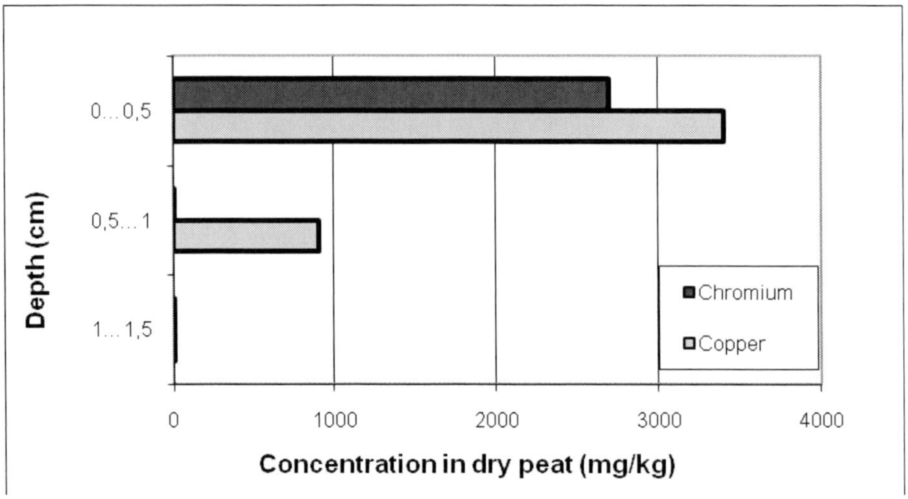

Figure 5. The adsorbed concentrations of chromium and copper in dry peat after the diffusion test.

It can be seen from the images taken with the SEM (figure 6.) that the metals are not distributed evenly on the sample surfaces and that the peat is not a homogenic material, although this does not show in the adsorption capacity, especially in large structures. The metals can be seen as the luminous white areas in the images.

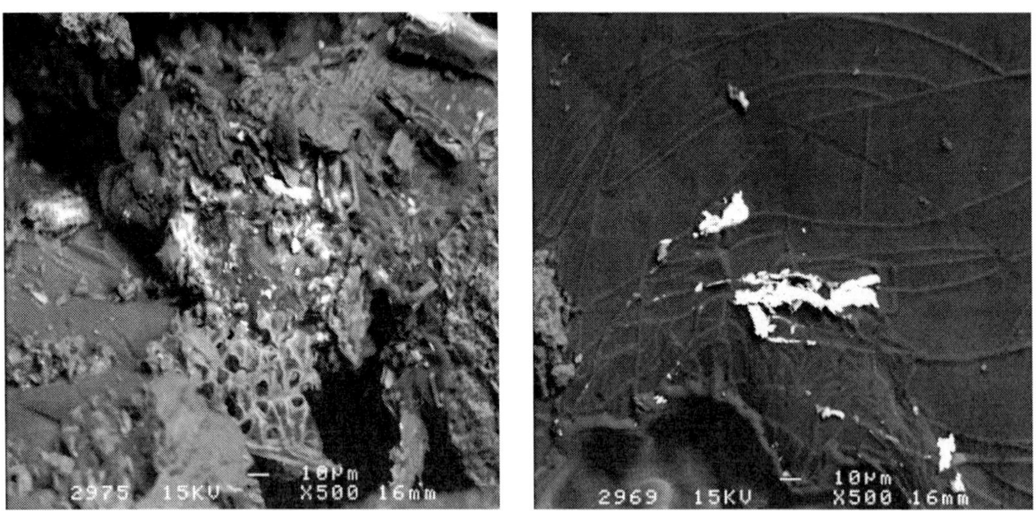

Figure 6.

3.3. The Migration in Landfill through Diffusion

To simulate the environment of a landfill, the migration of copper and chromium with diffusion was calculated for a 50 cm thick peat layer. The concentration in the landfill was assumed to stay at a steady level the whole time and the relative concentrations were calculated using Fick's second law.

$$\frac{\partial c}{\partial t} = D\frac{\partial^2 c}{\partial x^2} \tag{1}$$

where,

C is concentration [mg/l]
t time [s]
x diffusion distance [m]

There is an analytical solution for Fick's second law with the following boundary conditions:

$$\lim_{x\to\infty} C(x,t) = 0\;,\; C(0,t) = C_0\,.$$

$$c(x,t) = c_0 erfc\left(\frac{x}{x\sqrt{Dt}}\right) \tag{2}$$

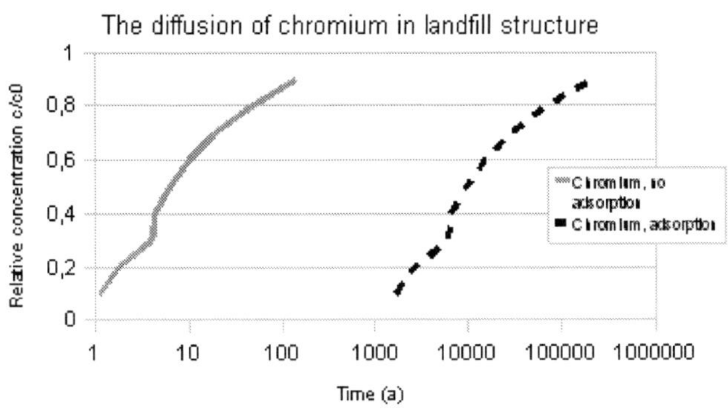

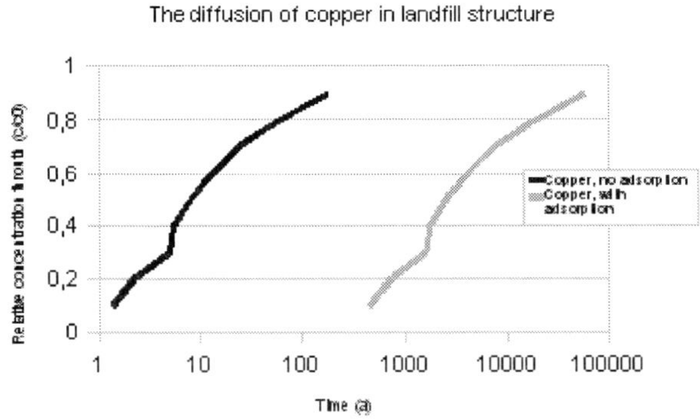

Figure 7. The relative concentrations of chromium and copper as a function of time at the bottom of a simulated hydraulic barrier layer (h=0.5m). The migration is assumed to happen only by diffusion and the relative concentrations were calculated with and without adsorption.

As can be seen from figure 7, the adsorption greatly affects the rate of diffusion. If the material was inert, the contaminants would diffuse through the peat layer in 140 years (chromium) and 180 years (copper) whereas with adsorptive material, the times would be 200,000 years and 60,000 years respectively. The adsorption capacity of the peat will not be filled in this time if the migration only happens by diffusion, since the migration is so slow (it would take about 500,000-600,000 years for the adsorption sites to be filled with the speed of diffusion). This is however only a calculatory result based on the empirical measurements and it depicts an ideal situation where there is no advection.

Conclusion

Peat adsorbed large amounts of heavy metals; approximately 2700 mg/kg of chromium and 3400 mg/kg of copper.

The adsorption affected the diffusion greatly; without adsorption the measured diffusion coefficient for chromium in peat was $6 \cdot 10^{-10}$ m^2/s and with adsorption the coefficient was $9.2 \cdot 10^{-13}$ m^2/s. For copper without the adsorption, the diffusion coefficient is $7 \cdot 10^{-10}$ m^2/s. With adsorption the value of the coefficient becomes almost a hundred times smaller; $3.4 \cdot 10^{-12}$ m^2/s.

The concentrations that leached through the sample were small during the test; the highest copper concentration was 0.8 mg/l and the chromium concentration only 0.01 mg/l.

Based on the results, the strong adsorption of the heavy metals on peat diminishes the influence of the diffusion on migration greatly, in comparison to a non-adsorptive material. Adsorptive materials would therefore increase the safety of the landfill structures.

References

Dean, S.A. & Tobin, J.M. (1999) Uptake of chromium cations and anions by milled peat, *Resources, Conservation and Recycling, vol. 27*, 151–156.

Directive 2006/21/EC of the European Parliament and of the Counsil of 15 March 2006 on the management of waste from extractive industries and amending Directive 2004/35/EC

Fityus, S.G., Smith, D.W. & Booker, J.R. (2006). Contaminant transport through an unsaturated soil liner beneath a landfill. *Canadian Geotechnical Journal 36 (2),* 330-354.

Gabaldon, C., Marzal, P. and Alvarez-Hornos, F.J. (2006) Modelling Cd(III) removal from aqueous solutions by adsorption on a highly mineralized peat: batch and fixed bed column experiments, *Journal of Chemical Technology and Biotechnology, 81 (7),*1107–1112.

Ho, Y.S. & McKay, G. (1998) The sorption of lead (II) ions on peat: research note. *Water Research, vol. 33 (2)*, 578–584.

Itakura, T., Airey, D.W. & Leo, C.J. (2003). The diffusion and sorption of volatile organic compounds through kaolinitic clayey soils. *Journal of Contaminant Hydrology, 65 (3-4)*, 219-243.

Koivula, M.P., Kujala, K., Rönkkömäki, H. & Mäkelä, M. (2009). Sorption of Pb(II), Cr(III), Cu(II), As(III) to peat, and utilization of the sorption properties in industrial waste landfill hydraulic barrier layers. *Journal of Hazardous Materials, 164*, 345-352.

Kotas, J. & Stasicka, Z. (2000). Chromium occurrence in the environment and methods of its speciation. *Environmental Pollution, 107*, 263-283.

Leroy, P., Revil, A. & Coelho, D. (2005). Diffusion of ionic species in bentonite. *Journal of colloid and Interface Science, 296 (1)*, 248-255.

Lorenzetti, R.J., Bartel-Hunt, S.L., Burns, S.E. & Smith, J.A. (2005). Hydraulic conductivities and effective diffusion coefficients of geosynthetic clay liners with organobentonite amendments. *Geotextiles and geomembranes 23 (5)*, 385-400.

Malusis, M.A., Shackelford, C.D. & Olsen, H.W. (2003). Flow and transport through clay membrane barriers. *Engineering Geology, 70 (3-4)*, 235-248.

Malusis, M.A. & Shackelford, C.D. (2002). Theory for reactive solute transport through clay membrane barriers. *Journal of Contaminant Hydrology, 59 (3-4)*, 291-316.

Myrand, D., Gillham, R.W., Sudicky, E.A., O'Hannesin, S.F. & Johnson, R.L. (1992) Diffusion of volatile organic compounds in natural clay deposits: Laboratory tests. *Journal of Contaminant Hydrology 10 (2),* 159-177.

Pourbaix, M. (1974). Atlas of electrochemical equilibria in aqueous solutions. 2nd English edition. Houston, Tex. National Association of Corrosion Engineers.

Rowe, R.K & Booker, J.R. (1987) An Efficient Analysis of Pollutant Migration through Soil. *Numerical Methods for Transient and Coupled Problems.* John Wiley & Sons Ltd, Chichester.

Sadhra, S.S., Wheatley, A.D. & Cross, H.J. (2007) Dietary exposure to copper in the European Union and its assessment for EU regulatory risk assessment. Science of The Total Environment 374 (2-3), 223-234.

Shackelford, C.D. & Daniel, D.E. (1991) Diffusion in saturated soil. 1. Background. *Journal of geotechnical engineering 117(3),* 467-484.

Wei, M & Tobin, J.M. (2003) Development of multimetal binding model and application to binary metal biosorption onto peat biomass,*Water Research, vol. 37*, 3967–3977.

INDEX

A

B

C

D

E

F

G

H

M

N

Q

R

S

T

U

V

W

X

Y

Z